Modernity and Technology

Modernity and Technology

edited by
Thomas J. Misa, Philip Brey, and Andrew Feenberg

The MIT Press
Cambridge, Massachusetts
London, England

This book was set in Sabon by UG/GGS Information Services.

Printed and bound in the United States of America.

Library of Congress Cataloging-in-Publication Data

Modernity and technology / edited by Thomas J. Misa, Philip Brey, and Andrew Feenberg.
  p. cm
  ISBN 0-262-13421-7 (hc. : alk. paper)
  1. Technology—Social aspects. I. Misa, Thomas J. II. Brey, Philip.
III. Feenberg, Andrew.

T14.5 .M63   2003
303.48′3—dc21   2002071754

10 9 8 7 6 5 4 3 2 1

Contents

Acknowledgments   vii

Workshop Participants   ix

1 The Compelling Tangle of Modernity and Technology   1
Thomas J. Misa

I Modernity Theory and Technology Studies

2 Theorizing Modernity and Technology   33
Philip Brey

3 Modernity Theory and Technology Studies: Reflections on Bridging the Gap   73
Andrew Feenberg

4 Critical Theory, Feminist Theory, and Technology Studies   105
Barbara L. Marshall

II Technologies of Modernity

5 Modernity under Construction: Building the Internet in Trinidad   139
Don Slater

6 Surveillance Technology and Surveillance Society   161
David Lyon

7 Infrastructure and Modernity: Force, Time, and Social Organization in the History of Sociotechnical Systems   185
Paul N. Edwards

8 Creativity of Technology: An Origin of Modernity? 227
Junichi Murata

III Changing Modernist Regimes

9 The Contested Rise of a Modernist Technology Politics 257
Johan Schot

10 Technology, Medicine, and Modernity: The Problem of Alternatives 279
David Hess

11 The Environmental Transformation of the Modern Order 303
Arthur P. J. Mol

12 Technology, Modernity, and Development: Creating Social Capabilities in a POLIS 327
Haider A. Khan

13 Modernity and Technology—An Afterword 359
Arie Rip

References 373
About the Authors 409
Index 413

Acknowledgments

The volume draws on an international workshop held at the University of Twente in the Netherlands in November 1999, which brought together a diverse group of scholars from the communities of modernity studies (philosophy, cultural studies, and social theory) and technology studies (history, sociology, anthropology); see <www.iit.edu/~misa/twente/>. This workshop was made possible by financial support from the U.S. National Science Foundation (grant SES-9900894), the University of Twente, and the Dutch Graduate School for Science, Technology and Modern Culture (WTMC). Our collective efforts to understand technology and modernity go back to a seminar series that Pieter Tijmes, Johan Schot, and Tom Misa organized at Twente in the spring of 1997.

At Twente we would especially like to thank Gerdien Linde-de Ruiter for a thousand small acts of kindness and assistance in planning and preparing the workshop, as well as the friendly and expert assistance of Femke Merkx in conducting the workshop. Many workshop participants freely shared their own ongoing research and reflections, giving our authors invaluable insights.

Workshop Participants

Hans Achterhuis   (University of Twente, Netherlands)
Michael Allen   (Georgia Institute of Technology, USA)
Anne Balsamo   (Xerox Palo Alto Research Center, USA)
Anne-Jorun Berg   (SINTEF Institute for Social Research in Industry, Norway)
Regina Lee Blaszczyk   (Boston University, USA)
Philip Brey   (University of Twente, Netherlands)
Nil Disco   (University of Twente, Netherlands)
Paul Edwards   (University of Michigan, USA)
Andrew Feenberg   (San Diego State University, USA)
Hans Harbers   (University of Groningen and University of Twente, Netherlands)
Mikael Hård   (Technical University Darmstadt, Germany)
Gabrielle Hecht   (University of Michigan, USA)
David Hess   (Rensselaer Polytechnic Institute, USA)
Thomas Hughes   (University of Pennsylvania, USA)
Haider Khan   (University of Denver USA)
Michiel Korthals   (Wageningen University, Netherlands)
Andrew Light   (New York University, USA)
David Lyon   (Queen's University, Canada)
Barbara Marshall   (Trent University, Canada)
Thomas Misa   (Illinois Institute of Technology, USA)
Joel Mokyr   (Northwestern University, USA)
Arthur Mol   (Wageningen University, Netherlands)
Junichi Murata   (University of Tokyo, Japan)
David Nye   (Odense University, Denmark)
Johan Schot   (University of Twente, Netherlands)
Don Slater   (London School of Economics, United Kingdom)
Pieter Tijmes   (University of Twente, Netherlands)
Ulrich Wengenroth   (München University of Technology, Germany)

Modernity and Technology

1

The Compelling Tangle of Modernity and Technology

Thomas J. Misa

Schiphol Airport, November 21, 1999. I'm checking in, heading home, answering questions. "Please step this way, I have a few things to ask you. . . . Did you pack your own bags this morning? Has a stranger given you anything to carry? Where were you staying in the Netherlands? I *do* need to see your passport." I decide to give straight answers, even if the smiling young woman—officially, I suppose, with the full power of the Dutch nation-state behind her—soon enough goes way beyond the script of ensuring safe travel. "How many days did you stay? What were you doing here?" Stay calm, I think. This is no concrete-and-barbed-wire interrogation, even if she still has my passport. I'm on friendly and familiar terrain. Schiphol is an unmistakably human-made space, beautiful in its way. Bright painted steel-framed ceilings high overhead, a wall of windows spotless as only the Dutch can make them, the quiet hum of air conditioning, the periodic clunk of baggage conveyors, the pleasant babble of a thousand people on their journeys. Five minutes ago I arrived on a sleek electric train, whose bulb-nosed profile still calls to mind the classic shape of a Boeing 747. So Claire's next question—I've sneaked a peak at her name tag—takes me off-guard. "This workshop you were at, I don't understand, what *exactly* do you mean by 'modern' and 'technology'?" Well, I say, look around you.

Is there anything more assertively modern and more thoroughly technological than an airport? Airports—we might equally think of harbors, subways, skyscrapers, automobiles, telephones, or the Internet—are deeply implicated in the social and cultural formations deemed "modern" by the founding fathers of social theory. Can you imagine an anthropologist of any "traditional" society doing his or her fieldwork on some exotic ritual in which 300 strangers willingly line up to be crowded into a narrow cylinder-shaped space, placed in seats so close their shoulders touch, and strapped down for hours on end? And they *pay* for this privilege!

Yet the airport ritual is a common experience of contemporary life, and more to the point, it embodies and enacts certain key features of modernity. It is not that airports are "new." Airports provide a techno-logically mediated instance of the increased interpersonal contact and communication that Emile Durkheim deemed characteristic of modern society. For his contemporary Max Weber, increasing rationalization characterized modern society. Weber's observations on German civil servants ring surprisingly true for airports.[1] Any sizable airport in the world has check-in counters, boarding passes, security and surveillance systems, indexical location schemes, English-language signs, and a high degree of time consciousness. Checking in at Portland, Oregon, one learns that Lagos, Nigeria, has failed its international safety inspection. The sign might as well say: you are entering a space of global standards.

As theorists of modernity, Marx and Engels shared with Weber a faith in the rationalization of society (in the sense of technological "progress" as well as growing social awareness of the process of change). Yet even though they misread the capacity of capitalism to avoid the cataclysms of revolution, Marx and Engels grasped the crucial point that modern economies, societies, and cultures are fundamentally about unremitting and unceasing *change*—in their memorable image, "all that is solid melts into air."[2] This insight historicizes the "great divide" that theorists from Francis Bacon forward to Bruno Latour (1993) have used to separate the modern world from the premodern world that it supposedly supplanted.[3] If you accept the divide and the terms used to describe it—traditional and modern, *Gemeinschaft* and *Gesellschaft*, lifeworld and system, Self and Net—you cannot help but put airports on the "modern" side.[4] "Those marvelous flights which furrow our skies" were among the soul-inspiring "tangible miracles of contemporary life" identified and celebrated by the Italian Futurists, the primordial theorists of aesthetic modernism.[5] Not bad for a painters' manifesto penned within a year of Louis Blériot's first cross-channel flight in 1909.

If one goal of this volume is to examine modernist icons such as airports, harbors, train stations, mechanical clocks, automobiles, pharmaceuticals, and surveillance and information technologies in the light of social theory, another goal is to consider them at the same time explicitly as technologies. In popular discourse technologies often appear as "black boxes," fixed entities that irresistibly change society and culture. However, the contributors to this volume want to understand them instead as embodiments of human desires and ambitions, as solutions to complex problems, and as interacting networks and systems. Social theories that assume static categories of "technology" and "society" or that presume technologies are always coercive structures are of scant help.[6] Technologies interact deeply with society and culture, but the interactions involve mutual influence, substantial uncertainty, and historical ambiguity, eliciting resistance, accommodation, acceptance, and even enthusiasm. In an effort to capture these fluid relations, we adopt the notion of co-construction.[7]

In compelling ways, airports combine transportation, production, and consumption, activities that we usually think of as being conducted in railroads, factories, and stores.[8] Think for a moment of your favorite airport not merely as a way of leaving town but as a rational factory with countercurrent flows of raw materials and products: departing and arriving passengers; food, beverages, and lavatory waste; jet fuel and pollution. Airports are in fact not only the location of electrical systems, ventilating systems, water systems, and communication systems, among others; they are also nodes in road and rail networks. Airports are created by, and in their day-to-day functioning depend on, the integration of these numerous systems. They are "systems of systems" or, as some theorists put it, second-order technological systems (Braun and Joerges 1994).

Solutions to the unique spatial problems of airports and other systems of systems often take novel forms and entail social and cultural changes. Sometimes what is important is a physical coupling of technologies; you can see this in the invention of jetways, which bridge the dangerous space between the check-in counter and the airplane's door, and which emerged at Amsterdam's Schiphol and Chicago's O'Hare airports around 1960. Equally important are the nonphysical couplings that occur through a welter of communication and control systems guiding the flow of passengers, ground traffic, and airplanes. One might say, on an abstract level, that airports process information.[9] Recently, as more and more airports have become display sites for luxury goods, they have displaced the shop windows of the metropolis and serve as a new site of modernism as consumption.

These transport, communication, and merchandising technologies have created a "modern" experience, and they serve as one long argument for a technological framing of modernity. Airport authorities, like railroad companies before them, seem to understand their culture-making power intuitively and act on it instrumentally. The experience they create is not always, as the founders of the modern movement in architecture had hoped, spiritually satisfying. In our own time, what better display of a banal and homogenized global economy is there than a quick stroll through the enticements of "airport culture"? How can you decide (even if you are merely going to Cincinnati) between Motorola cell phones, Komatsu earth-moving equipment, or Mannesmann engineering? Perhaps you try to escape the blare of CNN by retreating to an authentic "local" airport bar?

The impossibility of escaping this tangle of technology and modernity is our volume's point of departure.

Forget retreating to some mythical nontechnological past of small farms and happy peasants. Modern society—whether aspiring East or industrialized West, wealthy North or resentfully poor South—is constituted, in varied ways, through technological systems and networks. These systems and networks not only are the "connective tissues and the circulatory systems" of the modern economy,[10] they also constrain and enable social and cultural formations. Birthing babies, educating children, exercising citizenship, going to work, eating and drinking, visiting with distant friends and family, maintaining health or combating sickness, even dying—these human experiences are all mediated by technology. We cannot responsibly escape this condition of modernity, and we need ways to confront it constructively.

In this respect most existing approaches to the "problem of technology" leave much to be desired. Habermas's elegant opposition of "lifeworld" and "system," and the legion of philosophers, critics, and commentators who have followed his lead, takes you straight to dead ends or to despair. As humans we identify deeply with lifeworld, but as inhabitants of a modern world we are enmeshed in systems. As scholars and citizens we have no choice but to wrestle with the cultural formations and technological systems that together constitute modern society. "Our fate is worked out here as surely as on Heidegger's forest paths," as Andrew Feenberg phrases our contemporary dilemma (Feenberg 1999a: p. 197). Our volume takes up this pressing task.

Proposal One: The concepts "technology" and "modernity" have a complex and tangled history.

For more than a century "modernity" has been a key theoretical construct in interpreting and evaluating social and cultural formations. What it means to be "modern," however, is by no means clear. The term is bound up with overlapping and controversial notions about the imperatives of change and progress, of rationality and purposeful action, of universal norms and the promise of a better life.

Let us start at the present and dig down through the layers of sedimented meaning. In common speech, "modern" is often a synonym for the latest, and it is assumed inevitably the best, in a triumphant progression to the present. Contemporary designers, as Herbert Muschamp has recently observed, imaginatively draw a modernist veil over such varied products as computers, personal organizers, so-called designer drugs, cyber-prosthetics, and interior designs. "As expressions of The New, these products have inherited the myth of progress, modernity's defining legend."[11] The legend of progress through a parade of technologies, which has especially deep roots in American culture, forms a stock-in-trade for contemporary advertising.

The tie between modern technology and social progress was much in the minds of "modernists" in the early twentieth century. In Thomas Hughes's (1989) formulation, Americans invented modern technology in the early twentieth century, while European artists and architects, inspired by Americans' electric systems, automobile factories, and managerial organizations, theorized the "modern" movement. For Walter Gropius and Le Corbusier no less than for Frederick Taylor or Henry Ford, the values of order, regularity, system, and control constituted modernism. Inspired by the creative possibilities of new technologies such as electricity, automobiles, and mass-produced steel and glass, avant-garde artists and architects argued that modern forms were an authentic expression of the new machine age, and a necessary agent for progressive social change.

Among the well-known icons of modernism theorized by Europeans were the Futurists' city planning schemes and "dynamic" art, Le Corbusier's rational "machine for living," and the sleek rectilinear International Style architecture of Mies van der Rohe and Walter Gropius.

These early twentieth century modernists were "technological funda-mentalists" who embraced a messianic vision of societal transformation and spiritual redemption through the embrace of technology. In effect, they floated their aesthetic modernism on the deeper currents of socio-economic modernization (Banham 1986; Smith 1993; Trommler 1995).

Modernism in literature and poetry also drew on the technological dynamism of the age, especially the urban experience and the cinema, although its theorists were less likely to admit explicitly technical inspira-tion (Berman 1982; Tichi 1987; Charney and Schwartz 1995; Charney 1998; Harootunian 2000b).[12] Another expression of these mythic ideas was modernization theory in social science, which posited a deterministic link between technology, industrial growth, and desirable social and cultural changes (see later discussion).

Digging deeper, we can locate alternative and complementary concep-tions in the various revolutions of the seventeenth and eighteenth cen-turies that were deemed to have ushered in the modern age: the scientific revolution, the Enlightenment, the consumer revolution, and the indus-trial revolution.[13] For Francis Bacon in 1620, it was printing, gunpow-der, and the compass "which were unknown to the ancients" and which had "changed the appearance and state of the whole world."[14] Along with the physical embodiments of progress, rationality, and science in iconic technologies such as steam engines, laboratories, factories, and prisons, the habits of mind associated with mechanical metaphors are key interpretive notions. In this vein Lewis Mumford (1934) famously argued that the defining symbol of the industrial age was not the steam engine but the mechanical clock, while Otto Mayr (1986) contrasted continental Europeans' preoccupation with clock metaphors with British preferences for feedback mechanisms in politics and technolo-gies. Recently, a small scholarly industry has grown up relating science, standards, and state formation in early modern Europe.[15] Some, delving yet deeper, find a defining departure from traditional society in the acquisitive economy of the early modern town.[16] For that matter, declaring a "modern" period in history was a polemical act that defined who was "in" and who was "other."

On balance, the single most influential touchstone for modernity the-orists is the Enlightenment, with its affinity for rationality and social progress. Miles Ogborn, in *Spaces of Modernity*, writes: "[A]gainst the backdrop of the Enlightenment, modernity is associated with the release of the individual from the bonds of tradition, with the progressive differentiation of society, with the emergence of civil society, with political equality, with innovation and change. All of these accomplishments are associated with capitalism, industrialism, secularisation, urbanisation and rationalisation."[17] (In like measure, postmodern critics target these very same articles of faith.) In various ways, to conjure up "modernity" is to summon a noisy carnival of historical actors and images.

Technology also cannot be defined statically since its nature and meaning have shifted over time. In etymology, "technology" refers to a body of knowledge about the useful arts. It was this sense that prevailed, in the physical form of handbooks and written knowledge about the useful arts, from the Renaissance well into the industrial era. Even Jacob Bigelow, the Harvard professor whose *Elements of Technology* (1831 [1829]) is typically cited as introducing the term into popular English, used "technology" mostly in the sense of the useful arts or accumulated knowledge. "We traverse the ocean in security, because the arts [sic] have furnished us a more unfailing guide than the stars," he wrote, "We accomplish what the ancients only dreamt of in their fables; we ascend above the clouds, and penetrate into the abysses of the ocean." (In his chapters Bigelow described such "useful arts" as writing, printing, painting, sculpture, modeling, and casting as well as materials, machines, and processes.)[18] Technology, as a *set* of devices, a complex of industries, or as an abstract force in itself, had yet to appear.

Other modernist key words, including "scientist," "socialism," and "capitalism" were coined around the 1830s, and as Raymond Williams has observed, such loaded terms as "industry," "class," and "culture" emerged in the surrounding decades. Put another way, Karl Marx's famous observation that the culture of the working class was a product of modern technology and industry could not have been expressed, at least in English, before the mid-nineteenth century. The word "technology" took on something like its present meaning—abstract and culture-changing, systemic and symbolic—only after midcentury. "Technology" as Bigelow himself told his audience in 1865 at the newly founded and aptly named Massachusetts Institute of Technology (MIT), "in the present century and almost under our eyes . . . has advanced with greater strides than any other agent of civilization."[19]

Proposal Two: Technology may be *the* truly distinctive feature of modernity.[20]

This volume takes up the task of reintegrating the close empirical study of technology with broader theoretical reflections on modernity. The drive to professionalize, itself a characteristic of the modern era, helps account for the enormous gap between empirical studies of technology and theoretical reflections on modernity that has persisted for a generation or more. No such gap can be found in writings by the founding fathers of social theory and technology studies. Marx's scathing critique of the orthodox political economists of his day focused on their blind ignorance of the social processes of industrialization. And in Friedrich Engels, who for years actively managed and came to jointly own his father's Manchester cotton factory, Marx had an unusually well-informed critical source on industrial capitalism. Weber similarly argued for a historically and empirically grounded analysis of society. Lewis Mumford, a founding father of technology studies, was deeply informed by his philosophical commitment to organicism. For all these authors, theoretical reflections are bound up with empirical studies.

Oddly enough, the "modern society" that has emerged in the writings of social theorists and philosophers in the past several decades has been a theoretical construct that is surprisingly devoid of technology. Theorists of modernity frequently conjure a decontextualized image of scientific or technological rationality that has little relation to the complex, messy, collective, problem-solving activities of actual engineers and scientists.[21] Technology, abstractly, dominates humans. In representative formulations Heidegger writes of "enframing" (*Gestell*) and Horkheimer emphasizes "the domination of instrumental rationality." Ellul in his work floated the notion of a boundless, omnipotent, and deterministic "technique." And Habermas, as Feenberg (chapter 3 in this volume) writes, "has elaborated the most architectonically sophisticated theory of modernity without any reference at all to technology."

These theorists of modernity invariably posit "technology," where they deal with it at all, as an abstract, unitary, and totalizing entity, and typically counterpose it against traditional formulations (such as lifeworld, self, or focal practices). Heidegger followed such an abstract, macro-level conception of technology and concluded that the rationalization of modern society (inescapably) leads to humans being caught in technology's grip. "Agriculture is now the mechanized food industry, in essence the same as the manufacturing of corpses in gas chambers and extermination camps, the same as the blockade and starvation of nations, the same as the production of hydrogen bombs," he wrote in 1949. In the end, he famously despaired, "only a god can save us now" from this technology-driven juggernaut.[22]

Yet one central finding of this volume is that such despair, however elegantly arrived at, is certainly misplaced. Whether modernist or, as discussed later, postmodernist, these overaggregated approaches cannot help us discern the *varieties* of technologies we face and the *ambiguities* in the technologies that we might exploit.[23] Abstract, reified, and universalistic conceptions of technology obscure the significant differences between birth control and hydrogen bombs, and blind us to the ways different groups and cultures have appropriated the same technology and used it to different ends. To constructively confront technology and modernity, we must look more closely at individual technologies and inquire more carefully into social and cultural processes.

To be fair, empirical students of technology who have this detailed understanding have been instinctively antagonistic to the broad-scale interpretive schemes offered by social theory and philosophy, including reflections on modernity. The 1970s were something of a watershed. At more or less the same historical moment that postmodern theorists boldly asserted that information, media, and communication technologies had brought about a new, postmodern society, most empirical students of technology took hostile aim at all such "technological determinist" schemes.

In their detailed empirical studies, historians, sociologists, and many anthropologists of technology aimed to deconstruct the process by which a given technology supposedly imposed its logic on society. An early target was Marx's famous line in *The Poverty of Philosophy* (1847, chap. 2): "The hand-mill gives you society with the feudal lord; the steam-mill society with the industrial capitalist." In combating such technological determinist arguments, the empirical students' chosen method was to reconstruct in great detail the social and political choices that conditioned how technologies were invented, chosen, or deployed. The "logic of technology" invoked by modernist and postmodern theorists alike simply vanishes in these detailed micro-level accounts.[24]

A concise way of making the same point is to say that while philosophers and social theorists asserted the "technological shaping of society," historians and sociologists countered with the "social construction of technology." For years, these groups just talked past each other.[25] One can see, of course, that these rival positions are not logically opposed ones. Modern social and cultural formations are technologically shaped; try to think carefully about mobility or interpersonal relations or a rational society without considering the technologies of harbors, railroad stations, roads, telephones, and airports; and the communities of scientists and engineers that make them possible. At the same time, one must understand that technologies, in the modern era as in earlier ones, are socially constructed; they embody varied and even contradictory economic, social, professional, managerial, and military goals. In many ways designers, engineers, managers, financiers, and users of technology all influence the course of technological developments. The development of a technology is contested and controversial as well as constrained and constraining.

The central aim of this volume is to grasp both perspectives—the social construction of technology and the technological shaping of society—and to develop new intellectual frames by which to comprehend them. Indeed, we argue that theories of modernity at the macro level must engage the detail, ambiguity, and variety of technology evident at the micro level of empirical analysis. Theories of modernity that lack a reasonable and robust account of technology are hopelessly hollow. At the same time, we take seriously the criticism that empirical work on technology too often offers little more than instances of messy complexity without a larger aim in sight.[26] In proposing the *co*-construction of technology and modernity as our methodological point of departure, we emphatically reject the idea that either technology or modernity alone can be used as a template to "explain" the other. In different ways, the chapters in this volume problematize both "modernity" and "technology."

Proposal Three: Modernization theory missed what was modern about technology.

Some readers may inadvertently assume that we wish to revive the social-scientific "modernization theory" that was popular in the 1950s and 1960s. Quite the contrary. Advocates of modernization theory, under the sway of rationalistic and universalistic models, sought to define and measure a single path leading from traditional societies to modern ones. Modernization theorists with a flair for policy advice capitalized on the political context of the Cold War, as the two superpowers competed for the hearts and minds of the developing world (recall that Walt Rostow's famous *Stages of Economic Growth* [1960] was subtitled *A Non-Communist Manifesto*). Historical indexes of industrial production, education, literacy, and other "factors" deemed important in the successful industrialization and modernization of North America and western Europe were quickly transformed into policy targets for the developing world. Unfortunately, what appeared to work for England in the nineteenth century was often a disaster for many developing countries in Asia, Africa, and Latin America in the later twentieth century. Modernization theory can be a compelling object of study, but it offers few useful tools for understanding technology and modernity.[27]

As I noted earlier, the word "technology" took on its contemporary meaning—in the twin sense of a complex of industrial systems and a dynamic force bringing about social change—well into the industrial era. Leo Marx (1994) suggests that it was the railroad systems and the elaboration of other complex mechanical and industrial systems in the late nineteenth century that gave rise to something approximating our contemporary understanding of technology. Ruth Oldenziel (1999) also locates the emergence of our contemporary understanding of the term in the two decades before and after 1900, focusing on the male identity of the American engineering community. In these decades, it was continent-spanning railroads; electric lighting and communications; immense bridge, dam, and skyscraper constructions; and sprawling factory complexes like Henry Ford's that captured the public's imagination and seemed to change culture. In the middle of the twentieth century, synthetic chemicals, mass automobility, and atomic power ushered in a new era. Today, such heavily hyped visions as pervasive computing, wireless communication, genetic engineering, or nanotechnology capture the imagination and, at least for their visionary promoters, promise an endlessly better future. These culture-changing technologies have been at the core of modernity because their presence and their promoters' promises have seemingly offered proof of the modernist storyline that society is incessantly changing, ever progressing, transcending frontiers without an end in sight.

Yet, then as now, the symbol-making technologies, and the set of culture-changing expectations their promoters create, are only part of the modern story. Like the users of most technological systems, as travelers we hardly notice the dozens of technologies knitted together at an airport. They are unexamined black boxes whose internal characteristics we notice only when they fail.[28] This apparently smooth, silent functioning of networks of networks, or systems of systems, constitutes an infrastructure of daily life, choreographing the members of modern societies in an intricate routine. Technology, then, in its relations with modernity, is not only symbol making and culture changing but also, in the infrastructure of daily life, society constituting.[29]

Proposal Four: Postmodernism no less and no more than modernism is tangled up with technology.

For many writers, modernity refers to a specific historical period, beginning sometime during the succession of scientific, industrial, and political revolutions considered to usher in the modern age, and which lasted through at least the middle of the twentieth century. Some authors furthermore distinguish "classic," "high," "low," or "late" modernity (Harvey 1989; Lash and Friedman 1993; Scott 1998). Although their terminology is by no means clear, postmodern theorists argue that modern society has been superseded by a postmodern one. Postmodernism in architecture can be understood as a revolt from the formalism and minimalism of modernist, International Style architecture, and can be dated rather precisely with the publication in 1966 of Robert Venturi's *Complexity and Contradiction in Architecture.* While Mies van der Rohe preached that "less is more," Venturi's postmodern stance is that "less is a bore."

Postmodernism in social theory is similarly a revolt, from the project of Enlightenment. As Michel Foucault (2000: p.273) phrased the dilemma, "the Enlightenment's promise of attaining freedom through the exercise of reason has been turned upside down, resulting in a domination by reason itself, which increasingly usurps the place of freedom." But while postmodern skyscrapers literally stand next to modernist ones, modernist and postmodernist writings are not easily compared. Many postmoderns deliberately deploy alternative narrative forms—rejecting as a point of principle linear cause-and-effect relationships, formal logic, and rational argument. Writers informed by poststructuralist sympathies, while not adopting a specific postmodern theory of society, often utilize nontraditional writing styles—rejecting the objective third person and taking up multiple narrative voices.[30]

It is too little appreciated that most postmodern theorists repeat the modernist mistake of conceiving technology as a universalistic force. A defining distinction for many postmodern theorist-critics is that modern society has changed into a postmodern society with distinctive cultural forms. Yet looking closely at what brought about this cultural transformation, one finds a well-worn argument hinging on technology: post-Fordist manufacturing technology, media technology, communication technology, and especially computer and information technology. From this volume's viewpoint, these technologically determinist theories—common to many modernists and postmodernists alike—simply miss the theoretical salience of technology. It is in the details of technology, and not its macro-level abstractions, that one can escape the (various) traps that Heidegger, Ellul, Lyotard, Borgmann, and others have set for themselves.

Given our media-saturated culture, it is alarming to find so little empirical discussion of modern media technologies. An apparent exception to this pattern of neglect, Jürgen Habermas's media studies, turns out upon close inspection to be an analysis of an abstract concept of media. The gap between theories of media and empirical studies of media technology is all the more unfortunate in that Susan Douglas (1987, 1995, 1999), Lisa Gitelman (1999), and others have demonstrated that the *history* of media technologies really matters, not least in who dominated which media when—and where the media have served countervailing, even oppositional social formations. Even Foucault's famous reading of Bentham's Panopticon is hardly the last word on that historic technology.[31]

This volume is informed but not captured by the fractious debates in recent decades between modernists and postmodernists. All contributors take seriously the methodological problems raised by postmodernists (such as essentialism, foundationalism, and determinism), and many adopt poststructuralist sympathies. As noted earlier, however, the problematic of this volume—which departs from and extends this debate—is a focus on relating theories of modernity (and postmodernity) to empirical studies of technology. Until now, the work done on this problem has been suggestive but episodic.

Perhaps the most compelling use of postmodernist and modernist themes in technology studies is Sherry Turkle's exposition of rival computer aesthetics.[32] In the mid-1990s she found that users of IBM-DOS personal computers (PCs) tended to use modernist images in their effort to understand and relate to their machines (Turkle 1995). These users wanted detailed understanding and absolute control over their machines. Far from being irritated by the need to set dozens of parameters just to plug in a modem, they praised their machines' operational transparency and conceptual openness.

By contrast, users of Apple Macintoshes often used postmodernist images in describing their machines. Early Macs were literally factory-sealed beige boxes that not only frustrated users eager to know what was "going on inside" (you needed a special factory tool just to open them) but also discouraged reductive understanding and detailed control. Whereas PC users found satisfaction in controlling their machines directly, by typing inscrutable computer codes at the "command line," few Mac users ever experienced this level of their machines. Instead of plumbing their machines' conceptual depths, Mac users surfed the conceptual "surface" of their machines with mouse clicks, windows, and icons. (Needless to say, Turkle's neat dichotomy is considerably clouded by the rise of mouse-enabled and windows-savvy PCs as well as transparent iMacs that show off their insides, not to mention the "command line" to Unix within Macintosh's latest operating system.)

Contributors to this volume come from several disciplines and theoretical traditions, but we all share a conviction that comprehending technology and modernity is a compelling theoretical, practical, and political problem. In moving from the international workshop we held at the University of Twente, the Netherlands, in November 1999, to this volume, the editors have selected eight essays, commissioned four new ones, and asked each author to develop three levels of analysis. The volume as a whole, and nearly all the essays in it, suggests and exemplifies relations between theory, methodology, and empirical research. Our goal is not only to illuminate the co-construction of technology and modernity but also to develop ways of moving across various levels of understanding.

The papers in part I, "Modernity Theory and Technology Studies," are methodological pieces concerned with description and analysis. Philip Brey, Andrew Feenberg, and Barbara Marshall take up various disciplinary angles (respectively, technology studies, philosophy, and sociology). Each of their essays reflects on the interactions between technology and either modern socioeconomic structures or modern notions of culture, ideology, or identity. Brey and Feenberg have a predominantly methodological orientation in that they focus on the question of how to combine modernity theory and technology studies, and how to deal with different levels of analysis. Marshall exemplifies a way of integrating feminist and critical theory with technology studies, while raising methodological issues.

Philip Brey's chapter offers a wide-ranging interdisciplinary survey of theoretical and methodological issues in bringing together modernity studies (e.g. Marx, Weber, Habermas, Heidegger, Giddens, Beck, Latour, Castells) and technology studies (sociology and history of technology), including a perspective on postmodern theory (including Harvey, Jameson, Baudrillard, Lyotard). He develops the co-construction theme, which jointly problematizes modernity and technology, first by discussing disciplinary and philosophical obstacles to analyzing technology and modernity together, and then by developing methodological proposals for surmounting these obstacles. Feenberg aims similarly at "bridging the gap" by diagnosing the philosophical and methodological gaps and overlaps between technology studies and modernity theory. Using Thomas Kuhn and Karl Marx as exemplars of these two traditions, and gathering together complementary strands in their respective bodies of work, he then develops a synthetic "instrumentalization" theory, introduced in his recent *Questioning Technology* (Feenberg 1999a).

Barbara Marshall's chapter continues her work in combining critical theory with feminist theory (Marshall 1994, 2000). She surveys these theoretical constructs with an eye to developing methodological prescriptions for the empirical analysis of technology. Even more so than Feenberg and Brey, she combines her theoretical comments with detailed empirical discussion. Her illustrations of what she terms the "gender-technology-modernity nexus" include the feminist-inspired sexual assault evidence kit as a forensic technology and the pharmaceutical framing of erectile dysfunction with Viagra. "[T]here is no point at which technology and modernity are not joined in some way in the production of sexual bodies," she finds. While the conjunction of technology and human might call to mind Donna Haraway's postmodern notion of cyborg bodies, Marshall finds more compelling the "distinctively modernist framing shared by the scientists, pharmaceutical companies, physicians, and consumers."

Part II, "Technologies of Modernity," continues the methodological discussion with a focus on the co-construction theme. These essays, however, examine particular sociotechnical systems or technologies with prominent symbolic and material relations to modernity. These include the Internet, surveillance, infrastructures, and western technologies in China and Japan. Don Slater's essay deals with the Internet and its varied uses by Trinidadians to grapple with global modernity, while David Lyon's deals with the technologies of surveillance and their relation to modernist and postmodernist cultural formations. Paul Edwards, building on his work in computer history, proposes a wide-ranging interpretation of modernity and infrastructure technologies. Junichi Murata, drawing on contemporary Japanese philosophy, revisits and interprets the influx of western technologies into China and Japan during the late nineteenth and early twentieth centuries.

Don Slater's essay, drawing on his investigations of the Internet and Trinidad (Miller and Slater 2000), illustrates perfectly the co-construction theme. He insists that neither technology nor modernity can be taken as global, totalizing, or unitary. Even "context," if we assume it to be a fixed entity, may mislead: "the context of a technology is also partly a consequence of that technology; it is produced by the very 'thing' one is trying to put into it." As with much good ethnography, Slater's work challenges and problematizes our conceptual categories. He asks, for instance, what is constant about the categories "modernity" and "technology" when Trinidadians use the Internet—itself an amalgam of email, chat rooms, World Wide Web (WWW) sites, intranets, and e-commerce—to participate directly and actively in global modernity (by linking up and sharing technical capabilities with world-leading North American technology companies) but also use email to sustain traditional family functions (such as mothers nagging daughters, often at a significant physical distance, about staying out too late)?

Yet Slater's chapter is not principally an exercise in category crashing, since he proposes reconstructing the "big picture" from his theoretically aware ethnography. To this end, he offers four "dynamics"—objectification, mediation, normative freedom, and positioning—as methodological heuristics which would make sense of both the Internet and modernity "in a wide range of different places," and which might "allow us to ask intelligent questions about the similarities and differences in peoples' responses to new communication possibilities." His framework also has the valuable feature of highlighting actors' agency in and perceptions of "modernity."

"Modernity is in part constituted by surveillance practices and surveillance technologies," observes David Lyon in his chapter. Lyon is concerned to show the deep historical relation of surveillance technologies and modern societies, with their functions of taxing, policing, and providing welfare, as well as producing and consuming goods and services linked to the state's routine monitoring of individuals. Yet what most intrigues him about surveillance is the shift he discerns beginning in the 1960s with extensive computerization of surveillance in the capitalist workplace and modern nation-state. Equal in importance to new hardware, he emphasizes, are the practices of data matching between government departments as well as the outsourcing of government functions to private firms. These changes have brought about, not the centralized "Big Brother" that haunts the Orwellian imagination, but perhaps just as ominously, a decentralized network of databases that facilitates national and international flows of personal data.

For Lyon these shifts embody a shift away from classical modern society and toward a condition of "postmodernity . . . where some aspects of modernity have been inflated to such an extent that modernity becomes less recognizable as such." People are still under scrutiny, he observes, but less as citizens of the modern nation-state and more as workers and consumers, often in a globalized economy. (He notes that the Internet company Engage tracks the Web-surfing patterns of more than 30 million individuals.) Postmodernity as a social formation involves "widespread and deepening reliance on computers and telecommunications as enabling technologies, and an intensification of consumer enterprises and consumer cultures."

Finally, anticipating the normative bent of part III, Lyon considers how co-construction itself is a technology-making process. Consider privacy advocates. "The ad hoc practices of organizations as well as the self-conscious political stances of those who question and resist encroaching surveillance are inextricable elements of that co-construction process," he maintains. While lobbying the more-or-less centralized nation-state on privacy concerns has been an effective way of changing laws and thus altering surveillance technologies and practices, it is vastly more difficult to exert meaningful influence when confronting contemporary surveillance that is "networked, polycentric, and multidimensional." In this way Lyon, while aware of practical limits to effecting change in a polycentric world, echoes Feenberg's advocacy of democratic rationalization as a strategy for affecting sociotechnical change.[33]

In directing our attention from "new" to mature technologies, Paul Edwards subtly enlarges the co-construction theme with his consideration of infrastructures. The order, regularity, predictability, and stability of modernity, he argues, "fundamentally depend on" the presence and mostly silent functioning of mature technological systems—cars, roads, municipal water supplies, sewers, telephones, railroads, weather forecasting, even most routine uses of computers. At the same time, he writes, the "ideologies and discourses of modernism have helped define the purposes, goals, and characteristics of those infrastructures." Technology and modernity, to repeat the theme, are co-constructions.

Showing how these co-constructions occur, and developing a methodology for understanding these processes, are the goals of Edwards' chapter. Reviewing the SAGE early-warning military system in the 1950s and the ARPANET/Internet system beginning in the 1970s, and retelling the narratives from several perspectives, he shows how these infrastructures link varied scales of force, time, and social organization. Edwards, like Slater, finds that the "same" technology can seemingly possess contradictory characteristics, yet he goes beyond this telling observation by providing a method to grasp the varied effects of technologies. Edwards develops and illustrates a typology of scales ranging from the detailed micro, through intermediate meso, to the aggregated macro level. Then, to comprehend seemingly contradictory phenomena occurring at different scales, Edwards offers the concept of "mutual orientation."

Junichi Murata's chapter is deeper and more subtle than it may appear at initial approach. At first, his essay appears to be a straightforward discussion of the technology studies literature, essentialism in philosophy, and a comparison of modernization processes in Japan and China. One should appreciate, however, that Murata is seeking to engage technology studies with modern Japanese philosophy—an effort that Feenberg does for the critical theory tradition (see Feenberg 1986, 1995, 1999a, 2002). For both, as philosophers, the point is exploring the overlaps, contradictions, and extensions of the two (once-separate) literatures. To this end Murata offers an interpretation of the modern Japanese philosopher Nishida, focusing on his notions of the "otherness" of technology, an exposition of "reverse determination," and a discussion of the natural and human worlds.

By embedding his discussion of empirical cases of modernization within Japanese philosophy, Murata arrives at results that will be at once familiar and fresh. The "otherness" of technology is not a takeoff on the feminist Other, but rather an exploration of the unplanned, often unforeseeable, noninstrumental and nonrational aspects of technology. (These "creative" aspects of technology are also a concern in Khan's chapter.) The transformation of the Internet from a military tool to a commercial medium, or the reconception of automobiles as speed machines, are instances of creativity in the *use* of technology. Murata suggests that this feature should be called "creative" because "a new meaning for artifacts is realized." That the results may go against the original intent of designers and producers agrees solidly with the "user heuristic" in technology studies (see Fischer 1992; Borg 1999; Oudshoorn and Pinch forthcoming).

Murata further develops his ideas through examples drawn from the wrenching modernization that Japan experienced in the years following the Perry mission in 1853. On the one hand, Murata fully grasps that Japanese elites saw little option but to import and master such western technologies as telegraphs, railroads, and military equipment—and to adopt western institutions and ways of life that were consonant with industrialization and modernization. All the same, as hinted by the slogans "Japanese spirit and western technology" and "Enrich the country, strengthen the army," Japan thoroughly industrialized and modernized in these years, but did not clearly westernize. Murata provides tantalizing evidence for this proposition in a detailed comparison of western, Chinese, and Japanese mechanical clocks and the persistence of indigenous conceptions of time.

While the essays in parts I and II are mostly concerned with description and analysis of existing or historical conditions, the essays in part III, "Changing Modernist Regimes," shift attention to practical and political matters. These chapters provide a normative critique of modernity and technology as unitary, totalizing, and universal by suggesting alternative modes of developing technology or, indeed, alternative modernities.[34] In their discussion of alternatives, the chapters in this section also offer original and substantial critiques of technology policy, medicine, environmental technology, and international development.

In his chapter, Johan Schot accepts the broad framework of modernization as a way to analyze the various structural changes in western societies since the eighteenth century, but uses it to criticize the "modernist technology politics" that developed during this era. While a classic modernist account might point at "progress" in dealing with social conflict about technology, or the increasing acceptance of technical rationality, or the emergence of an autonomous technical realm, Schot instead emphasizes that a modernist technology politics emerged under a continual cloud of contestation. In the early industrial era, there was little separation of "technical" criteria from broader social and cultural considerations; the Luddites in England, in his view, knew precisely that certain machines embodied a dangerous worldview and moved to destroy them. By comparison, he finds the discourse of technology dissent much impoverished by the early twentieth century. In a classic "men versus machines" set piece—a two-year conflict over mechanizing the unloading of grain at Rotterdam Harbor—the working-class critics of the machines failed to find alternatives to mechanization while their socialist leaders even supported job-destroying mechanization.[35] Schot's diagnosis is that by accepting the terms of modernist technology politics, which has split a unified problem into separate "promotion" and "regulation" realms that are frequently are in conflict, we have lost the ability to have reasoned discussion and dialogue on these vital matters.

Drawing on Beck and Giddens, Schot is guardedly optimistic about the prospects of a "reflexive modernization." He presents a case study of recent Dutch attempts to integrate technical and political decision making, which should be compared with Hughes's (1998) discussion of Boston's Central Artery project and Khan's discussion of a positive feedback loop innovation structure (see chapter 12). Critics of the grandiose expansion schemes of Schiphol Airport in the late twentieth century have succeeded in slowing and shifting the airport's grand march into the future by preemptively buying needed land and forcing the airport to build a state-of-the-art train station. However, a persistent divide between "promoters" and "critics" has poisoned dialogue between the parties, led to substantial mistrust, and left both sides disillusioned and angry. To go beyond this modernist stalemate, Schot proposes a fundamental reform of design processes using the criteria of anticipation, reflexivity, and social learning.[36]

David Hess develops a broad three-part framework to explore the practical, political, and theoretical implications of the rise of "complementary and alternative" medical therapies. His long-term viewpoint, informed by the anthropology-inspired frameworks of cultural ecology, cultural values, and political economy, and his claim that the political economy of technology and modernity needs to be situated far beyond the past 500 years, will challenge readers presuming a "recent" view on modernity. He contrasts orthodox allopathic, science-based medicine with alternative therapies such as acupuncture, herbal remedies, and chiropractic, and finds a conceptually puzzling (if practically popular) result. In a simple narrative of the "triumph" of modern science in medicine, bitter conflict "ought" to occur between alternative therapies, which are often based on belief systems antithetical to "modern" reductionist science, and orthodox medical practices. But something more intriguing has happened. While orthodox medical doctors have been "surprisingly" open to alternative therapies—after all, the therapies frequently work, even if their underlying biomedical mechanisms are unclear—at the same time the mainstreaming of alternative therapies has brought about their acceptance through insurance payments and licensing rights.

Hess departs from a conventional technology-studies framing of these issues (for instance as alternative rationalities) with his insistence on a "broader terrain of shifts in environmental consciousness and disease ecology." As with Mol's chapter that follows, Hess takes Beck's "risk society" seriously. He also considers the understudied notion of "technological pluralism," a parallel to medical pluralism, as a way to conceptualize the tensions between the local and the global, patterns of domination and resistance, and relations between "normal" and "alternative" technologies.[37] Yet these analytical or methodological observations, important in themselves, are a means for Hess to spotlight the "deep normative question about the kind of global material-social world that should be co-constructed." These problems, he concludes, "require both empirical research and normative debate." He proposes sustainability, equality, and community as "three major criteria that provide viable points of reference for a general discussion of technological and social redesign."

Arthur Mol's chapter on ecological modernization focuses attention on how modernity is understood and deployed by *actors* in the realm of politics and policy making. In the environmental field, the understandings of modernity and the actions based on them are changing, he reports. For decades, environmentalism was antimodern; to be an environmentalist was to be against capitalism, industrialism, modern science and technology, and the bureaucratic nation-state. In the past 15 years, however, the landscape of "green" philosophical positions has become far more complex and decidedly less hostile toward modernity. Mol surveys four positions in these environmental debates—neo-Marxists, demodernization or counterproductivity theorists, postmodernists, and reflexive modernization advocates—but his underlying concern is to situate the rise of "ecological modernization" as both a philosophy and a set of political and policy-making strategies (see Mol 1995; Hajer 1996).

As a theory, ecological modernization has deep affinities with Beck's and Giddens' writings on reflexive modernization and the risk society (see Beck 1992; Beck et al. 1994). Both ecological and reflexive modernization are philosophies positing the imperative of fundamental change in society (to deal with the environmental and risk crises, respectively) absent a requirement for radical social transformation. Specifically, the advocates of ecological modernization view the "institutions of modernity, not only as the main causes of environmental problems but also as the principal instruments of ecological reform." Scientific researchers, technology developers, industrial corporations, and the nation-state are at once part of the problem and part of the solution, and are themselves changing in response to environmental problems. For example, in the growing "autonomy" of the environmental sector (where environmental functions are institutionalized within and across governments, businesses, and nongovernment organizations, or NGOs), itself a classic symptom of modernization, the role of the nation-state is transformed. And while changes in the content of the sciences, for instance "soft chemistry," have thus far been more speculative than practical, changes in business have been substantial, meaningful, and fundamental. Businesses, especially in the European chemical sector Mol reviews, are using environmental criteria to shape their business strategies, to prioritize technological choices, and to relate to environmentally conscious consumers.

Yet Mol's objective is not to lay out a "global" theory, as Beck is sometimes criticized for attempting. Rather, he uses ecological modernization to evaluate both sectoral and regional variations in business and society. Surveying how this highlights environment-induced transformations in modern social practices and institutions, Mol enumerates five key heuristics of ecological modernization—for example, the contributions of science and technology to environmental reform, the increasing importance of market dynamics and economic agents, and the transformations of the modern "environmental state," along with the rise of new ideologies, practices, and discourses for the environmental movement. These heuristics, while valuable in framing research for analysts, are also used by policy actors as "normative paths for change."

Focusing principally on Europe, he reviews and evaluates changes in the chemical industry over the past 15 years as the industry has introduced new environmental management functions and activities, new products, and new relationships among its members and with governments and NGOs. As Mol observes, from the viewpoint of ecological modernization, "all ways out of the ecological crisis will lead further into modernity."

The entwined "path dependence" of modernity and technology is also a central concern for Haider Khan. Khan's chapter begins with a careful and critical discussion of "methodological aspects of connecting theories of modernity with empirical approaches in the context of technology and development." Yet for Khan, as for Schot, the principal concern is to identify the limits imposed by a modernist framing of technology and development and to explore a rigorous conceptual model for moving forward and beyond the modernist impasse. Indeed, both of their chapters use Beck and Giddens' rather abstract notion of reflexive modernization as an entry point for their real-world discussions.

Khan's principal aim is to critique the modernist framing of development policy and to develop his own, holistic model (a POLIS). Targeting the modernist framing of development, especially the influential national innovation systems (NIS) model, which has focused narrowly on economic measures, he advocates instead a multidimensional "capabilities" approach (drawing on the work of Amartya Sen and Martha Nussbaum), insisting that human capabilities should be understood as enhancing a "complex" notion of freedom and that technology developments are central in realizing human capabilities. While "technology as freedom" is too often loosely theorized,[38] Khan's specified criteria and constructive stance make it clear that he is seeking a fundamental change in development policies and practices. In a stance that resonates with Feenberg's democratic rationalization and Mol's ecological modernization, Khan is embracing technology as a powerful means to enhance societal development.

Stories of development schemes gone awry are, alas, common enough. Instead, Khan critically analyzes one of the "success" stories, Taiwan, and especially its top-down, modernist, national-innovation-system model of development. Even though the country succeeded beyond its planners' wildest dreams in the worldwide export of computer components, Khan nonetheless finds Taiwan lacking in a range of fundamental human capabilities. The corporate economy boomed while democracy, among other human capabilities, languished. The path-dependent paradox is that the country's "very success in exports may have forced the Taiwanese companies to seek a closure that largely excludes their domestic constituencies." In the end, Khan inquires into the conditions at the national, regional, and city levels that might bring about an alternative POLIS model of development that is "cognizant of the complex interactions among technology, economy and polity . . . [and] emphasizes the teleological desideratum of equalizing social capabilities as the end of development."

Still at Schiphol Airport, my fellow travelers have long gone to their gates. "OK, yes, I can see the point about technology and modernity now. I've been working here at Schiphol only a few months, and it is quite a place. Sounds like a nice workshop. Is there a chance you can send me the papers?" Yes, of course, I tell Claire. At long last she explains that she is a Ph.D. student in medieval history, and that she works part-time as a security guard in the airport to make ends meet. Working in both a premodern field of history and a thoroughly modern airport, she is making her own journey through the compelling tangle of modernity and technology. Finally I see the point of her questions. "Can you give me your address?" I ask. "I'll send you the essays. But I need to go now, my flight home is leaving soon." "Of course. Have a pleasant journey. Here's your passport."

Acknowledgments

I am grateful for specific comments on this essay from Mikael Hård, Nil Disco, Henk van den Belt, and the MIT Press reviewer of this volume. More generally, this essay serves as a long-term reflection on a Mumford-cluster seminar on technology and modernity at the University of Twente that I had the good fortune to co-organize with Johan Schot and Pieter Tijmes during the spring of 1997. I appreciate feedback on earlier versions from the Center for Science and Technology Studies at Trondheim, Norway, and the Humanities Colloquium at Illinois Institute of Technology.

Notes

1. For studies of technology and modernization processes directly inspired by Durkheim and Weber, respectively, see Fischer (1992) and Hård (1994). For

studies adopting a loosely theorized "agents of modernity" approach, see Rose (1995), Tobey (1996), and Kline (2000).

2. For expositions of modernity *as* change, see Berman (1982), Lash and Friedman (1993), and Charney (1998).

3. A parallel argument is made by Adas (1989), who observes that Europeans' perceptions of cultural superiority over African, Indian, and Chinese peoples were a product of the technical superiority they believed opened up in the course of industrialization.

4. Similar binary opposites figure prominently in recent discussions on the latest manifestation of "global modernity," i.e., globalization: Jihad and McWorld (Benjamin Barber), Lexus and olive tree (Thomas Friedman). Of course, one need not accept any "great divide" and the modernist assumptions it entails. In fields as diverse as science studies, history of technology, and the "new" (post-Chandler) business history, scholars in the past two decades or so have adopted a determinedly skeptical approach to the very core of the modernist paradigm: facts and rationality. The solid "facts" of science, technology, and capitalist business, it turns out, are not so solid and indeed are shot through with contingencies and compromises. For this reason, these scholars tend to reject (or ignore) any formulation (like Habermas's) separating system and lifeworld, science and society, rationality and practice. For examples of the "new" business history, see Scranton (1997) and Sabel and Zeitlin (1997). The field of technology studies is addressed later as well as by Brey and Feenberg in this volume.

5. Umberto Boccioni et al., "Manifesto of the Futurist Painters," at <www. futurism.org.uk/manifestos/manifesto 02.htm> (13 July 2002). On "technological fundamentalism," see Trommler (1995) and Todd (2001).

6. For criticism of social theorists' approaches to technology that are essentialist, reified, or deterministic, see Feenberg (chapter 3) and Brey (chapter 2). For social theories with a more interactive and fluid conception of technology and society, see Bourdieu's notion of *dispositif* and Giddens' notion of the duality of agency and structure: Bourdieu and Wacquant (1992) and Giddens (1979, 1984). I thank Mikael Hård for the latter suggestion.

7. For recent studies exploring the co-construction of technology and modern culture in a variety of settings, see Mayr (1986), Overy (1990), Nye (1990), Nolan (1994), Misa (1995), Edwards (1996), Alder (1997), Brooks (1997), Charney (1998), Hecht (1998), Schatzberg (1999), Gitelman (1999), Slaton (2001), and Allen (2001).

8. For the historical evolution and multifunctionality of Schiphol Airport, see Mom et al. (1999).

9. As if to underscore its role as an information processor, Chicago's O'Hare Airport just retired a signature artifact of the mid twentieth century, a three-bladed DC-6 propeller whose springiness and surface texture you could physically engage. The airport filled the space occupied by the propeller with a bank of pay-by-the-minute Internet-linked computer workstations.

10. See Edwards' essay in this volume.

11. Herbert Muschamp, "A Happy, Scary New Day for Design," *New York Times* (15 Oct. 2000). <www.nytimes.com/2000/10/15/arts/15MUSC.html> (17 Oct. 2000).

12. The modernity of cities, city life, and city planning, from St. Petersburg to New York, and from Brasília to Chandigarh, is a prominent theme of Berman (1982), Ward and Zunz (1997), Scott (1998), Driver and Gilbert (1999), and many other authors.

13. There are varied approaches to modernity in the seventeenth and eighteenth centuries; for science, see Whitney (1986), Toulmin (1990), and Iliffe (2000); on consumption, see Clunas (1999); and for the economy, see de Vries and Woude (1996).

14. Francis Bacon, *Novum Organum* (1620), aphorism 129, cited in Eisenstein (1983: p. 12).

15. On science, standards, and state formation, see Porter (1995), Wise (1995), Alder (1997), and Scott (1998).

16. Conceptions of modernity can be located much earlier in human history; see Hess's chapter in this volume.

17. Miles Ogborn, *Spaces of Modernity* (p. 10), quoted in Porter (2000: p. 488, note 10).

18. I consulted Bigelow's second edition of 1831 (Jacob Bigelow, *Elements of Technology*, Boston: Hilliard, Gray, Little and Wilkins, 1831; 2nd ed., p. 4.) First printed in 1829.

19. For discussions of Bigelow and "technology," see Segal (1985: pp. 74–97, quote on p. 81) and Oldenziel (1999: pp. 9–26). Oldenziel argues forcefully that technology took on its modern sense, as an abstract and gender-bound concept, only in the years after 1865. She cites (p. 195, note 8), for instance, the founding of institutes and colleges of "technology" e.g., Massachusetts (1861), Stevens (1870), Georgia (1885), Clarkson (1896), Carnegie-Mellon (1912), and California (1920).

20. I think this volume comes close to operationalizing Leo Marx's call (in an exchange with Mel Kranzberg in *Technology and Culture*, vol. 33 [1992]: 407), "Why not start with the intuitively compelling idea that technology may be *the* truly distinctive feature of modernity? . . . The aim would be to understand all of the ways that technological knowledge, processes, and behaviors in fact distinguish modernity from other ages—other societies and cultures."

21. For a recent evaluation of technological rationality by a well-informed historian of technology, see Constant (2000).

22. Heidegger, quoted in Feenberg (2000a: p. 297, note 3).

23. Discerning these varieties in technologies and exploiting their ambiguities for alternative social formations is the goal of Feenberg's "subversive" or "democratic" rationalization; see Feenberg (1995). Douglas Kellner (2000: p. 236)

also appreciates the pressing need for theoretical approaches that can discern "some of the more positive, but also more ambiguous and enduring features of modernity" and technology.

24. For analysis of technological determinism, see MacKenzie (1984), Sherwood (1985), Misa (1988), Adler (1990), Smith and Marx (1994), and Edgerton (1998). For a well-regarded exemplar attacking technological determinism, see Noble (1984). For Marx's "handmill" quote <www.marxists.org/archive/marx/works/1847/poverty-philosophy/ch02.htm#s2> (23 April 2002).

25. For apposite instances of Thomas Kuhn's incommensurability thesis, compare Russell (1986) with Pinch and Bijker (1986) and Winner (1993) with Pinch (1999). In criticizing social constructivism, Langdon Winner (2001: p. 15) states that "the scholarly community in STS is so inward looking that it seems not to notice the glaring disconnect between its own favored theories and the visions of run-away technology that prevail in society at large."

26. "Putting on boots" is how one of my Dutch philosopher colleagues refers to doing empirical work, which is something like wading through the muck in a cow barn; for his work, he prefers a book-lined study.

27. For historical critiques of modernization theory and development, see Adas (1989: pp. 402–418), Moon (1998), Scott (1998), and Engerman (2000).

28. When I wrote these lines some months before September 11, 2001, I had in mind such "failures" as lost baggage and missed connections. The multiple system failures evident on that day (airport security at Boston, Newark, and Dulles; the tracking systems that lost American Airlines flight 77 en route to the Pentagon; the faulty antihijacking transponders), which substantially contributed to the success of the attacks, have indeed forced the scrutiny of many technologies and practices taken for granted. Conversely, we have heard far less about the striking successes of the air traffic control system, which quickly and effectively shut down U.S. airspace in short order on that date, or the stairwells of the World Trade Center towers that enabled thousands to save themselves. Personally, I can no longer forgive the Futurists' architectural dictum that "the stairs—now useless—must be abolished."

29. See Edwards (chapter 7, this volume).

30. For this distinction I am indebted to Barb Marshall, who in her contribution to this volume tries to sort out postmodernism and poststructuralism.

31. On Benthamite reforms in London, see Hamlin (1998) and Linebaugh (1992: pp. 371–401).

32. For other suggestive modern and postmodern readings of technologies, see Rosen (1993) on the global bicycle industry, Duncombe (1997) on IBM and SONY, Marshall (chapter 4, this volume) on sexual technologies, and Lyon (chapter 6, this volume) on surveillance.

33. Feenberg (1995) suggests the notion of a "subversive" or democratic rationalization to encourage would-be reformers to engage rationalization processes, including technological change, and to strive to bend them toward

more democratic outcomes. Rationalization, for Feenberg, can favor dominant power structures and systems, but approached critically, rationalization processes can also enhance nondominant values.

34. The generous notion of "alternative modernities" (see Feenberg 1995; Lash and Friedman 1993; Lash 1999) is more complex than it first appears. While Feenberg uses "alternative" in the specific sense of *opposition to* the dominant (capitalist) rationalization processes (and in favor of a "subversive" rationalization), Lash uses "alternative" in the sense of *neither* the "high" modernists' embrace of the Enlightenment tradition *nor* the postmoderns' rejection of the Enlightenment (and in favor of a "reflexive" rationalization). Yet adopting a loose notion of "alternative" drains modernity of its emancipatory and universalizing potential; this has been a contentious issue within feminist scholarship, human-rights debates, and development thinking, as well as in environmental reform (see the chapters by Marshall and Mol in this volume). From a different angle, Harootunian (2000a: p. 163, note 4) firmly declares his opposition to "more fashionable descriptions, such as 'alternative modernities,' 'divergent modernities,' 'competing modernities,' and 'retroactive modernities,' that imply the existence of an 'original' that was formulated in the 'West' followed by a series of 'copies' and lesser inflections." He demonstrates (Harootunian 2000b) that intellectuals in interwar Japan, just as their western counterparts, wrestled with the cultural implications of modern life, including capitalism, cities, and industrialization.

All the same, modernity as a totalizing force should not be overdrawn. Historians studying the industrial revolution have piled up a huge literature on alternative paths to industrial revolution (Sabel and Zeitlin 1985, 1997). Essays in Hård and Jamison (1998) show empirically that in the formative years of the early and mid twentieth century, Europeans did not experience modernity as a single phenomenon; rather, each country absorbed and reinterpreted a global notion of modernity in a nation-specific tradition of discourse.

35. For a detailed treatment of the Rotterdam conflict, see van Lente (1998a,b).

36. On reflexivity and social learning in technology, see Rip et al. (1995: chaps. 2, 7–10).

37. Hess hints at the prescriptive sense of "normal" (rather than in the "mundane but deadly" sense used by organizational sociologist Charles Perrow [1984]). A similar point has been argued by John Staudenmaier (1985: p. 200): "A technological style can be defined as a set of congruent technologies that become 'normal' (accepted as ordinary and at the same time as normative) within a given culture. They are congruent in the sense that all of them embody the same set of overarching values within their various technical domains. For example, it can be argued that the United States, beginning with the U.S. Ordnance Department's 1816 commitment to the philosophical ideal of standardization and interchangeability, gradually adopted a set of normal technologies that incorporate that ideal. From this point of view many distinct technological developments—the machine tool tradition, the growth of standardized and centrally

controlled rail systems, the centralization and standardization of corporate research and development, the use of consumer advertising to program individual buying habits, the increasing centralization and complexity of electricity and communications networks, etc.—can be interpreted as participating in a single style, embodying a specific set of values within a specific world view."

38. For a loose formulation of "technology as freedom," see Tobey (1996). Sen (1999) provides a more rigorous formulation of "development as freedom."

I
Modernity Theory and Technology Studies

2

Theorizing Modernity and Technology

Philip Brey

The Need for Integrated Studies of Modernity and Technology

Technology made modernity possible. It has been the engine of modernity, shaping it and propelling it forward. The Renaissance was made possible by major fourteenth- and fifteenth-century inventions like the mechanical clock, the full-rigged ship, fixed-viewpoint perspective, global maps, and the printing press. The emergence of industrial society in the eighteenth century was the result of an industrial revolution that was made possible by technological innovations in metallurgy, chemical technology, and mechanical engineering. The recent emergence of an information society is also the product of a largely technological revolution, in information technology. Technology has catalyzed the transition to modernity and catalyzed major transitions within it. More than that, technologies are and continue to be an integral part of the infrastructure of modernity, being deeply implicated in its institutions, organizing and reorganizing the industrial system of production, the capitalist economic system, surveillance and military power; and shaping cultural symbols, categories, and practices (see Lyon and Edwards, chapters 6 and 7 in this volume).

If modernity is shaped by technology, then the converse also holds: technology is a creation of modernity. The common wisdom of technology studies, that technology is socially shaped or even socially constructed, that it is "society made durable," implies that a full understanding of modern technology and its evolution requires a conception of modernity within which modern technology can be explained as one of its products. If this holds for technology at large, it certainly also holds for particular technologies, technical artifacts, and systems. These are also products of modernity and bear the imprint, not only of the behaviors of actors immediately involved in their construction, but also of the larger sociocultural and economic conditions within which they are developed. To ignore this larger context is to leave out part of the story that can be told about that technology. It would be like staging Wagner's *Parsifal* with only the actors on stage, without any settings, costumes, or props.

In the current specialized academic landscape, modernity is the object of study of modernity theory, and technology is studied in technology studies. Few works exist that bridge these two fields and that study technology with extensive reference to modernity, or modernity with extensive reference to technology, or that concentrate on both by studying the way in which evolutions within modernity intersect with technological changes. In modernity theory, technology is often treated as a "black box" that is discussed, if at all, in abstract and often essentialist and technological determinist terms. In technology studies, the black box of technology is opened, and technologies and their development are studied in great empirical detail, yet technology studies generate their own black box, which is society. The larger sociocultural and economic context in which actors operate is either treated as a background phenomenon to which some hand-waving references are made, or it is not treated at all—a black box returned to sender, address unknown.

Undoubtedly, part of the reason that modernity theory has not adequately come to grips with technology has been the lack of empirically informed accounts of technology. It is only in the past few decades that major progress has been made in our understanding of technology and technological change, with the establishment of technology studies as a mature field of study. The same reason cannot be given for the lack of reference to modernity theory in technology studies because modernity theory has been around a lot longer than technology studies. Here, this lack of reference is more likely explained by the abstract and totalizing character of many theories of modernity; their often inadequate accounts of technology; the speculative, untested character of many of their claims; and the difficulty of connecting the microlevel concepts of technology studies to the macrolevel categories of modernity theory.

These criticisms do not apply equally to all theories of modernity. There is a world of difference between the abstract, totalizing theories of modernity of classical critical theory, Marxism, and phenomenology, and many recent theories of modernity, such as those of David Harvey and Manuel Castells, that are empirically rich and mindful of heterogeneity and difference. So if the sociocultural and economic context that is modernity ought to be considered in technology studies, then technology studies should work to appropriate more adequate theories of modernity, or start developing its own.[1]

It is time, then, to bridge the disciplinary gaps that now separate modernity theory and technology studies and to work at empirically informed and theoretically sophisticated accounts of technology, modernity, and their mutual shaping. In this essay, I contribute to this task through an analysis of the problems and misunderstanding that now beset modernity theory and technology studies in their respective treatment (or nontreatment) of technology and modernity.

A key conclusion is that the major obstacle to a future synthesis of modernity theory and technology studies is that technology studies mostly operate at the micro (and meso) level, whereas modernity theory operates at the macrolevel, and it is difficult to link the two. I analyze the micro-macro problem and ways in which it may be overcome in technology studies and modernity theory. The next two sections provide basic expositions of concepts, themes, and approaches in modernity theory and technology studies. Their aim is to introduce these fields to readers insufficiently familiar with them, as well as to set the stage for the analysis that follows.

Modernity Theory: Understanding the Modern Condition

Structure and Aims

Modernity is the historical condition that characterizes modern societies, cultures, and human agents. Theories of modernity aim to describe and analyze this historical condition. A distinction can be made between cultural and epistemological theories of modernity, most of which are found in the humanities, and institutional theories, which are common in social theory—although in both traditions many theories of modernity can be found that blend cultural, institutional, and epistemological aspects.

Cultural and epistemological theories of modernity focus on the distinction between premodern and modern cultural forms and modes of knowledge. These theories usually place the transition from traditional society to modernity in the Renaissance period, in fifteenth- and sixteenth-century Europe. The transition to modernity, in this conception, is characterized by the emergence of the notion of an autonomous subject, the transition from an organic to a mechanistic world picture, and the embrace of humanistic values and objective scientific inquiry. Some theories date the transition to modernity later than this, as late as the eighteenth century, during which Enlightenment thought had culminated in a genuine project of modernity, with universal pretensions to progress, and with fully developed conceptions of objective science, universal morality and law, and autonomous art (e.g., Habermas 1983). The cultural-epistemological approach to modernity dominates in philosophy, with Hegel, Nietzsche, and Heidegger as early proponents, and is also well represented in cultural history and cultural studies.

Many studies in the humanities that analyze modernity as a cultural phenomenon also focus on modernism, which is a phenomenon distinct from modernity. Modernism, or aesthetic modernism, as it is also called, was a cultural movement that began in the mid-nineteenth century as a reaction against the European realist tradition, in which works of art were intended to "mirror" external nature or society, without any additions or subtractions by the artist. Modernist artists, in often quite different ways, rejected this realism and held that it is the form of works of art, rather than their content, that guarantees authenticity and liberates art from tradition. Modernism has been very influential in literature, in the visual arts, and in architecture, with movements as diverse as naturalism, expressionism, surrealism, and functionalism being collected under it.

The emergence of modernism has often been explained by reference to major social transformations in nineteenth- and early twentieth-century modernity. David Harvey, for instance, has argued that modernism was a cultural response to a crisis in the experience of space and time, which was the result of processes of time-space compression under late nineteenth-century capitalism (Harvey 1989, chap. 8). The label "modernism" is also used in a broader sense, in which it does not refer to an aesthetic movement, but to the culture and ideology of modernity at large (e.g., Bell 1976). "Modernism," in this sense, stands for positivism, rationalism, the belief in linear progress and universal truth, the rational planning of ideal social orders, and the standardization of knowledge and production. When used in this latter sense, the notion of modernism becomes almost interchangeable with the notion of modernity construed as a cultural or epistemological condition (see Berman 1982).

Institutional theories of modernity focus on the social and institutional structure of modern societies, and tend to locate the transition to modernity in the eighteenth century, with the rise of industrial society in Europe. Institutional theories of modernity are as old as social theory itself, with early proponents like Weber, Marx, and Durkheim outlining key structural features of modern societies and theorizing major transitions from traditional to modern society. Modernity, in the institutional conception, is a mode of social life or organization rather than a cultural or epistemological condition. It is characterized by institutional structures and processes, such as industrialism, capitalism, rationalization, and reflexivity. It is with this institutional meaning of modernity that one can correlate the notion of modernization, which is the transformation of traditional societies into industrial societies. Modernity used to describe a condition that emerged in eighteenth-century European societies, but today it characterizes industrial societies around the globe.[2]

In my discussion of modernity theory, I give special emphasis to the social theory tradition, with the understanding that much of this work analyzes not only institutional aspects of modernity but cultural and epistemological dimensions as well. Indeed, it is quite common to see these aspects combined in social theories of modernity, even if institutional features receive the most emphasis. This blending of traditions has been particularly strong in critical theory, with authors like Habermas, Marcuse, and Adorno referring to Hegel and Heidegger as liberally as to Marx and Weber. However, it is also quite visible in more recent theories of modernity, such as those of Giddens, Harvey, Wagner, and Castells, as well as in the early institutional theories of modernity developed by Weber and Marx.

Theories of modernity in the social theory tradition present an account of the distinct structural features that characterize modern societies and the way these features came into being. Typically, they contain most or all of the following elements:

• They draw the boundaries of modernity as a historical period, contrasting it with a premodern period and sometimes also with a postmodern period.

• They describe and analyze the special features of modernity, with an emphasis on institutional, cultural, or epistemological dimensions. They almost invariably do this through macrolevel or "abstract" analysis. However, they may contain various elaborations, case studies, or illustrations of the macro theory.

• They (optionally) describe the dynamics of modernity, delineating (1) the historic changes that led to modern society, (2) various epochs within modernity (e.g., early, high, and late modernity; classical and reflexive modernity), and (3) the transitions between these epochs.

• Some theories of modernity also contain normative evaluations or critiques of the condition of modernity. Some propose visions of an alternative society or speculate how present modernity may transform itself into another type of social formation.

Next to grand theories of modernity, such as those of Marx, Weber, Habermas, and Giddens, one can find studies of particular eras within modernity, of major transitions and developments within the modern era, and of particular features or structures of modernity. Theories of particular eras within modernity attempt to characterize a particular historical epoch and to analyze the transitions that led to it (Wagner 1994). Many contemporary social theorists focus on late modernity as a historical epoch emerging in the second half of the twentieth century, and attempt to characterize its special features. Thus, one finds theories of "reflexive modernity" (Beck et al. 1994), "the risk society" (Beck 1992), "postindustrial society" (Bell 1976; Touraine 1971), "the information age" and "the information society" (Castells 1996; Schiller 1981), "the global age" (Albrow 1996), and many others. Akin to these theories, one finds theories of postmodernity, which hypothesize that we have already left (late) modernity and have recently entered a new postmodern era (e.g., Jameson 1991; Harvey 1989).

Besides theories of particular eras in modernity, there are many studies of major sociocultural, technological, or economic transitions within modernity. These range from studies of the scientific revolution and the industrial revolution to studies of the control revolution (a revolution in technologies of control that is claimed by Beniger [1989] to have paved the way for the information society) or the emergence of Fordism, to theories of the historical development of the modern subject and of new modern forms of power (e.g., Foucault 1977). Not all these works explicitly situate the developments they analyze within the wider context of modern social institutions and culture. Finally, one can find studies that are concerned with particular aspects or structures of modernity, such as modern identity (Lash and Friedman 1993; Giddens 1991), capitalism (Sayer 1991), pornography (Hunt 1993), consumer culture (Slater 1997), and gender (McGaw 1989; Marshall 1994; Felski 1995).

Not every work in social theory is a work in modernity theory. For it to qualify as such, it would have to be centrally concerned with major institutional, cultural, or epistemological aspects of or transformations within modernity, such as capitalism, the autonomous self, modern technology, and the Enlightenment. Alternatively, for phenomena that are not inherently tied to modernity or at least do not define it, such as pornography, adolescence, or the automobile, it would study these in relation to the larger institutional, cultural, and epistemological context of modernity. Thus, an analysis of adolescence would be a study in modernity theory if it explicitly considered the historical, cultural, and institutional constructions of adolescence in the modern era and changes in these constructions over time, but not if it treated adolescence in a largely ahistorical way (e.g., as a set of locally enacted constructions with little historical continuity), or if it studied its historical treatment in a particular country or setting without reference to its relation to modern social institutions and culture.[3]

Modernity and Social Theory

Theories of modernity have always held a prominent position in social theory. What follows is a brief review. Any such review will have to start with Karl Marx and Max Weber, who are often identified as the fathers of modernity theory. They are both known for their theories of the transition between feudal and industrial society, and their theories of (capitalist) industrial society. They are hence early proponents of institutional theories of modernity and of the transition of the premodern to the modern period.

In Marx's historical materialist conception of modernity, the difference between the modern and the premodern era is characterized by qualitative differences in the economic structure. The economic structure of a society is made up of production relations and it changes when the development of the productive forces (means of production and labor power) results in greater productive power. According to Marx, the transition from feudal to capitalist society was caused by large increases in productive power in feudal society. These increases caused changes in production relations, and hence in the economic structure. The resulting economic structure was capitalist in the late nineteenth century, but Marx of course envisioned a transition to a post-class socialist society, a transition that would occur when further increases in production power made a socialist state possible. He hence envisioned an early, capitalist, and a late, socialist state of modernity. Both are characterized by an industrial system of production, but their social form and culture are significantly different.

Weber (1958[1905]) did not see the transition from feudal to industrial society as caused by the development of productive power. Instead, he held that the capitalist economic system that made industrial society possible was an outgrowth of the Protestant work ethic, which demanded hard work and the accumulation of wealth. Because capitalism is profit based, it demanded *rationalization* so that results could be calculated and so that efficiency and effectiveness could be increased. In this way, rationalization became the distinguishing characteristic of modern industrial societies. The rationalization of society is the widespread acceptance of rules, efficiency, and practical results as the right way to approach human affairs and the construction of a social organization around this notion. According to Weber, rationalization has a dual face. On the one hand, it has enabled the liberation of humanity from traditional constraints and has led to increased reason and freedom. On the other hand, it has also produced a new oppression, the "iron cage" of modern bureaucratic organizational forms that limit human potential.[4]

Weber's notion of rationalization as the hallmark of modernity has been very influential in modernity theory. It has been particularly influential in critical theory, particularly with members of the Frankfurt school such as Adorno, Horkheimer, Marcuse, and Habermas, who built on Weberian notions as well as Marxist ideas in formulating their sweeping critiques of modern society (see, e.g., Marcuse 1964 and Horkheimer and Adorno 1972). Jürgen Habermas, without doubt the most influential scholar in the critical theory tradition, has advanced a theory of modernity with strong Weberian and Marxist influences, in which he analyzes modernity as an "unfinished project" (Habermas 1983). He theorizes an early phase of modernity and a later phase. Early modernity witnessed the rise of the "bourgeois public sphere," which mediated between the state and the public sphere. In late modernity, the state and private corporations took over vital functions of the public sphere, as a result of which the public sphere became a sphere of domination (Habermas 1989).

Although he is critical of late modernity, Habermas sees an emancipatory potential in early modernity, with its still-intact bourgeois public sphere. He hence sees modernity as an "unfinished project" and has attempted to redeem some elements of modernity (the Enlightenment ideal of a rational society, the modern differentiation of cultural spheres with autonomous criteria of value, the ideal of democracy) while criticizing others (the dominant role of scientific-technological rationality, the culture of experts and specialists). Central in this undertaking has been his distinction between two types of rationality: purposive or instrumental rationality, which is a means for exchange and control and which is based on a subject-object relationship, and communicative or social rationality, which is geared toward understanding and is based on a subject-subject relationship that is the basis for communicative action. Habermas claims that there has been a one-sided emphasis since the Enlightenment on instrumental, scientific-technological rationality, which has stifled possibilities for expression. The result has been a colonization of the lifeworld by an amalgamated system of economy and state, technology and science, that carries out its functional laws in all spheres of life. Habermas regards communicative action as a means to put boundaries on this system and to develop the lifeworld as a sphere of enlightened social integration and cultural expression.

Looking beyond critical theory, one cannot escape the powerful analysis of modernity in the work of Anthony Giddens (1990, 1991, 1994b). Giddens analyzes modernity as resting on four major institutions: industrialism, capitalism, surveillance, and military power. These and other institutions in modernity moreover exhibit an extreme dynamism and globalizing scope. To account for this dynamism, Giddens identifies three developments. The first is the separation of time and space, through new time- and space-organizing devices and techniques, from each other and from the contextual features of local places to which they were tied. Time and space become separate, empty parameters that can be used as structuring principles for large-scale social and technical systems. The second development is the disembedding of social life, the removal of social relations and institutions from local contexts by disembedding mechanisms, such as money, timetables, organization charts, and systems of expert knowledge. Disembedding mechanisms define social relations and guide social interactions without reference to the peculiarities of place. The third development is the reflexive appropriation of knowledge, which is the production of systematic knowledge about social life that is then reflexively applied to social activity. Jointly, these developments create a social dynamic of displacement, impersonality, and risk. These can be overcome through reembedding (the manufacture of familiarity), trust (in the reliability of disembedding mechanisms), and intimacy (the establishment of relationships of trust with others based on mutual processes of self-disclosure).

Risk, trust, and the reflexive appropriation of knowledge are also central themes in Ulrich Beck's theory of (late) modernity (Beck 1992). Beck distinguishes two stages of modernization, the first of which is simple modernization: the transformation of agrarian society into industrial society. The second stage, which began in the second half of the twentieth century, is that of reflexive modernization. This is a process in which modern society confronts itself with the negative consequences of (simple) modernization and moves from a conflict structure based on the distribution of goods to a model based on the distribution of risks. Our current society is the risk society, in which risks are manufactured by institutions and can be distributed in different ways. The distribution of risk occurs with major social transformations at the backdrop, transformations in which traditional social forms such as family and gender roles, which continued to play an important role in industrial society, are in the risk society undergoing radical change, leading to a progressive "individualization of inequality."

The idea that modernity has recently entered a new phase is pervasive in contemporary social theory, even among those authors that stop short of claiming that we have entered or are entering a phase of postmodernity. Intensifying globalization, the expansion and intensification of social reflexivity, the proliferation of nontraditional social forms, the fragmentation of authority, the fusion of political power and expertise, the transition to a post-Fordist economy that is no longer focused on mass production and consumption and in which the production of signs and spaces becomes paramount—all have been mentioned as recent developments that point to a new stage of reflexive or radicalized modernity (e.g., Lash and Urry 1994; Beck et al. 1994; Giddens 1990, 1994b; Albrow 1996; Lipietz 1987), with most authors identifying the late 1970s as a transition period. Many authors point specifically to the revolution in information technology in claiming that we have entered an information age (or, equivalently, a postindustrial age) in which an economy based on information, not goods, has become the organizing principle of society (e.g., Bell 1976; see Webster 1995 for an overview).

In the transition from an industrial to an information society, the economic system is transformed, and along with it the occupational structure, the structure of organizations, and social structure and culture at large. According to Manuel Castells, who has presented the most comprehensive theory of the information society to date, the basic unit of economic organization in the information age is the network, made up of subjects and organizations, and continually modified as networks adapt to their (market) environments. Castells argues that contemporary society is characterized by a bipolar opposition between the Net (the abstract universalism of global networks) and the Self (the strategies by which people try to affirm their identities), which is the source of new forms of social struggle (Castells 1996, 1997, 1998).

Modernity and Postmodernity

Not all scholars agree that modernity is still the condition that we are in. Theorists of postmodernity claim that we have recently entered an era of postmodernity, which follows modernity. Postmodernity is often considered, like modernity, to be a historical condition. Most theorists who consider postmodernity in this way place the transition from modern to postmodern society somewhere in the 1960s or 1970s, although some hold that we are still in the middle of a transition phase. They hold that changes in society over the past century accumulated during these decades to produce a society whose institutional, cultural, or epistemological condition is sufficiently different from that of modern society to warrant the new label.

Many postmodern theorists point only to cultural changes to support this claim. Some, however, emphasize technological and economic changes and see changes in cultural and social forms as resulting from them. David Harvey emphasizes the 1970s transition from a Fordist economy of mass production and consumption to a global post-Fordist regime characterized by greater product differentiation, intensified rates of technological and organizational innovation, and more flexible use of labor power (Harvey 1989). Frederick Jameson has theorized a transition to "late capitalism," which is global and in which all realms of personal and social life and spheres of knowledge are turned into commodities. He claims that late capitalism comes with its own cultural logic, which is postmodernism (Jameson 1991). Lash and Urry (1994) point to the shift from an economy of goods to an economy of signs and spaces, as does Jean Baudrillard (1995), who claims that information technology, mass media, and cybernetics have effected a transition from an era of industrial production to an era of simulation, in which models, signs, and codes determine new social orders. The culture of postmodernity is often characterized by consumerism, commodification, the simulation of knowledge and experience; the blurring if not disappearance of the distinction between representation and reality; and an orientation on the present that erases both past history and a sense of a significantly different future. The cultural shifts also include a decline in epistemic and political authority, the fragmentation of experience and personal identity, and the emergence of a disorienting postmodern hyperspace.

Not all postmodern theorists hold postmodernity to be a historical condition, however. For some, like Jean-François Lyotard, postmodernity is rather a cultural or epistemological form that is not essentially tied to a particular historical period. Lyotard holds that within contemporary society, one can find both modern and postmodern forms existing together.[5] The characteristic of postmoderns like Lyotard is that they resist the modern form. For Lyotard, modernity is equivalent to reason, the Enlightenment, totalizing and universalizing thought, and grand historical narratives. It is equivalent to what I identified earlier as modernism in a broad sense, that is, the culture and ideology of modernity. Lyotard criticizes the modern form of knowledge and calls for new kinds of knowledge that do not impose a grid on reality, but that emphasize difference. Lyotard's cultural critique is also a critique of scholarly method. He argues that postmodern scholars should not do theory. They are also not to produce new grand narratives of society, but should deconstruct and criticize modernist claims for universalistic knowledge by doing local, microlevel studies that emphasize heterogeneity and plurality (Lyotard 1984a). He rejects the old methodology of social theory, along with any and all of its theoretical claims. This call for a postmodernization of the social sciences and humanities has been echoed by Richard Rorty, Jacques Derrida, and Zygmunt Bauman, and can be seen in the profusion of postmodern case studies and analyses that uncover difference and heterogeneity and celebrate cultural "others."

Postmodern theorists thus range from writers like Jameson and Harvey, who study postmodernity as a historical era, to those like Lyotard and Rorty, who criticize modernist ideology and develop and employ postmodern methodologies for the humanities and social sciences. As a critique of modernist thought, postmodernism is moreover an intellectual orientation that is different from, even if it overlaps with, *aesthetic* postmodernism, which has emerged in literature, architecture, and the visual arts since the 1960s and 1970s as a response to aesthetic modernism. Critics of (academic) postmodernism, which include Habermas and Giddens, criticize both the hypothesized transition from modernity to postmodernity and the intellectual attitude of postmodern scholars. Giddens, for example, claims that in spite of the discontinuities cited by postmodernists, the major institutions of modernity as it existed in the nineteenth century and the early twentieth century—industrialism, capitalism, surveillance, and military power—are still in place, and he therefore only wants to go as far as to theorize a late or "radicalized" stage of modernity (Giddens 1990). He and Habermas have both criticized postmodernism's antitheoretical attitude, its epistemological and moral relativism, its irrationalism, and its laissez-faire attitude to politics (Giddens 1990; Habermas 1987). Similar debates exist within postmodern theory, with Harvey (1989) theorizing a transition to postmodernity while criticizing postmodernist thought, and Lyotard (1984b) criticizing Jameson's "totalizing dogmas" and defense of master narratives.[6]

Technology Studies: New Visions of Technology

Technology Studies as a Field

"Technology studies" is the name for a loosely knit multidisciplinary field with a wide variety of contributing disciplines, such as sociology, history, cultural studies, anthropology, policy studies, urban studies, and economics. Technology studies are concerned with the empirical study of the development of technical artifacts, systems, and techniques and their relation to society. Technology studies are part of science and technology studies, or STS, a larger field that emerged in the 1970s and that is based on studies of science and technology and their relation to society that are both empirically informed and on sound theoretical footing. STS is today an established discipline, with departments and programs around the world, as well as specialized conferences and journals.[7]

A full review of theories and approaches in technology studies is well beyond the scope of this paper and is complicated because of the relative youth of the field and the diversity of its topics and approaches. In what follows, I focus on two subfields of technology studies that are at the core of many STS departments and programs. They are *social studies of technology*, which look at social and cultural aspects of technology, and the *history of technology*, which studies the historical development of technologies and their relation to society.[8] In discussing the history of technology, moreover, I focus on contextual approaches, which are dominant in STS, and which look at the historical development of technologies in relation to their social context, instead of taking an internalist approach that focuses on purely scientific and technological contexts only.[9] This selective choice means that I ignore, among other studies, the important work that has been done in two distinct fields: economics of technology and philosophy of technology.[10]

Contemporary technology studies, with their focus on social, cultural, and historical dimensions, cover a wide variety of topics. Scholars rarely consider "Technology-with-a-capital-T." Instead, they examine specific technologies, such as genetic engineering or nuclear technology; specific engineering fields and approaches, such as mechanical engineering or cold fusion research; specific techniques, such as rapid prototyping or cerebral angiography; and technical artifacts, machines, materials, and built structures, such as ceramic vases, Van de Graaff generators, polystyrene, and the Eiffel Tower. In addition, many scholars study large technological systems, such as railroad systems or early warning systems in missile defense, and the processes of technological change, such as the development of the bicycle in the nineteenth century or the invention and development of electric lighting.

Technology studies analyze these technological entities in relation to their social context. Roughly, this is done in one of three ways. In one set of studies, the focus is on the shaping of the technology itself and the role of societal processes. How did the technology come into existence? What (social) factors played a role in this process? What modifications has it undergone since it first came into being, and why did these occur? In other studies, the focus is on how a technology has shaped society, or, alternatively, on the social changes that accompanied the introduction and use of the technology. In yet other studies, these processes are considered together, emphasizing how a technology and its social context co-evolve, or co-construct each other. A significant proportion of work that takes up this co-construction theme even denies that there is a meaningful distinction between technology and society, and attempts to study "sociotechnology," which consists of dynamic seamless webs of entities that are only labeled as technological or social after they have fully evolved (Bijker and Law 1992; Latour 1987; Callon 1987). There is also a fourth category of studies in technology, which historian John Staudenmaier (1985: p. 17) calls "externalist," that do not focus on technology per se but only on contextual aspects, such as engineer-ing communities, technological support networks, or public images of technology.

The core of contemporary technology studies consists of social studies of technology and the history of technology, both of which have been influenced by New Left critiques of science and technology. I discuss these two subfields in order. In social studies of technology, the research focus is on the social contexts in which technologies are developed and used, such as engineering labs, factories, and homes. The research examines how elements in these contexts interact with each other and with the technology in question. Such elements include individual agents and social groups, along with their behaviors, interactions, identities, and statuses (gender, race, class), as well as organizational structures, institutional settings, and cultural contexts.

Contemporary social studies of technology are in large part an outgrowth of social studies of science. The specific tradition of which it is an outgrowth is sometimes called sociology of scientific knowledge (SSK). The SSK approach to the sociology of science, which is the dominant approach today, holds that scientific knowledge itself, and not just the social and institutional context of scientific inquiry, ought to be the key focus of the sociology of science. SSK holds that scientific knowledge is not a rational process exempt from social influences, but a social process, and that scientific truth is not objectively given but socially constructed. This SSK approach deviates from what was the dominant approach in the sociology of science until the late 1970s: the Mertonian approach, named after Robert K. Merton, which focused only on the institutional context of scientific inquiry while assuming that scientific inquiry itself is by and large rational and objective. SSK also distinguishes itself from traditional (positivist) philosophy of science and epistemology, which also holds scientific inquiry and truth to be rational and objective. Instead, it takes its inspiration from philosopher of science Thomas Kuhn's work on the structure of scientific revolutions, which is critical of images of science as a rational and cumulative process (Kuhn 1962).[11]

It was a founding principle of SSK that "nature" and "rationality" and "truth" in science do not explain the process of scientific inquiry, but are themselves contingent social constructs that must be explained. This central principle was extended in the early 1980s, when some SSKers began to publish work in social studies of technology. The principle is modified to read: the working of machines does not provide an explanation of technological and social change, but is itself something that must be explained, at least in part by investigating social agents, their interactions, and their beliefs about technology.[12] Technology is regarded, in part or wholly, as a social construction that must be explained by reference to social processes, and within which no appeal can be made to objective standards of truth, efficiency or technological rationality.

Although some contemporary work in (contextual) history of technology finds inspiration in social studies of technology, the history of technology is itself a much older field (Cutcliffe and Post 1989; Westrum 1991; Staudenmaier 1985; Fox 1999). Yet, although there has always been an interest in the social context of technology in the history of technology, approaches that put this social context at center stage have only recently come to dominate. A typical study in a contextual history of technology considers how a particular technology, such as electric power transmission, the internal combustion engine, or the personal computer, evolved historically and how the technology came to reflect the contexts in which it has been developed and used. The investigation is often bounded in time (a particular historical era or development stage of the technology) and space (a particular geographical area or setting). Contextual elements that such historians consider may include organizational, policy, and legal settings, including relevant individual actors, social groups, and organizations (engineers, firms, industries, government bodies, activist groups) and their discourses and behaviors. In sociohistorical studies of technology, in which social studies of technology intersect with the history of technology, the development of technologies is studied with special reference to their social contexts and uses (see Bijker 1995b for a review).

Most studies in social studies and the history of technology are case studies that consider particular settings or events in which technologies are developed and used.[13] Others are what John Staudenmaier (1985: p. 206) calls "expanded studies," which look more broadly at several types of technologies or several types of settings or historical episodes. Yet other studies are primarily theoretical or methodological, focusing on such issues as technological determinism or the interpretive flexibility of technological artifacts, or on methodological issues within technology studies. Most studies operate at a micro or mesolevel of analysis, focusing on individual actors, social groups and organizations, and their interactions, rather than on the macrolevel of institutions and cultural frameworks. The research methods are diverse and include textual analysis, discourse analysis, participant observation, ethnomethodology, and quantitative analysis.

Theoretical Claims of Technology Studies

The strong empirical orientation of most work in social studies and the history of technology is visible, not only in its case analyses, but also in its theoretical and methodological assumptions, which have often been inspired by, or modified as a result of, these case studies. As a consequence of this, there has been a fair amount of agreement on a number of theoretical assumptions. I will try to characterize some of these assumptions, along with some others that are also salient but more controversial.

One of the most central theoretical assumptions in technology studies is the assumption that *technology is socially shaped*. Technological change is conditioned by social factors, and technological designs and functions are the outcome of social processes rather than of internal standards of scientific-technological rationality; technology is society made durable.[14] The social shaping thesis denies the technological determinist idea that technological change follows a fixed, linear path, which can be explained by reference to some inner technological "logic," or perhaps through economic laws. Instead, technological change is radically underdetermined by such constraint and involves technological controversies, disagreements, and difficulties that engage different actors or relevant social groups in strategies to shape technology according to their own insights.

Some scholars may discern technological or scientific constraints on technological change, but others point out that such constraints, if they exist at all, are themselves also socially shaped—for example, expectations of growth within the business, engineering, or user communities. Also, while some scholars recognize separate stages in the development of technology (e.g., invention, development, innovation), others, particularly in social studies of technology, analyze technological change as an entirely contingent and messy process, in which heterogeneous factors affect technological outcomes, and in which the process of invention continues after technologies leave the laboratory or factory. These scholars emphasize that users, regulators, and others also affect the design and operation of technologies and the way in which technologies are interpreted and used (Bijker 1992; Lie and Sørenson 1996; Oudshoorn and Pinch forthcoming). In contrast to a linear-path model of technological change, proposals have been made for a variation and selection model, according to which technological change is multidirectional: there are always multiple varieties of particular design concepts, of which some die, and others, which have a good fit with social context, survive (e.g., Pinch and Bijker 1987; Ziman 2000).

The social-shaping thesis implies a weak constructivist claim that technological configurations are variable and strongly conditioned by social factors. Social constructivist approaches go beyond this claim to arrive at the strong constructivist claim that technological change can be entirely analyzed as the result of processes of social negotiation and interpretation, and that the properties of technologies are not objective, but are effectively read into the technologies by social groups. Social constructivism is hence a contemporary form of idealism, denying the possibility or desirability of a reference to any "real" structures or forces beyond the representations of social groups. Whether a certain technology works or is efficient or user-friendly, and the nature of its functions, powers, and effects is not a pregiven, but the outcome of social processes or negotiation and interpretation.[15]

Those social-shaping theorists who do not embrace social constructivism also recognize that the meaning or use of technologies is not pregiven. Most theorists agree that technology has interpretive flexibility, meaning that technologies can be interpreted and used in different ways (Pinch and Bijker 1987). When social negotiations surrounding technological change come to a close, interpretive flexibility is held to diminish because the technology stabilizes, along with concomitant (co-produced) meanings and social relations. Stabilization implies the embedding of the technology in a stable network consisting of humans and other technologies, and the acceptance of a dominant view on how to interpret and use the technology. Stabilization of a technology implies that its contents are "black-boxed" and are no longer a subject of controversy. Its stabilized properties come to determine the way that the technology functions in society. Yet, black boxes can be reopened. The history of technology shows how technologies such as the telephone, the Internet, or the automobile take on particular functions or societal roles that may vary from time to time and place to place.

The flip side of the claim that technology is socially shaped is the claim that *society is technologically shaped*, meaning that technologies shape their social contexts. This goes considerably beyond the claim that new technologies may open up new possibilities that change society, or that technologies may have side effects. Obviously, the steam engine changed society by making new types of industrial production possible, and the printing press effected change by making written information more available and easier to distribute. Obviously, also, technologies may have side effects such as environmental pollution or unemployment. The technological-shaping thesis refers not just to such recognized functions and side effects of technologies, but to the multiplicity of functions, meanings, and effects that always, often quite subtly, accompany the use of a technology. Technologies become part of the fabric of society, part of its social structure and culture, transforming it in the process. The idea of society as a network of social relations is false, because society is made up of sociotechnical networks, consisting of arrangements of linked human and nonhuman actors.

The notion of a sociotechnical network is a central notion in *actor-network theory* (ANT), which is a third influential approach to technology studies, next to the social-shaping and social-construction approaches. It studies the stabilization processes of technical and scientific objects as these result from the building of actor networks, which are networks of human actors and natural and technical phenomena. Actor-network theorists employ a principle of generalized symmetry, according to which any element (social, natural, or technical) in a heterogeneous network of entities that participate in the stabilization of a technology has a similar explanatory role (Callon 1987; Latour 1987; Callon and Latour 1992). Social constructivism is criticized by ANT for giving special preference to social elements, such as social groups and interpretation processes, on which its explanations are based, whereas natural or technical elements, such as natural forces and technical devices, are prohibited from being explanatory elements. Actor-network theory allows technical devices and natural forces to be actors (or "actants") in networks through which technical or scientific objects are stabilized. By an analysis of actor networks, any entity can be shown to be a post hoc construction, but entities are not normally *socially* constructed because stabilization is not the result of social factors alone.

The notion that society is technologically shaped means, according to most scholars in technology studies, that technology seriously affects social roles and relations; political arrangements; organizational structures; and cultural beliefs, symbols, and experiences. Technology scholars have claimed that technical artifacts sometimes have built-in political consequences (Winner 1980), that they may contain gender biases (Wajcman 1991; Bray 1997), that they may subtly guide the behavior of their users (Sclove 1995; Latour 1992), that they may presuppose certain types of users and may fail to accommodate nonstandard users (Akrich 1992) and that they may modify fundamental cultural categories used in human thought (Turkle 1984, 1995).

Latour (1992), for example, discusses how mundane artifacts, such as seat belts and hotel keys, may direct their users toward certain behaviors. Hotel keys in Europe often have heavy weights attached to compel hotel guests to bring their key to the reception desk upon leaving their room. Winner (1980) argues that nuclear power plants require centralized, hierarchical managerial control for their proper operation. They cannot be safely operated in an egalitarian manner, unlike, for example, solar energy technology. In this way, nuclear plants shape society by requiring a particular mode of social organization for their operation. Sclove (1995) points out that modern sofas with two or three separate seat cushions define distinct personal spaces, and thus work to both respect and perpetuate the emphasis of modern western culture on individuality and privacy, in contrast to, for example, Japanese futon sofa-beds. Finally, Turkle (1984) discusses how computers and computer-operated

toys affect conceptions of life. Because computer toys are capable of behaviors that inanimate objects are not normally capable of, they lead children to reassess the traditional dividing lines between "alive" and "not
alive" and hence to develop a different concept of "alive." Most authors
would not want to claim that technologies have inherent power to effect
such changes. Rather, it is technologies in use, technologies that are already embedded in a social context and that have been assigned an interpretation, that may generate such consequences.

To conclude, the major insights of technology studies have been that
technologies are socially shaped and at the same time society is shaped
by technology, or, alternatively, that society and technology co-construct
each other. They are not separate structures or forces, but are deeply interwoven. Moreover, technological change is not a linear process but
proceeds by variation and selection, and technologies have interpretive
flexibility, implying that their meanings and functions and even (according to social constructivists) their contents are continually open to rene-
gotiation by users and others.

Technology Studies and Modernity Theory: Mutual Criticism

The Treatment of Technology in Modernity Theory

It is difficult to overlook the pervasive role of technology in the making
of modernity. As argued earlier, technology is a central means by which
modernity is made possible. It is a catalyst for change and a necessary
condition for the functioning of modern institutions. However, it is
more than that. What can be learned from technology studies is that the
institutions and culture of modernity are not just shaped or influenced
by technology, they are also formed by it. The social systems of modernity are sociotechnical systems, with technology an integral part of the
workings of social institutions. Social institutions are societal structures
that regulate and coordinate behavior and in this way determine how
certain societal needs are met. In the modern age, however, their regulative functions are no longer a direct outcome of collective actions, since
most collective actions have become thoroughly mediated and shaped by
modern technologies, which function as co-actors. For example, collective acts of voting are now thoroughly mediated by voting technologies

that help determine whether people get to vote at all, how votes are defined, and whether votes are counted. Modern culture is, likewise, a technological culture, in which technologies are not just material substrates of existing cultural patterns, but also have a major role in defining, shaping, and transforming cultural forms. Information technology, for example, is transforming basic cultural concepts and experiences such as those of time, space, reality, privacy, and community and is also effecting fundamental shifts in cultural practices.

If this analysis of the role of technology in modernity is anywhere near correct, then it is surprising, to say the least, to find that technology is not a central topic in the vast literature in modernity theory. Indeed, of the many hundreds of books that bear the word "modernity" in the title, fewer than a handful also refer to technology or one of its major synonyms or metonyms (e.g., technological, computers, biotechnology, industrial).[16] Many of the major works in modernity theory make only passing reference to technology. For example, technology is referenced only once in the recent edited volume, *Theories of Modernity and Postmodernity*; it is not mentioned at all in Zygmunt Bauman's *Intimations of Postmodernity*; and there are only four or five brief references to it in Alain Touraine's *Critique of Modernity* (see Turner 1990; Bauman 1992; Touraine 1995).

What can explain this apparent neglect of technology in modernity theory? It is not denial that technology has an important role in the constitution of modernity, for most authors would agree that its role is pivotal. A better explanation is that the dominant dimensions along which modernity has traditionally been analyzed (institutional, cultural, and epistemological) have not allowed technology to play a major identifiable role, but have instead assigned it the status of a background condition. Technology is often analyzed as a mere catalyst of institutional, cultural, and epistemological change, or as a mere means through which institutions, cultural forms, and knowledge structures are realized.

In institutional analyses, modernity is analyzed as being constituted by institutions and their transformations. Technology is not usually recognized as an institution itself; it is not seen as a separate regulative framework such as capitalism, government, or the family, but rather as one of the means through which these frameworks operate. More often than not, institutions such as capitalism, industrialism, or military power are discussed without specific reference to the technologies that sustain them. The role of technology in transforming these institutions (e.g., in the transition to an information society) is more difficult to ignore. However, here one often finds technology subsumed as part of a broader phenomenon, such as rationalization (Weber), productive forces (Marx), or disembedding mechanisms (Giddens), of which it is only a part. Even in Marxist theory, which assigns an important role to production technology in the making of modernity, this technology still only serves as an external constraint on economic structure, which ultimately determines the social forms of society.

In most cultural and epistemological theories of modernity, technology is either analyzed as a mere catalyst of cultural and epistemological changes, or it is robbed of its materiality and reduced to knowledge, language, or ideas. In Heidegger's critique of modernity, in which technology "enframes" us and turns the world into "standing reserves," technology turns out not to be defined as a material process or as a mode of action, but as a particular mode of thinking (Heidegger 1977). The same idealism is also visible in much of critical theory, in spite of its greater emphasis on social institutions. There, technology is often identified with technological or formal rationality, which is a mode of thinking that characterizes not only modern technology but also modern thought and economic and social processes. Habermas, moreover, has defined technology as "technological knowledge and ideas about technology" (Habermas 1987: p. 228). Finally, in postmodern theory, technology is often reduced to language, signs, or modes of knowledge, along with everything else.

When technology is referred to in modernity theory without being reduced to something else, still other problems emerge. One is the level of abstraction at which technology is discussed: technology is usually treated as a monolith, as a macroscopic entity, Technology-with-a-capital-T, about which broad generalizations are made that are supposed to apply equally to nuclear technology and dental technology, to vacuum cleaners and gene splicers. This abstract, undifferentiated treatment leads to vagueness, obscures differences between technologies, and fails to distinguish the varied ingredients that make up technology (knowledge, artifacts, systems, actions) and the way these relate to their context.

Giddens, for example, employs the notion of an "expert system," which is a key mechanism for decontextualizing social relations. He defines expert systems as "systems of technical accomplishment or professional expertise that organize large areas of the material and social environments in which we live today" (Giddens 1990: p. 27). He discusses few examples of expert systems, but makes it clear that virtually any system in which the knowledge of experts is integrated and that contains relevant safety measures qualifies as an expert system, including automobiles, intersections, buildings, and railroad systems. Moreover, Giddens goes into hardly any detail on the way in which expert systems decontextualize social relations.

A monolithic treatment of technology easily leads to essentialism and reification. In an essentialist conception, technology has fixed, context-independent properties that apply to all technologies. As Andrew Feenberg (1999a: pp. viii–ix) has argued, technological essentialism usually construes technology's essence as its instrumental rationality and its functionalism, which reduces everything to functions and raw materials. This essentialism often correlates with a reified conception of technology, according to which it is a "thing," with static properties, that interacts with other "things," such as the economy and the state. Essentialism and reification, in turn, have a tendency to promote technological determinism, in which technology develops according to an internal logic, uninfluenced by social factors, and operates as an autonomous force in society, generating social consequences that are unavoidable.[17] Technological determinism is evident in dystopian critiques of modernity, such as those of Heidegger, Marcuse, and Ellul, in which technology engulfs humanity and rationalizes society and culture. In many other theories of modernity, it is also present, albeit in a more subtle way. Marx's thesis that the productive forces determine or constrain production relations has often been interpreted as a form of technological determinism. Daniel Bell (1976) presents a similar view in characterizing the transition to a postindustrial society as the result of economic changes that are due to increased productivity, which is conditioned by information technology. Baudrillard (1995) construes the transition from modernity to postmodernity in technological determinist terms by analyzing it as the result of information technology and media, whose models and codes yield a new social order. James Beniger's (1989) detailed historical study of the making of the information society is also built on technological determinist principles, with technological change being a cause of social change, while itself remaining relatively independent of social influences.

In conclusion, the treatment of technology in modernity theory is problematic in several respects. Often, technology is not assigned a major role in modernity; it is subsumed under broader or narrower phenomena or one-dimensional phenomena; its treatment is often abstract, leading to vagueness, overgeneralization, detachment from context, and a failure to discern detailed mechanisms of change. In addition, technology is often reified and essentialized, and the conceptions of it are often deterministic. There is also the problem that modernity theory's sweeping generalizations about technologies do not normally rest on micro-level elaborations of the macro theory or on case studies. Modernity theory's generalizations, it will be clear by now, tend to go against many key ideas of technology studies (the social character of technology and its interpretive flexibility, the path dependence of technological change, etc.). Moreover, when theories of modernity provide inadequate accounts of technology and its role in modernity, their accounts of social institutions, culture, and the dynamics of modernity suffer as well. There are theories that avoid many of the problems listed (e.g., Castells 1996), but they are exceptions to the rule.

The Treatment of Modernity in Technology Studies

Modernity theory must provide an account of technology because of its major role in the shaping of modernity. Technology studies, on the contrary, do not seem to require a consideration of modernity in their analyses of technology. It is not obvious that a historical study of the telephone or an analysis of the development and advertisement of fluorescent lighting must refer to macroscopic structures and events such as disembedding mechanisms and changes in capitalist production modes. And in fact, most work in technology studies does not refer to such macro structures but instead remains at the micro (or meso) level. It studies actors (individuals, social groups, organizational units); their values, beliefs, and interests; their relations and (inter)actions; and the way in which these shape or are shaped by specific technologies. Case studies and extended studies based on this approach contain rich descriptions of complex dynamics that lead to social and technological outcomes. However, the aim of many of these studies is not just to describe what happens, but also to explain why it happened. For example, in analyzing the history of the Penny Farthing bicycle, Pinch and Bijker (1987: p. 24) do not want to only describe various bicycle models and the social groups involved in their manufacture and use; they want to understand the factors that determine what models are successful and the reasons social groups assign certain meanings to a model. I argue that microlevel accounts cannot fully explain technological and social change unless they are linked with macrolevel accounts.

The main reason for this is that a sufficiently rich account of actors and their relationships, beliefs, and behaviors requires an analysis of the wider sociocultural and economic context in which these actors are operating. This broader analysis is needed to explain why actors have certain attitudes, values, beliefs, or relationships, and it may even be necessary to infer their very existence. For example, an understanding of why certain types of men were attracted to high-wheeled bicycles in late nineteenth-century England, and perhaps also the identification of social groups with this attraction, is likely to require an account of masculine culture in late nineteenth-century England. The failure to look at this cultural context would result in superficial and possibly also unreliable descriptions of actors. More generally, to base explanations of technological and social change merely on observations of actors and their behaviors would be to subscribe to a form of methodological individualism, a questionable form of reductionism that holds that social explanations can be reduced to facts about individuals and hence that no reference to supraindividual social structures is required (Lukes 1994).

Granted, the actors in technology studies also include more complex entities, such as social groups and organizations, and nonhuman actors such as machines, but these are still particular actors to which agency is attributed, frequently along with beliefs and attitudes. If the actions, beliefs, and attitudes of these actors are not related to wider sociocultural contexts, then any explanation is likely to fall short. This is a recurring problem in most approaches in technology studies that emphasize an actor perspective, including social-shaping and social-constructivist approaches and the actor-network approach of Bruno Latour, Michel Callon, John Law, and their associates. This latter approach does relate the properties of individual actors to a wider context, which is the network of actors in which they are operating, and holds that this network defines these properties. However, the networks are limited in scope, usually containing only the actors thought to have a direct role in the development or functioning of a particular technology. Actor-network studies rarely provide sufficiently complete accounts of the networks that shape the behaviors or attitudes of the other actors in the network (e.g., engineers, corporations, or politicians), who therefore tend to be analyzed in a methodological individualist way.[18]

There is also another reason microlevel approaches have only limited explanatory power. As Paul Edwards points out (chapter 7, this volume), a major distinguishing feature of modern societies is their reliance on infrastructures, large sociotechnical systems such as information and communications networks, energy infrastructures, and banking and finance institutions. As Edwards argues, these infrastructures mediate among the actors that are studied in microlevel analysis. In this sense they function as disembedding mechanisms, defining social relations and guiding social interactions over large distances of time and space (Brey 1998). However, these infrastructures themselves are best studied at the macrolevel. Microlevel approaches that ignore infrastructures run the risk of providing an insufficient account of the relations between actors in modernity (whereas accounts of social relations in premodern societies can more easily remain at the microlevel because they are not usually mediated by infrastructures). The recent transition to a post-Fordist, global economy has heightened the inadequacy of microlevel analyses by fragmenting industrial production and marketing and reorganizing it on a global scale (Rosen 1993).

Social constructivists, while acknowledging the need to consider the societal context in which actors operate, have sometimes objected to an appeal to social theory because of its "realism," which would be incompatible with (strong) social constructivism (see, e.g., Pickering 1995; Elam 1994) However, there is no inconsistency in invoking categories of social theory in social constructivist analyses. Social constructivist explanations proceed by deconstructing entities in terms of the activity of other entities, specifically social groups. These entities are often not deconstructed themselves for pragmatic reasons because deconstruction has to stop somewhere. For instance, Bijker's (1992) social constructivist analysis of fluorescent lamps refers to the involvement of General Electric as a "real" entity. As Bijker (1993) later explained, his primary interest had been the social construction of fluorescent lamps, and not the social construction of General Electric. Because of this specific interest, it was excusable to present some parts of the sociotechnical world as fixed and as undeconstructed entities that function in the explanation of the development of fluorescent lamps, even though these entities are social constructions as well. But if reference can be made to General Electric in social explanation, then surely reference can be make to Fordism, disembedding mechanisms, and other socially constructed entities of social theory.[19]

Another criticism of modernity theories, and a reason cited for avoiding them, is their alleged tendency to totalization, universalization, functionalism, rationalism, panopticism, and determinism, not just in their treatment of technology, but in their treatment of society as a whole. This mirrors the criticism by postmodernists of macrolevel metanarratives. Tom Misa has argued, for instance, that macrolevel theories tend to "impute rationality on actors' behalfs or posit functionality for their actions, and to be order-driven," and that these tendencies quickly lead to "technological, economic or ecological determinism." Microlevel studies, instead, focus on "historical contingency and variety of experience" and are "disorder-respecting" (Misa 1994: p. 119). While the former tendencies are clearly visible in the majority of theories of modernity, I hold that they are not inherent to macrotheorizing. The macrostructures postulated in macrotheories inevitably impose constraints on the actions of individuals, but this does not mean that they must also determine these actions. Moreover, macrostructures can be defined as contingent, heterogeneous, and context dependent, such as Castells' networks.

A final objection to macrotheories is that they are often speculative and not elaborated or tested empirically. While there are good exceptions (again, Castells), these virtuoso performances confirm the general rule. My point is that this category of theorizing should not be rejected, but instead that these theories should be developed, tested, and refined. I conclude, tentatively, that there are no good reasons for scholars in technology studies to avoid macrotheories of modernity and that there are good reasons to employ them. Working toward integrated studies of modernity and technology involves, then, developing and testing macrotheories and working to bridge the micro-macro gap that now often separates modernity theory from technology studies. These two tasks are the topic of the next section.

Modernity, Technology, and Micro-Macro Linkages

The Problem of Micro and Macro

In large part, the problem of connecting the topics of modernity and technology, and of connecting modernity theory with technology studies, is the problem of connecting the macro with the micro. Modernity theory typically employs a macrolevel of analysis, analyzing macrolevel phenomena, such as late modernity and globalization, in terms of other macrolevel phenomena, such as time-space disembedding and the gradual decline in Western global hegemony. Much work in technology studies operates at the micro level, analyzing microlevel entities such as fluorescent lighting or the advertising of a new daylight fluorescent lamp by reference to other microlevel entities such as "the influence of Ward Harrison of the incandescent lamp department of General Electric" or "the writing of a report on daylight lighting by the Electrical Testing Committees." In addition, one could claim that modernity theory typically employs a structure perspective, focusing on social structures and their properties, whereas technology studies often employ an actor perspective.

I assume that there is a mutual need in technology studies and modernity theory to bridge the gap between the micro and the macro, and between structure and actor perspectives. Nevertheless, the problem of micro and macro (not to mention the problem of structure and agency) remains one of the great unsolved problems in social science. In spite of the attention this problem has generated, there is still no recipe, no method, and few inspiring exemplars on how to connect macrolevel and microlevel analyses. In the discussion that follows, I try to advance this general issue by looking more analytically at the problem. I argue that progress on the micro-macro problem has been hampered by a failure to recognize the multiplicity of levels of analysis (micro and macro being coarse distinctions only) and a failure to distinguish two distinct dimensions within the micro-macro distinction: size and level of abstraction. I then outline four principal ways in which levels of analysis may map onto each other, and conclude by drawing implications for an integration of modernity theory with technology studies.

Size and Level of Abstraction

What makes a phenomenon studied in the social sciences or humanities a macro phenomenon? And what makes a concept a microlevel concept? Considerable confusion exists over this matter. Sometimes it is held that macroanalysis is distinct from microanalysis because it focuses on larger things. Social systems are large and individuals and their actions are small; therefore social systems are the subject of macroanalysis and individuals the subject of microanalysis. Another claim sometimes made about the micro-macro distinction is that macrolevel phenomena and the concepts that refer to them are abstract and general, whereas microlevel phenomena tend to be concrete and specific.

Thus there are at least two parameters along which macroanalyses are distinguished from microanalyses: the size of the units of analysis, and their level of abstraction.[20] Very few attempts exist in the literature to further define or operationalize these parameters, or to study their interrelationships. It is usually assumed that they tend to interrelate; that the units of macrolevel analysis are typically, if not invariably, large, abstract, and general, whereas things in microlevel analysis tend to be small, concrete, and specific. Yet there are many exceptions to this rule. For example, the modern self is both smaller and more abstract than protest marches during the inauguration of George W. Bush. The modern self is a smaller unit of analysis because protest marches involve many modern selves. It is more abstract because it refers to a general type of phenomenon, whereas the protest marches denote a specific type of phenomenon. Other units of analysis do not seem to have a definite size at all. For example, reflexivity is a property that can apply to both large things (e.g., social systems) and small things (e.g., the knowledge processes of organizational units).

To understand the connections between the micro and the macro, we must therefore first better understand the parameters by which these notions are defined—the notions of level of abstraction and size. I discuss these in order.

Level of Abstraction   What does it mean to say that one phenomenon is more abstract than another? Principally, I want to argue, this means that the phenomenon is more general. For example, rationalization is a more general process than the standardization of testing in aviation schools (a form of rationalization), and that is why it is more abstract. Starting from an abstract phenomenon, one can arrive at more concrete phenomena by introducing additional properties that bound it. Starting with the abstract phenomenon of industrial society, one can arrive at the more specific and therefore more concrete phenomenon of late nineteenth-century British industrial society by adding properties that specify time period and nationality. Likewise, starting from the notion of a parent, one can arrive at the somewhat more concrete notion of a mother (a female parent) by adding a gender property (female). Conversely, when one starts with a concrete phenomenon, one can arrive at a more abstract one by removing properties from it. One can go from late nineteenth-century British industrial society to industrial society and from mother to parent by subtracting properties, that is, by generalizing.

In this way it is possible to construct hierarchies of entities that range from abstract to concrete, with the more concrete entities being species (subtypes or instances) of the more abstract entities. For example, one can construct a hierarchy going from transportation vehicle to bicycle to Penny Farthing bicycle to the Bayliss-Thomson Ordinary Penny Farthing bicycle to the specific Bayliss-Thomson Ordinary Penny Farthing bicycle of which I have a picture. Notice, however, that concretization is not just a matter of adding adjectives (or abstraction a matter of subtracting them). The relation between more abstract and more concrete phenomena is not always linguistically transparent, and conceptual analysis, if not empirical investigation, may be needed to observe the relation (e.g., that a mother is a type of parent, or that a standard-setting body in health care is a type of bureaucratic organization).

Size Units of analysis can often be ordered according to their size. For example, a social system is obviously larger than a social group in that system, and a social group is larger than an individual in that group. The reference to size here does not imply a reference to absolute metric or numerical properties. Rather, size is used here in a relative or comparative sense. Phenomenon *a* is larger than phenomenon *b* if *a* can contain *b*, or *b* is a part of *a*. For example, there are part-whole relations between the economy and the individuals participating in it because an analysis of economic processes ultimately reveals individuals engaged in economic behavior. This is why economic systems are larger than individuals. Large units of analysis are larger than small units of analysis because they are able to stand in a part-whole relation or a relation of (partial) containment to these smaller units (Castells 1996: pp. 174–179). Because parts may have parts themselves, hierarchies can be constructed of units of analysis that range from large to small. For example, a social system may include a market system that includes organizations that include organizational units that include divisions that include employees. Likewise, the British railway system includes train stations that include platforms and station staff. Also, items may be parts of multiple wholes. For example, pay-per-view virtual museums may be part of the post-Fordist economy, but they are also part of postmodern culture. Notice that part-whole relations between units of analysis, which refer to their size, are clearly different from the types of genus-species relations discussed earlier, which refer to level of abstraction. For instance, Internet advertising is a species of advertising, but a part of the post-Fordist economy.

Levels of Analysis and Their Interrelationships
What the distinction between size and level of abstraction shows is that the micro-macro distinction encompasses at least two hierarchies: a hierarchy from abstract to concrete and one from large to small. Things can be simultaneously small and abstract (the modern self) or large and

concrete (the locations of capital cities across the globe in the year 2001). In practice, however, there are correlations between these two hierarchies. What should also be clear from the discussion is that the distinction between two levels of analysis (macro and micro) or even three levels (macro, meso, and micro) is a gross oversimplification. Going from abstract to concrete or from large to small, many levels may be encountered in between. So what is commonly called the "macrolevel" in fact relates to multiple levels of analysis that may range from very large or abstract phenomena such as modernity, western culture, and industrial society to significantly smaller or more concrete entities such as the Internet economy, gender in late nineteenth-century France, and the Kansai region in Japan. Similarly, microlevel phenomena may range from larger and more abstract entities such as advertising agencies, hackers and local area networks, to smaller or more concrete entities such as Bill Gates, Mary's filing of a petition, and the software error in Fred's computer.

The terminology of micro and macro is therefore too coarse because it does not distinguish between size and level of abstraction, and it does not discriminate the different levels and hierarchies that exist within macro- and microlevel analyses. The consequence of this is that it becomes difficult to see how various kinds of micro- and macrolevel analyses may be integrated with each other. Yet, arriving at an adequate integration of levels of analysis is the major problem faced by theories of modernity and technology. How do you get from a discussion of late modernity, rationalization, and the state to a consideration of the development and use of specific technologies? Conversely, how do you get from talk about Pentium computers, hacker culture, and virtual communities to talk about globalization, the modern self, and post-Fordist economies? Any adequate study of modernity and technology requires such a bridging of the micro and the macro, of the abstract and general and the concrete and empirical, of the large and diffuse and the small and singular.

The major question, then, for theories of modernity and technology, is how the gaps that exist between different levels of analysis can be bridged. My distinction between size and level of abstraction indicates that gaps between levels occur in two ways: because the higher level refers to more abstract phenomena than the lower level (e.g., bureaucratic organizations versus standard-setting bodies in health care), or because it refers to larger phenomena (e.g., social systems versus markets). But my discussion also suggests how these gaps may be bridged: by identifying genus-species relationships (for phenomena at different levels of abstraction) and part-whole relationships (for phenomena of different sizes). For instance, an analysis of standard-setting bodies in health care may be linked to an analysis of bureaucratic organizations by identifying standard-setting bodies as species or instances of bureaucratic organizations. Similarly, an analysis of markets may be linked to an analysis of social systems by identifying markets as subunits within social systems. Such matches provide the conceptual links that are necessary to connect discourses that would otherwise remain disconnected.

However, in most studies in the social sciences and humanities that involve the linking of levels of analysis, the aim of such linking is not merely to connect disparate discourses. Most studies have a more specific aim; for instance, explaining events at the micro level, or analyzing the structure of macrolevel phenomena. Most studies center on a specific macro- or microlevel phenomenon that the study aims to analyze (e.g., late industrial society, or changes in the design of the bicycle in the late twentieth century). Links created to levels of analyses that are either higher or lower than that of the analysandum are hence asymmetrical: the higher- or lower-level entities are invoked to explain or analyze the analysandum.

Four Types of Interlevel Analysis

When something is analyzed in terms of phenomena at another level, these phenomena may be from a lower or a higher level, and may differ in their level of abstraction and their size. This implies that there are four ways in which analysis may bridge levels. I call these "decomposition" (the analysis of a larger unit in terms of smaller units), "subsumption" (the analysis of a smaller unit by reference to larger units), "deduction" (the analysis of a more concrete unit by analyzing it as a subclass of a more general phenomenon) and "specification" (concretization; analyzing a more abstract unit by analyzing one or more of its more concrete forms). I discuss these in order.

• In *decompositional analysis* (or reductive analysis), a large phenomenon is analyzed in terms of (much) smaller phenomena. For example, the behavior of markets (at the macro level) is analyzed as the product of the behavior of individuals (at the micro level).

• *Subsumptive analysis* is the opposite of decompositional analysis. With it, one tries to account for smaller phenomena by (partially) subsuming them under a larger (structural, functional, or causal) pattern of which they are a part. For example, given the macroevent of a transition from Fordism to post-Fordism, in which the bicycle firm Raleigh (a microentity) is one of the players, there is a modest expectation that Raleigh will invest in product differentiation, since product differentiation is part of the transition to post-Fordism (see Rosen 1993).

• In *deductive analysis*, a phenomenon is identified as a species or token of a more general phenomenon, and knowledge about this more general phenomenon is subsequently applied to the more specific phenomenon. That is, one deduces features from the general to the specific. For example, a regulatory agency in health care is identified as a bureaucratic organization, and one's theory of bureaucratic organizations is subsequently applied to it.

• In *specificatory analysis*, finally, a phenomenon is studied by identifying and studying one or more subtypes or tokens of it. Case analysis, when used to elaborate a more abstract analysis, is one type of specificatory analysis, one that makes reference to tokens. An example of specificatory analysis is Castells' analysis of East Asian business networks (a meso unit). Castells analyzes them by distinguishing various kinds (at meso- and microlevels of analysis) and studying their similarities and differences (Castells 1996: pp. 174–179).

Implications for Studies of Technology and Modernity

This perspective on levels of analysis can be used to show how modernity theory can incorporate lower-level analyses in technology studies and how technology studies can make better use of higher-level analyses in modernity theory. To begin with the former, macrolevel modernity theory can benefit from microlevel work in technology studies by using such work to elaborate its macrolevel descriptions in a way that makes the overall account more concrete and empirical. Such elaboration can proceed through a process of decomposition and specification in which macro units are decomposed into smaller parts and concretized through the identification of species or subtypes. For example, in an elaboration of the notion of the bureaucratic organization, decomposition would specify the components of bureaucratic organizations, and specification would aim to distinguish various sorts of bureaucratic organizations. Both types of analysis may be repeated to arrive at levels of analysis that refer to ever smaller and more concrete phenomena. Such elaboration makes macrotheories both more easily testable and more capable of informing microlevel analyses. Such elaboration ultimately makes it easy to link up with the lower-level analyses of technology studies.

The incorporation of modernity theory into studies of technology can be similarly clarified. Here, the required types of analysis are subsumption and deduction. To illustrate, Paul Rosen (1993) has attempted to use David Harvey's theory of the shift from a Fordist mode of production to flexible accumulation in the late 1960s and early 1970s in an explanation of the constant shifts in the design of mountain bikes. Connecting this latter fact to Harvey's theory requires deduction (e.g., identifying it as a species of product differentiation, a process mentioned in Harvey's theory) and subsumption (e.g., identifying accompanying advertisements as part of the dialectic of fashion and function in post-Fordist economies). To make an adequate connection, Rosen has to do a good deal of level building, analyzing the cycle industry and advertising at various levels. This not only involves bottom-up construction (building up levels from his microlevel analyses of mountain bike design, firms, and advertisements) but also top-down construction (elaborating Harvey's theory). This makes it possible for him to have the two analyses meet halfway.

I conclude that integrated analyses of technology and modernity, which build on macrotheories of modernity and microtheories of technology, are possible, although they require hard work. Analysts have to work at level building, engaging often in decomposition, subsumption, deduction, and specification. This, I believe, is the responsibility of both modernity theorists and scholars in technology studies. It is a joint project that can begin to abolish the boundaries between two now all-too-separate fields.

Notes

1. For some recent attempts in technology studies to appropriate (and update) existing theories of modernity, see Feenberg (1995, 1999a), Rosen (1993), and Slevin (2000). For attempts at a theory of modernity from within technology studies, see Latour (1993) and Law (1994).

2. For further discussion of the notions of modernity, modernism, and modernization see Featherstone (1991), Turner (1990), and Harvey (1989).

3. What is and is not a defining aspect of modernity is, of course, a matter of debate. Thus, whereas some would consider gender to be just a social form within modernity, others have argued that it a major constitutive force, and that our very conceptions of the modern are the result of a deeply gendered ontology (e.g., Felski 1995; Marshall, chapter 4, this volume).

4. For an account of Marx's theory of modernity, see Antonio (2001). For Weber, see Scaff (1989) and Turner (1993). Sayer (1991) and Giddens (1973) treat Marx's and Weber's accounts jointly.

5. Compare Lyotard and Thébaud (1985: p. 9): "Postmodern is not to be taken in a periodizing sense." At other times, Lyotard seems to endorse an epochal conception of postmodernity in which postmodernity is the cultural condition that has resulted from the information technology revolution (cf. Lyotard 1984: p. 3).

6. For reviews of postmodern theory, see Best and Kellner (1991) and Smart (2000).

7. For surveys of STS as an academic field, see Jasanoff et al. (1995), Cutcliffe and Mitcham (2001), and Cutcliffe (2000).

8. See MacKenzie and Wajcman (1999) for a representative anthology of social studies of technology, and see its introduction for a survey. See Fox (1999) for a review of themes and approaches in the history of technology.

9. See Staudenmaier (1985) for a review of the contextual approach and its history.

10. See the respective reviews by Dosi et al. (1988) and Mitcham (1994). Achterhuis (2001) surveys contemporary American philosophy of technology.

11. Bloor (1976) is a seminal work in SSK. Other important works include those by Latour and Woolgar (1979, 1986) and Latour (1987).

12. See Woolgar (1991, 1996) and Bijker (1993) for accounts of the turn to technology in social studies of science. The early classic that marked the beginning of contemporary social studies of technology is still Pinch and Bijker (1987).

13. Staudenmaier (1985: p. 201) has surveyed this for the history of technology. My own review of issues from the year 2000 of the journals *Social Studies of Science* and *Science, Technology and Human Values* confirms that the same is true for social studies of technology.

14. For one of the original statements of this position, see MacKenzie and Wajcman (1985).

15. Pinch and Bijker (1987) give a classical statement of social constructivism in technology studies, specifically of the influential social construction of technology (SCOT) approach. For a recent review of social constructivist approaches, see Pinch (1999).

16. Based on a search on title words at Amazon.com, January 2001.

17. See Smith and Marx (1994) for a historical and Winner (1977) for a philosophical critique of technological determinism.

18. For critiques of the lack of attention of (constructivist) technology studies to sociocultural contexts, see Rosen (1993) and Winner (1993).

19. Along the same lines, the rejection by actor-network theory of social theory because it maintains artificial distinctions between society, nature, and technology is also overstated because these distinctions are not evident in many concepts in social theory. Such notions as disembedding mechanisms (Giddens), rationalization (Weber), and the Net (Castells) are all defined as sociotechnical phenomena.

20. Time scale is an often-mentioned third parameter (see Edwards, chapter 7 in this volume). It is often claimed that macro-analysis typically analyzes processes stretching over years or even centuries, whereas micro-analysis covers shorter time spans, ranging from minutes to months.

21. My claim that large units of analysis may have smaller units of analysis as parts does not imply the *reductionist* claim that larger units of analysis are wholly composed of smaller units of analysis and therefore can be analyzed without remainder in terms of these smaller units and their relation to one another. I am skeptical about this.

3

Modernity Theory and Technology Studies: Reflections on Bridging the Gap

Andrew Feenberg

Posing the Problem

Theories of modernity and technology studies have both made great strides in recent years, but remain quite disconnected despite the obvious overlap in their concerns. How can one expect to understand modernity without an adequate account of the technological developments that make it possible, and how can one study specific technologies without a theory of the larger society in which they develop? These questions have not even been posed, much less answered persuasively, by most leading contributors to the fields. The basic issue I would like to address is the why and wherefore of this peculiar mutual ignorance.[1]

In the first half of this chapter I review the positions of some of the major figures in each field. After posing the problem briefly in this section, I sketch the background to the current impasse in the original contributions of Marx and Kuhn, and then consider the obstacles each field places in the way of encountering the other. In the second half of the chapter I propose one possible resolution of the dilemma, bridging the gap between the two fields through a synthesis of their main contributions. Both modernity theory and technology studies employ hermeneutic approaches that I elaborate further in a loosely Heideggerian account of innovation. In the concluding sections I summarize my own instrumentalization theory and show how it can be applied to the computerization of society.

Modernity theory relies on the key notion of rationalization to explain the uniqueness of modern societies. Rationalization refers to the generalization of technical rationality as a cultural form, specifically the introduction of calculation and control into social processes, with a consequent increase in efficiency. Rationalization also reduces the normative and qualitative richness of the traditional social world, exposing social reality to technical manipulation. Modernity theories often claim that this reduction impoverishes our relation to the world. But, the theorists argue, impoverished though it may be, technical rationality gives power over nature, supports large-scale organization, and eliminates many spatial constraints on social interaction. This view of modernity is characteristic of a normative style of cultural critique that is anathema to contemporary technology studies. Albert Borgmann's theory of the "device paradigm" is a well-known example of this approach (Borgmann 1984; Higgs et al. 2000).

Rationalization depends on a broad pattern of modern development described as the "differentiation" of society. This notion has obvious applications to the separation of property and political power, offices and persons, religion and the state, and so on. However, a rationality differentiated from society as such appears to lie beyond the reach of social study. If technology is a product of such a rationality, it too would escape sociocultural determination.

Technology studies reject this whole approach. They point out the social complexity of technology, the multiple actors involved with its creation, and the consequent richness of the values embedded in design. The principles of symmetry embraced by technology studies undergird rigorous case studies that persuasively refute the very idea of pure rationality. Thus modernity theory goes wrong when it claims that all of society operates under values somehow specific to a science and technology differentiated from other spheres. However, if technology and society are not substantial "things" belonging to separate spheres, it makes no sense to claim that technology dominates society and transforms its values. Rather, technology is a social phenomenon through and through, no more and no less significant than any other social phenomenon.

Technology studies lose part of the truth when they emphasize only the social complexity and embeddedness of technology and minimize the distinctive emphasis on top-down control that accompanies technical rationalization. This trend depends on the differentiation of institutions such as corporations that wield technical rationality in the interest of control. Limited though that differentiation may be, it nevertheless makes it possible to grasp any concrete value or thing as a manipulable variable, and this includes human beings themselves. Where traditional craft work expressed the vocational investment of the whole personality, the modern organization of work separates occupations from personal character and growth, the better to expose the worker to external controls (deskilling). Similarly, whereas traditional architecture combined historical and aesthetic expression with stability and durability, today strictly "utilitarian" construction is the rule. True, other social values fill the vacuum left by the differentiation of the technical sphere— e.g., profit—but this differentiation process is a real characteristic of modernity, and it has immense social consequences.

Is it possible to find some truth in both these positions or are they mutually exclusive, as they certainly appear to be at first sight? I believe a synthesis is possible, but only if the concept of technical rationality is revised to free it from implicit positivistic assumptions. It is this implicit positivism that leads modernity theory into the error of assuming that differentiation imposes a purely rational form on social processes, when in fact, as technology studies demonstrate, technology is social through and through. Science and technology studies could thus help us to avoid hypostatizing rationality as a substantial reality responsive only to its own logic.

We must also find a way to preserve modernity theory's insight into the distinctiveness of modernity and its problems. We need to explain how rationality operates as such even as it is intertwined with society through internal relations that determine its concrete realizations. This technology, that market, will always be socially specific and inexplicable in the terms of a philosophically purified concept of reason.[2] In the next section I sketch the background to the two very different ways of understanding rationality in modernity theory and technology studies.

Science of Society and History of Science

The writings of Karl Marx are surely the single most influential source of theories of modernity. His thought is usually identified with a universalistic faith in progress. At its core there is an intuition he shared with his century, the notion that a "great divide" forever separates pre-modern from modern societies. All later contrasts of *Gesellschaft* versus *Gemeinschaft*, organic versus mechanical solidarity, traditional versus post-traditional society, and so on, owe something to Marx's canonical formulation of this idea in texts such as the *Communist Manifesto* and *Capital*.[3] After World War II, modernization theory emerged as the chief competitor to Marxism, but it shared Marx's progressive universalism.

The sense of radical discontinuity in these texts involves more than a theory of society. Marx's notion of what Max Weber later called "rationalization" covers not only the changes in economic and technical systems Weber identified, but a new form of individuality freed from ideology and religion. This new form of individuality is plain to see in the nineteenth-century novels contemporary with Marx's work, and he assumes its generalization to the lower classes under the conditions of modern capitalism. Modern workers have no fixed abode and are not subject to the paternalistic authority of nobles and clerics. As the tectonic plates of culture are thrown into movement by the market, workers are freed from naïve faith in their "betters" and acquire a rational appreciation of the gaps between ideals and realities. Under these conditions, they gain mental independence and become, in Engels' phrase, "free outlaw[s]" (Engels 1970: p. 23). Marx's social theory is thus founded not just on cognitive hypotheses but on the existential irony of this modern individual. Its method is therefore fundamentally hermeneutic and demystifying as well as analytical. This duality explains the contrast between the method in Marx's critique of ideology and that in his positive economic theory. It shows up in various guises in modernity theory and is especially clear in Habermas who, as we will see later in this chapter, employs both hermeneutic and analytical methods to study modern society.

If there is any one figure who has played a comparable role for contemporary technology studies, it is Thomas Kuhn. It is true that the case for Kuhn as a founding father is less clear. Many students of science and technology, particularly historians, avoided the positivistic errors Kuhn criticized in his famous book, *The Structure of Scientific Revolutions* (Kuhn 1970). However, Kuhn's overwhelming success lent philosophical legitimacy to these trends and encouraged others to follow their lead. Nonpositivist historiographic methods triumphed in science studies and subsequently influenced the new wave of technology studies that grew out of science studies in the 1980s. Unlike Marx, Kuhn is perhaps less a source than a symbol of a radically new approach.[4]

Of course neither Marx nor Kuhn are followed slavishly by contemporary scholars, but we should not be surprised to find that many of their background assumptions are still at work in the most up-to-date contributions to modernity theory and technology studies. I would like to begin by considering several such assumptions that may help to explain the gap between these two fields.

Like all modern historians and social theorists, Kuhn writes somewhere in the long shadow cast by Marx, as can be deduced from the place of "revolution" in the title of his major book, but his view of historical discontinuities is quite different from Marx's. Kuhn did not reject the idea of radical discontinuities in history, which, on the contrary, continue to shape his vision of the past. But where Marx took for granted the existence of a rationality gradient underlying the concept of modernity, Kuhn deconstructed the idea of a universal standard of rationality that was more or less identical with scientific reason and capable of transcending particular cultures and ordering them in a developmental sequence. The demystifying impulse is still present, but it is directed at the belief in a "great divide" that characterizes modernity itself. Now the ironic glance turns back on itself, undermining the cognitive self-assurance implied in the stance of the naïve ironist.

Kuhn's method had momentous consequences for the wider reception of science studies in the academic world. He showed that there is no one continuous scientific tradition, but a succession of different traditions, each with its own basic assumptions and standards of truth, its own "paradigms." The illusion of continuity arises from glossing over the complexities and ambiguities of scientific change and reconstructing it as an upwardly linear progression leading to the present. If we go back to the decisive moments in the scientific revolution and examine what actually occurred from the standpoint of the participants, their competing positions, their arguments and experimental results, we will discover that the case for continuity is by no means so clear.

This practice-oriented approach is neatly captured in Latour's suggestion that science resembles a Janus looking back on its past in an entirely different spirit from that in which it looks forward to the future (Latour 1987: p. 12). Science, Latour suggests, is a sum of results that "hold" under certain conditions, such as repeated experimental tests. While the backward glance shows nature confirming the results of science, the forward glance presents a very different picture in which the results that hold are called "nature." Looking backward, one can say that the conditions of truth were met because the hypotheses of science were true. Looking forward, one must say rather that meeting the conditions defines what scientists will use for truth. The backward glance tells of an evolutionary progress of knowledge about the way things are, independent of science; the forward glance tells of the sheer contingency of the process in which science decides on the way things are.

I doubt if Kuhn would have appreciated this Nietzschean twist to his original contribution, from which he unfortunately retreated in subsequent writings. Kuhn himself never challenges the notion of modernity or the material progress associated with it. But the point is really not so much to offer an interpretation of Kuhn as of his significance on the maps of theory. He certainly had no intention of commenting on issues beyond his field, the history of science, but a critique of Marx is implied in his notion of scientific revolution insofar as the latter did believe that his own work was scientific and, more deeply, that rationality characterizes the institutions and forms of modernity. Thus just because Kuhn undermines the pretensions of science to access transhistorical truths, his work also undercuts Marxism and the modernity theory which inherited many Marxist assumptions. From that standpoint, it is clear that Kuhn is in some sense the nemesis of Marx and the harbinger of what has come to be called "postmodernism." To the extent that many contributions to technology studies reflect Kuhn's methodological innovations, they too bear a certain elective affinity for postmodernism, or at least for a "nonmodern" critique of Marx's heritage.

The implicit conflict came to the surface in various formulations of postmodernism, but it still seemed a mere disagreement between abstract epistemological positions. Philosophers engaged in heated debates over the nature of truth, but these debates had only a few echoes in social theory, such as Habermas's critique of Foucault. Things have changed now that the conflict has emerged inside the ill-matched couple we are considering here, modernity theory and technology studies. Since no fully coherent account of modernity is possible without an approach to technology, and vice versa, the philosophical disagreement now appears as a tension between fields. It is no longer just a matter of one's position on the great question of realism versus relativism, but concerns basic analytical categories and research methods.

Consider the implications of technology studies for the notion of progress. If Kuhnian relativism has the power to dissolve the self-certainty of science and technology, then what becomes of the notion of a rationalized society? In most modernity theories, rationalization appears as a spontaneous consequence of the pursuit of efficiency once customary and ideological obstructions are removed. Technology studies, on the contrary, show that efficiency is not a uniquely constraining objective of design and development, but that many social forces play a role. The thesis of "underdetermination" holds that there is no one rational solution to technical problems, and this opens the technical sphere to these various influences. Technical development is not an arrow seeking its target, but a tree branching out in many directions. But if the criteria of progress themselves are in flux, societies cannot be located along a single continuum from the "less" to the "more" advanced. Like Kuhn's theory of scientific revolutions, but on the scale of society as a whole, constructivist technology studies complicate the notion of progress at the risk of dissolving it altogether.

In Latour's account, a contingent scientific-technical rationality can only gain a grip on society at large through the social practices by which it is actively "exported" out of the laboratory and into the farms, streets, and factories (Latour 1987: pp. 249ff.). The constructivist theorists export their relativistic method as they trace the movements of their object of study. They dissolve all the stable patterns of progress into contingent outcomes of "scaling up" or controversies. Institutional or cultural phenomena no longer have stable identities, but must be grasped through the process of their construction in the arguments and debates of the day. This approach ends up eliminating the very categories of modernity theory, such as universal and particular, reason and tradition, culture and class, which are transformed from explanations into explananda. One can neither rise above the level of case histories nor talk meaningfully about the essence and future of modernity under these conditions.

Modernity theory suffers disaster on its own ground once it encounters the new technology studies approach. If no fixed path of technical evolution guides social development toward higher stages, if social change can take different paths leading to different types of modern society, then the old certainties of modernity theory collapse. One can no longer be sure if such essential dimensions of modernity as rationalization and democratization are actually universal, progressive tendencies of modern societies or just local consequences of the peculiar path of recent western development. Unless it squarely faces these difficulties, modernity theory must become so abstract that this kind of objection no longer troubles it, with a consequent loss of usefulness, or cease to be a theory at all and transform itself into a descriptive and analytical study of specific cases. Here are two examples that show the depth of the problems.

System or Practice

Modernity as Differentiation

Modernity theory on the whole either continues to ignore technology or acknowledges it in an outmoded deterministic framework. Most revealing is the extreme but instructive case of Jürgen Habermas. Habermas is one of the major social theorists of our time. His influence is widespread and the rigor of his thought admirable. Yet he has elaborated the most architectonically sophisticated theory of modernity without any reference at all to technology. This blissful indifference to what should surely be a focal concern of any adequate theory of modernity requires explanation, especially since Habermas is strongly influenced by Marx, for whom technology is of central importance.

Habermas's approach is based to a considerable extent on Weberian rationalization theory. According to Weber, modernity consists essentially in the differentiation of the various "cultural spheres." The state, the market, religion, law, art, science, technology each become distinct social domains with their own logic and institutional identity. Under these conditions, science and technology take on their familiar post-traditional form as independent disciplines. Scientific-technical rationality is purified of religious and customary elements. Similarly, markets and administrations are liberated from the mixture of religious prejudices and family ties that bound them in the past. They emerge as what Habermas calls "systems" governed by an internal logic of equivalent exchange. Such systems organize an ever-increasing share of daily life in modern societies (Habermas 1984–87). Where formerly individuals discussed how to act together for their mutual benefit or to maintain customary rituals and roles, we moderns coordinate our actions with minimal communication through the quasi-automatic functioning of markets and administrations.

According to Habermas, the spread of such differentiated systems is the foundation of a complex modern society. But differentiation also releases everyday communicative interaction from the overwhelming burden of coordinating all social action. The communicative sphere, which Habermas calls the "lifeworld," now emerges as a domain in its own right as well. This lifeworld includes the family, the public sphere, education, and all the various contexts in which individuals are shaped as relatively autonomous members of society. It too, according to Habermas, is subject to a specific rationalization consisting in the emergence of democratic institutions and personal freedoms. However contestable this account of modernity, something significant is captured in it. Modern societies really are different from traditional ones, and the difference seems closely related to the impersonal functioning of institutions such as markets and administrations and the increase in personal and political freedom that results from new possibilities of communication.

At first Habermas argued that system rationalization threatened to create technocratic intrusions into the lifeworld of communicative interaction, and this reference to *techno*-cracy seemed to link his theory to the theme of technology (Habermas 1970; Feenberg 1995: chap. 4). However, his mature formulation of the theory ignores technology and focuses exclusively on the spread of markets and administration. The arbitrariness of this exclusion appears clearly in the following summary of Habermas's theory: "Because we are as fundamentally language-using as tool-using animals, the representation of reason as essentially instrumental and strategic is fatally one-sided. On the other hand, it is indeed the case that those types of rationality have achieved a certain dominance in our culture. The subsystems in which they are centrally institutionalized, the economy and government administration, have increasingly come to pervade other areas of life and make them over in their own image and likeness. The resultant 'monetarization' and 'bureaucratization' of life is what Habermas refers to as the 'colonization of the life world'" (McCarthy 1991: p. 52). What became of the "tool-using" animal of the first sentence of this passage? Are its only tools money and power? How is it possible to elide technological tools in a society such as ours? The failure of Habermasian critical theorists even to pose much less respond to these questions indicates a fatal weakness in their approach. There is worse to come.

Habermas's reformulation of Weber's differentiation theory neutralizes rational systems by identifying them with nonsocial rationality as such. This has conservative political implications. In many of Habermas's formulations, for example when he considers workers' control, it seems that radical demands would be irrational if they treated systems as socially constructed and hence transformable barriers to full freedom (Habermas 1986: pp. 45, 91, 187). He thus offers no concrete suggestions, at least in *The Theory of Communicative Action*, for reforming markets and administrations, and instead suggests limiting the range of their social influence.

In the case of science and technology, this puzzling retreat from a social account is carried to the point of caricature. Habermas claims that science and technology are based quite simply on a nonsocial "objectivating attitude" toward the natural world (Habermas 1984–87: Vol. I, p. 238). This would seem to leave no room at all for the social dimension of science and technology, which has been shown over and over to shape the formulation of concepts and designs. Clearly, if scientists and technologists stand in a purely objective relation to nature, there can be no *philosophical* interest in studying the social background of their insights. In Habermas's view, it is difficult to see how a properly differentiated rationality could incorporate social values and attitudes except as sources of error or extrinsic goals governing "use." This implies, too, a problematic methodological dualism in which phenomenological accounts of the lifeworld coexist with objectivistic systems-theoretic explanations of "systems" such as markets and administrations. No doubt there are objects best analyzed by these different methods, but which method is suited to analyzing the interactions between them? Habermas has little to say on this score beyond his account of the boundary shifts that preoccupy him.

The effect of this approach is to liberate social theory from all the details of sociological and historical study of actual instances of rationality. No matter what story sociologists and historians have to tell about a particular market, administration, or, a fortiori, technology, this is incidental to the philosophically abstracted forms of differentiated rationality. The real issue is not whether this or that contingent happening might have led to different practical results, for all that matters to social theory is the range of rational systems, the extent of their intrusions into the proper terrain of communicative action (Feenberg 1999a: chap. 7).

Could it be that the most important differentiation for Habermas is the one that separates social theory from certain sociological and historical disciplines, the material of which he feels he must ignore to pursue his own path as a philosopher? When the results are compared with earlier theories of modernity, it becomes clear what a tremendous price he pays to win a space for philosophy. Marx had a concrete critique of the revolutionary institutions of his epoch, the market and the factory system, and later modernization theory foresaw a host of social and political consequences of economic development. But Habermas's complaints about the boundaries of welfare state administration seem quite remote from the main sources of social development today, the response to environmental crisis, the revolutions in global markets, planetary inequalities, the growth of the Internet, and other technologies that are transforming the world. In his work the theory of modernity is no longer concerned with these material issues, but operates at a higher level, a level where, unfortunately, very little is going on.

Of course some social theorists have made contributions to the theory of modernity that do touch on technology in an interesting way, sometimes under the influence of other aspects of Habermas's theory.[5] Ulrich Beck has proposed a theory of "reflexive modernity" in which the role of technology is explicitly recognized and discussed in terms of transformations in the nature of rationality. Beck starts out from the same concept of differentiation as Habermas, but he considers it to be only a stage he calls "simple modernity." Simple modernity creates a technology that is both extremely powerful and totally fragmented. The uncontrolled interactions between the reified fragments have catastrophic consequences.[6] Beck argues that today a "risk society" is emerging and is especially noticeable in the environmental domain. "Risk society . . . arises in the continuity of autonomized modernization processes which are blind and deaf to their own effects and threats. Cumulatively and latently, the latter produce threats which call into question and eventually destroy the foundations of industrial society" (Beck 1994: pp. 5–6).

The risk society is inherently reflexive in the sense that its consequences contradict its premises. As it becomes conscious of the threat it poses for its own survival, reflexivity becomes self-reflection, leading to new kinds of political intervention aimed at transforming industrialism. Beck places his hope for an alternative modernity in a radical mixing of the differentiated spheres that overcomes their isolation and hence their tendency to blunder into unforeseen crises. "The rigid theory of simple modernity, which conceives of system codes as exclusive and assigns each code to one and only one subsystem, blocks out the horizon of future possibilities. . . . This reservoir is discovered and opened up only when code combinations, code alloys and code syntheses are imagined, understood, invented and tried out" (Beck 1994: p. 32).[7]

This revision of modernity theory is daring and suggestive, but it still rests on a notion of differentiation that would surely be contested by most contemporary students of science and technology. Their major goal has been to show that "differentiation" (Latour calls something similar "purification") is an illusion, that the various forms of modern rationality belong to the continuum of daily practice rather than to a separate sphere (Latour 1991: p. 81).

Yet the main phenomena identified by the theory of modernity do certainly exist and require explanation. We have reached a puzzling impasse in the interdisciplinary relationship around this problem. Practice-oriented accounts of particular cases cannot be generalized to explain the systemic character of modernity, while differentiation theory appears to be invalidated by what we have learned about the social character of rationality from science and technology studies. A large part of the reason for this impasse, I believe, is the continuing power of disciplinary boundaries which, even where they do not become a theoretical foundation as in Habermas, still divide theorists and researchers. Far from weakening, these boundaries have become still more rigid in the wake of the sharp empiricist turn in science and technology studies, and the growing skepticism in these fields with regard to the theory of modernity in all its forms (see Misa, chapter 1, this volume). I turn now to two examples from technology studies to illustrate this point.

The Logic of Symmetry

The constructivist "principle of symmetry" is supposed to ensure that the study of technological controversies is not biased by knowledge of the outcome (Bloor 1976: p. 7). Typically, the bias appears in popular understanding as an "asymmetrical" evaluation of the two sides of the controversy, ascribing "reason" to the winners and "prejudice," "emotion," "stubbornness," "venality," or some other irrational motive to the losers. A similar bias is also presupposed by such basic concepts of modernity theory as rationalization and ideology. These concepts appear to be cancelled by the principle of symmetry.

Social constructivists' main concern is to achieve a balanced view of controversies in which rationality is not awarded as a prize to one side only, but recognized wherever it appears, and in which nontechnical motives and methods are not dismissed as distortions, but are taken into account right alongside technical ones as normal aspects of technological debate. The losers often have excellent reasons for their beliefs, and the winners sometimes prevail at least in part through dramatic demonstrations or social advantage as well as rational arguments. The principle of symmetry orients the researcher toward an even-handed evaluation by contrast with the inevitable prejudice in favor of the winners that colors the backward glance of methodologically unsophisticated observers.

However, there is a risk in such even-handedness where technology is concerned: if the outcome cannot be invoked to judge the parties to the controversy, and if all their various motives and rhetorical assets are evaluated without prejudice, how are we to criticize mistakes and assign responsibility? Consider, for example, the analysis of the *Challenger* accident by Harry Collins and Trevor Pinch (Collins and Pinch 1998: chap. 2). Recall that several engineers at Morton Thiokol, the company that designed the space shuttles, at first refused to endorse a cold-weather liftoff. They feared that the O-rings sealing the sections of the launcher would not perform well at low temperatures. In the event they were proven right, but management overruled them and the launch went ahead, with disastrous results. The standard account of this controversy is asymmetrical, opposing reason—the engineers—to politics—the managers.

Collins and Pinch think otherwise. They show that the O-rings were simply one among many known problems in the *Challenger*'s design. Since no solid evidence was available to justify canceling the fateful flight, it was reasonable to go forward and not a heedless flaunting of a prescient warning. Scheduling needs as well as engineering considerations influenced the decision, not because of managerial irresponsibility, but as a way of resolving a deadlocked engineering controversy. It appears that no one is to blame for the tragic accident that followed, at least in the sense that this is a case where normally cautious people would in the normal course of events have made the same bad decision.

However, the evidence Collins and Pinch offer could have supported a rather different conclusion had they evaluated it in a broader context. Their symmetrical account obscures the asymmetrical treatment of different types of evidence within the technical community they study. It is clear from their presentation that the controversy at Morton Thiokol was irresolvable because of the systematic demand for quantitative data and the denigration of observation, even that of an experienced engineer. Can an analysis of the incident abstain from criticizing this bias?

Roger Boisjoly, the engineer who was most vociferous in arguing for the dangers of a cold-weather launch, based his warnings on the evidence of his eyes. This did not meet what Collins and Pinch prissily define as "prevailing technical standards" (Collins and Pinch 1998: p. 55). The fact that Boisjoly was probably right cannot be dismissed as a mere accident. Rather, it says something about the limitations of a certain paradigm of knowledge, and suggests the existence of an ideological bias masked by the principle of symmetry. Could it be that Boisjoly's observations were dismissed—and quantitative data demanded—mainly to keep the National Aeronautics and Space Administration (NASA) on schedule? Or put another way, would the need for quantitative data have seemed compelling in the absence of that pressure? By identifying this case with every other known risk in the design, without regarding Boisjoly's observations as a legitimate reason for extra caution, Collins and Pinch appear to surrender critical reason to so-called "prevailing technical standards."[8]

Now, I cannot claim to have made an independent study of the case, and Collins and Pinch may well have stronger reasons for their views than those that appear in their exposition. However, we know from experience that quantitative measures are all too easily manipulated to get the answer demanded by the powers that be. For example, quantitative studies were long thought to "prove" the irrelevance of classroom size to learning outcomes, contrary to the testimony of professional teachers. This "proof" was very convenient for state legislators anxious to cut budgets, but it resulted in an educational disaster that, like the *Challenger* accident, could not be denied. Similar abuses of cost-benefit analysis are all too familiar. How can critical reason be brought to bear on cases such as these without applying sociological notions such as "ideology," which presuppose asymmetry?

A similar problem regarding the supposed opposition of local and global analyses bedevils science studies. Science studies scholars sometimes claim that a purely local analysis extended to ever-wider reaches suffices in the study of society without the need for empirically "ungrounded" global categories. This is to be sure a puzzling dichotomy. If the local analysis is sufficiently extended, does it not become nonlocal, indeed global? Why not just generalize from local examples to macro categories and theories, as modernity theory does?

For Bruno Latour, the analysis of contingent contests for power within specific networks suffices, and the introduction of terms such as "culture," "society," or "nature" would simply mask the activities that establish these categories in the first place. "If I do not speak of 'culture,' that is because this word is reserved for only one of the units carved out by Westerners to define man. But forces can only be distributed between the 'human' and the 'nonhuman' locally and to reinforce certain networks" (Latour 1984: pp. 222–223, my translation).[9] Latour continues in this passage to similarly reduce the terms "society" and "nature" to local actions.

This "symmetry of humans and nonhumans" eliminates any fundamental difference between them. The "social" and the "natural" are to be understood now in the same terms. Attributions of social and natural status are contingent outcomes of processes operating at a more fundamental level. Then the distinctions we make between the social or natural status assigned to such things as a student protest in Paris and a dieoff of fish in the Mississippi, a politician's representation of American farmers and a scientist's representation of nuclear forces, are all products of the network to which we belong, not presuppositions of it.

This stance appears to have conservative political implications since in any conflictual situation the stronger party establishes the definition of the basic terms, "culture," "nature," and "society," and the defeated cannot appeal to an objective "essence" to validate their claims *quand même*. John Law's well-known network analysis of Portuguese navigation is thus widely criticized for ignoring the fate of the conquered peoples incorporated into the colonial network. And Hans Radder argues that actor-network theory contains an implicit bias toward the victors (Law 1987; Radder 1996: pp. 111–112).

Underlying Latour's difficulty with resistance is the strict operationalism that works as an Ockham's razor, stripping away generations of accumulated sociological and political conceptualization. If nature and society are exhaustively defined by the procedures through which they emerge as objects, it is unclear how unsuccessful competitors for the defining role can gain any grip on reality at all, even the feeble grip of ethical exigency. For example, the aspiring citizens of an aristocratic society may appeal to "natural" equality against the caste distinctions imposed by the "collective" to which they belong. But if nature is defined by the collective, not simply ideologically or theoretically but really, how can an appeal to nature be invoked oppositionally to sanction demands for change? Or consider demands for justice for the weak and dominated. The concept of justice stands here for an alternative organization of society, haunting the actual society as its better self. What can ground the appeal to such transcendent principles if the very meaning of society is defined by the forces that effectively organize and dominate it?

I have argued elsewhere that without a global social theory, it is difficult to establish what I call the "symmetry of program and antiprogram," i.e., the equal analytical value of the principal actors' intentions, more or less successfully realized in the structure of the network, and those of the weaker parties they dominate (Feenberg 1999a: chap. 5). In particular, the symmetry of humans and nonhumans blocks access to the central insight of modernity theory, the extension of technical control from nature to humans themselves. I concluded that although the empiricist preference for the local sounds innocent enough, in excluding all explanations based on the traditional categories of social theory, such as class, culture, ideology, and nature, truly rigorous localism blocks even-handed study of social conflict.

Latour's recent book on political ecology attempts to address criticisms like these (Latour 1999a). He faces up to the challenge of explaining oppositional agency, that is, resistance to the dominant definition of the network in which the subject is enrolled. Political morality requires that he find a place for such resistance in his theory. However, consistency requires that he do this without reintroducing a transcendent nature or morality. The following is a necessarily abbreviated account of his provocative central argument.

The operational reduction of society and nature in earlier presentations of his theory seemed paradoxically to eliminate the contingency of the phenomena he described. The case resembles artistic production. A musical composition depends on the composer's decisions, which might have been different, yet once it has been completed, the composition is perfectly self-defined. There is no higher authority to which one might appeal against it. Beethoven's Fifth is a necessary product of the contingencies of its creation. Similarly, Latourian networks define themselves as necessary in the course of their self-creation, with no higher authority able to cast doubt on that definition. The contrary hypothesis, that nature is not simply what the collective takes it to be, and that society overflows the bounds imposed on it by those with influence and power, would seem to violate Latour's operationalism. Yet without some such hypothesis, one inevitably ends up in the most uncritical conformism. Can Latour accept such a hypothesis without his theory cracking open at the seams?

Latour finds a way of having his operationalist cake and eating it too. He argues that the necessary conditions of opposition can be met without positing transcendent principles. The solution is again operational: look not to the transcendent *objects* but to the contestatory *procedures* by which they are given a chance to emerge within the collective. These procedures can prevent premature totalizations or closures that ignore the weak and violate human rights. In sum, Latour substitutes a democratic doctrine of legitimate debate for nature and morality as the ultimate ground of resistance (Latour 1999a: pp. 156, 172–173).

However, there is an ambiguity about this solution. Latour's claim might be interpreted as an antitechnocratic constitutional principle: "Thou shalt not interrupt the collective conversation with authoritative findings." He might be saying that this is all that philosophy can persuasively claim without prejudging the content of democratic discourse. In the terms of contemporary political philosophy, this would imply a distinction between the right and the good, the one universally valid, the other contentious and rationally undecidable. That interpretation still leaves open the possibility that ordinary actors could legitimately bring forward appeals to a transcendent nature and society. But this does not seem to satisfy Latour. He wants to expel the transcendent objects not only from theory but from practice as well. This is a consequence of ontologizing the network, treating it as the actual foundation of the objects it contains. Short of proposing a double discourse, a true one for the theorist and a false one for the masses, Latour is obliged to introduce his theoretical innovations into the collective conversation as an alternative to the outmoded discourse of transcendence.

These theoretical innovations consist of techniques of local analysis that trace the co-emergence of society and nature in the processes of social, scientific, and technological development. Since these processes are historical, what we call "nature" now develops and changes much as "society" does. Pasteur's discovery of lactic acid yeast was a great event, not only in Pasteur's life, but also in the life of the yeast. Latour refers to Whitehead's process philosophy for a metaphysical sanction for the effacement of the difference between nature and society to make room for a third term out of which both emerge (Latour 1994: p. 212). This is interesting and provocative as philosophy, but can these philosophical innovations become generally available to ordinary people as a substitute for the now disqualified appeal to transcendent grounds for resistance? That promises to be difficult, requiring that common sense itself become Latourian! Presumably, the traditional appeal to a preexisting "nature" (e.g., natural equality) would give way in a Latourian society to an appeal for a favorable evolution of nature itself. If I have understood him, Latour is confident something like this will occur (Latour 1999a: pp. 32–33), but that seems quite unlikely. I conclude that his attempt to evade the conformist implications of his position shows more good will than practical plausibility.

Now, there is no intrinsic reason why science studies should seek to explode the entire framework of social theory, and not all current approaches lead to such radical consequences. Yet the tendency to do so is influential in science studies circles. I call attention to it because it takes to the limit a consequence of certain original methodological choices applied to technology and through technology to modern social life. The results, I have argued, are intriguing but ultimately unsatisfactory.

Splitting the Difference

Interpretation and Worldhood

I now want to suggest one of several possible lines of argument leading to a partial resolution of the conflict between modernity theory and technology studies. The key point on which I focus is the role of interpretation in these two disciplines. Where society is not studied as a realm of causal interactions governed by law, it is usually considered to be a realm of meaning, engaging interacting subjects of some kind, for example, subjects of consciousness or language. Interpretative understanding of society is thus an alternative to deterministic accounts, and hermeneutics appears as an explanatory model better suited to society than the nomological approach imitated from physical science.

The place of interpretation in technology studies should be obvious from the Kuhnian critique of the "myth of the given." Data do not speak unambiguously, but must be interpreted, and interpretation calls into play the very theories the data are supposed to verify. This hermeneutic circularity has social ontological implications when a similar approach is applied to technology. Technologies serve needs while also contributing to the emergence of the very needs they serve; human beings make technologies that in turn shape what it means to be human.

These circular relationships are familiar from hermeneutics. The famous "hermeneutic circle" describes the paradoxical nature of interpretative understanding: we can only understand what, to some degree, we already understand. A completely unfamiliar object would remain impenetrable. However, this circularity is not vicious since we can bootstrap our way to fuller understanding, starting from a minimal "preunderstanding," "like using the pieces of a puzzle for its own understanding" (Palmer 1969: p. 25).

Pinch and Bijker's analysis of the bicycle highlights the role of "interpretative flexibility" in the evolution of design (Pinch and Bijker 1987). At its origin, the bicycle had two different meanings for two different social groups. That difference in interpretation of a largely overlapping assemblage of parts yielded designs with distinctive social significance and consequences. Pinch and Bijker conclude that "different interpretations by social groups of the content of artifacts lead by means of different chains of problems and solutions to different further developments" (Pinch and Bijker 1987: p. 42). This means that there is no stable, pregiven telos of technological development because goals are variables, not constants, and technical devices themselves have no self-evident purpose. Clearly, we are a long way here from the old deterministic conception of technology in which changes in design follow from the technical logic of innovation. Meaning is now central.

Interpretation plays an equally important role for modernity theorists such as Habermas and Heidegger. Both thinkers rely on a contrast between scientific-technical rationality and the phenomenological approach to the articulation of human experience. They see the everyday "lifeworld" as an original realm within which human identity and the meaning of the real are first and most profoundly encountered. Interpretation rather than law prevails in the study of this realm.

For Heidegger, worlds are realms of meaning and corresponding practices rather than collections of objects as in conventional usage. A world is "disclosed" according to Heidegger in the sense that the orientation of the subject opens up a coherent perspective on reality. Heideggerian worlds thus more nearly resemble our metaphoric concept of a "world of the theatre," or a "Chinese world" than the literal meaning. Here interpretation is no specialized intellectual activity, but the very basis of our existence as human beings (Spinosa et al. 1997: p. 17).

In his later work Heidegger developed a radical critique of technology for its power to "deworld," that is, to strip objects of their inherent potentialities and reduce them to mere raw materials. This turn in Heidegger's analysis seems to cancel its hermeneutic import since the message of technology is always the same, what Heidegger calls "enframing" (Heidegger 1977). Although his theory of technology is unremittingly negative, some of his followers have attempted to modify it in interesting ways.

The early Heidegger's concept of the lifeworld has been applied by Charles Spinosa, Fernando Flores, and Hubert Dreyfus in a recent book (*Disclosing New Worlds*). As we will see, their major focus is on leadership rather than technology, but this turns out to be a correctable error of emphasis. The authors' starting point in any case is the notion of disclosure that lies at the center of Heidegger's thought. They take up Heidegger's basic concepts in the context of a theory of history. The problem to which the book is addressed is how disclosive activities actually change the world we live in, opening us to new or different perspectives and reorganizing our practices around a different sense of what is real and important. The book reviews three main types of history-making disclosive practices that correspond to three main types of historical actors.

"Articulations" refocus a community on its core values and practices. This is primarily the task of political leaders. As an example, the authors cite John Kennedy's ability to generate enthusiasm for the space race around such themes as the new frontier. "Cross-appropriations" weave together values and practices from diverse domains of social life in new patterns that alter the structure of our world. This is the work of successful social movements, such as MADD (Mothers Against Drunk Driving), which transported ideas about responsible behavior from the domain of work to the domain of leisure. Finally, and most significantly, "reconfiguration" is the process by which a marginal practice is transformed into a dominant one. Entrepreneurs are the agents of reconfiguration, which they accomplish by introducing new products that suggest a new style of life. The focus of *Disclosing New Worlds* is not on the products but on the entrepreneurs. Yet the authors write explicitly, "it is the product or service, not the virtuous life-style of the entrepreneur, that makes the world change . . . " (Spinosa et al. 1997: p. 45).

Although technology studies are not mentioned, the examples illustrate nicely the theme of interpretative flexibility. The Gillette company's successful introduction of the disposable razor is a textbook case. The traditional straight razor belonged to a world in which men cared for and cherished finely made objects. Gillette sensed the possibility of a redefinition of the masculine relation to objects in terms of control and disposability and furthered that change with a new type of razor. In other words, Gillette did not just serve a preexisting need for sharper razors. "The entrepreneurial question was, what did his annoyance at the dullness mean? Did it mean that he just wanted a better-crafted straight-edge razor that kept its edge longer? Or did he want a new way of dealing with things? We shall argue that genuine entrepreneurs are sensitive to the historical questions, not the pragmatic ones, and that what is interesting about their innovations is that they change the style of our practices as a whole in some domain" (Spinosa et al. 1997: pp. 42–43). Style is a very general feature of worlds that is relevant to the design of artifacts. In this case the change in style involved the transition from a respectful to a controlling attitude toward objects.

We find more precise tools for discussing the reconfigurative work of artifacts in the notions of "actors" and "scripts" in technology studies (Akrich 1992; Latour 1992). In particular, the multiplicity of actors identified in many case histories offers a useful corrective to the book's implicit individualism. The bias toward the heroic disclosive power of poets, philosophers, and statesmen, who are presumed to be in touch with "Being," has been noted in Heidegger and his followers before.

Perhaps the overemphasis on entrepreneurs is a modest expression of that bias. In any case, the failure to deal adequately with technology confirms the tendency of modernity theories to abstract from the world of things. This time there is a difference: for once a theory lends itself to a shift in emphasis to take technology into account because in fact technology is already there at its core. "A *world*, for Heidegger," the authors write, "is a totality of interrelated pieces of *equipment*, each used to carry out a specific task such as hammering in a nail. These tasks are undertaken so as to achieve certain *purposes*, such as building a house. Finally, this activity enables those performing it to have *identities*, such as being a carpenter" (Spinosa et al. 1997: p. 17).

Instrumentalization Theory

We now have two complementary premises drawn from the two theoretical traditions we are attempting to reconcile. On the one hand, the evolution of technologies depends on the interpretative practices of their users. On the other hand, human beings are essentially interpreters shaped by world-disclosing technologies. Human beings and their technologies are involved in a co-construction without origin. Modernity theory asks how this process operates when it is mediated by differentiated technical disciplines and aims at the human control of human beings. Technology studies keeps us focused on the essentially social nature of the technical rationality deployed in those disciplines. The hermeneutic perspective builds a bridge between these different perspectives.

A synthesis must enable us to understand the central role of technology in modern life as both technically rational in form and rich in socially specific content. This then is the program: to explain the social and cultural impact of technical rationality without losing track of the concrete social embodiment of actual devices and systems. Here is where the concept of world disclosure can be helpful, on the condition that the analysis be pursued not just in terms of the question of style, but more specifically in terms of the practical constitution of technical objects and subjects.

I have proposed what I call "instrumentalization theory" to effect such a synthesis (Feenberg 1999a, chap. 9). Instrumentalization theory holds that disclosing new worlds involves a complementary process of deworlding inherent in technical action. The materials engaged in technical processes always already belong to a world that must be shattered if they are to be released for technical employment. The specific deworlding effect of technical action touches not only the object but also the subject. The technical actor stands in an insulated, external position with respect to his or her objects. We thus distinguish technical manipulation from the reciprocal relations of everyday communication. Philosophical models of instrumental rationality are generally based on this aspect of the technical. It is, for example, highlighted in Habermas's system/lifeworld distinction and Heidegger's critique of enframing.

Most modernity theory identifies deworlding with the essence of technology, without regard for the complexity of its disclosive dimension. I suspect that this identification is due to two features of the modern technical sphere. On the one hand, technical disciplines themselves incorporate social factors only in a stripped-down, abstract form. The most humane of values, for example compassion for the sick, is expressed technically in objective specifications such as a medical treatment protocol. The fact that the protocol can be followed without compassion suggests that the objective specifications are really self-sufficient, forming a closed universe from which values are excluded. On the other hand, modern technology has been structured around the extension of impersonal domination to human beings and nature, in profound indifference to their needs and interests. This line of technical development depends on severely restricting the range of social considerations that can be brought to bear on design. Thus deworlding looms especially large in the worlds disclosed in modern societies. These worlds differ from those of premodern societies in that they do not cover over the traces of their founding violence.

In demonstrating the contingency of technical development, technology studies encourage us to believe in the possibility of other ways of designing and using technology that show more respect for human and natural needs. However, an alternative technology is apparently unimaginable from the external perspective of modernity theorists, who are generally innocent of any involvement with the messy and complex process of actual technical development. The theorists simply fail to recognize that the deworlding associated with technology is necessarily and simultaneously entry into another world. The problems of our society are not due to deworlding as such, but to the flaws and limitations of the disclosure it supports under the social limitations of the existing form of modernity.

The duality of technical processes is reflected in the split between modernity theory and technology studies, each of which emphasizes one half of the process. Deworlding is a salient feature of modern societies, which are constantly engaged in disassembling natural objects and traditional ways of doing things and substituting new technically rational ways. An exclusive focus on the negative aspect of this process yields the dystopian critique we associate with thinkers like the later Heidegger. However, deworlding is only the other side of a process of disclosure that must be understood in social terms. Technology studies emphasize this aspect of the process. The antinomy results from the inherently dialectical character of technical action, which is unilaterally misunderstood in each case.

Instrumentalization theory characterizes this dialectic at two levels. Deworlding consists of a process of functionalization in which objects are torn out of their original contexts and exposed to analysis and manipulation while subjects are positioned for distanced control. Modern societies are unique in deworlding human beings in order to subject them to technical action—we call it "management"—and in theoretically prolonging the basic gesture of deworlding in technical disciplines that become the basis for complex technical networks. Disclosure involves a complementary process of realization, which qualifies functionalization by orienting it toward a new world containing those same objects and subjects. The two processes are analytically distinguishable but are essentially joined in practice.[10]

Terminal Subjects

I want to conclude these reflections with an example with which I am personally familiar and which I hope will illustrate the fruitfulness of a synthesis of modernity theory and technology studies. I have been involved with the evolution of communication by computer since the early 1980s, both as an active participant in innovation and as a researcher.

I came to this technology with a background in modernity theory, specifically Heidegger and Marcuse, whose student I was, but it quickly became apparent that they offered little guidance in understanding computerization. Their theories emphasized the role of technologies in dominating nature and human beings. Heidegger dismissed the computer as the pure type of modernity's machinery of control. Its deworlding power reaches language itself, which is reduced to the mere position of a switch (Heidegger 1998: p. 140).

However, what we were witnessing in the early 1980s was something quite different: the contested emergence of the new communication practices of online community. Subsequently, we have seen cultural critics inspired by modernity theory recycle the old approach for this new application, denouncing, for example, the supposed degradation of human communication on the Internet. Albert Borgmann argues that computer networks deworld the person, reducing human beings to a flow of data the "user" can easily control (Borgmann 1992: p. 108). The "terminal" subject is basically an asocial monster despite the appearance of interaction online. That reaction presupposes that computers actually are a communication medium, if an inferior one, which was precisely the issue 20 years ago. The prior question that must therefore be posed concerns the emergence of the medium itself. Most recently the debate over computerization has involved higher education, where proposals for automated online learning have met determined faculty resistance in the name of human values. Meanwhile, actual online education is emerging as a new kind of communicative practice (Feenberg 2001: chap. 5).

The pattern of these debates is suggestive. Approaches based on modernity theory are uniformly negative and fail to explain the experience of participants in computer communication. This experience can be analyzed in terms of instrumentalization theory. The computer reduces a full-blown person to a "user" in order to incorporate him or her into the network. Users are decontextualized in the sense that they are stripped of body and community in front of the terminal and positioned as detached technical subjects. At the same time, a highly simplified world is disclosed to the user. This world is open to the initiatives of rational consumers, who are asked to exercise choice there. Positioning and initiative as described here are correlated as primary and second instrumentalizations, interventions that deworld and disclose.[11]

The poverty of this world appears to be a function of the very radical deworlding involved in computing. However, we will see that this is not the correct explanation of what actually occurs. Nevertheless, the critique is not entirely artificial; there are types of online activity that confirm it and certain powerful actors do seek enhanced control through computerization. However, modernity theorists overlook the struggles and innovations of those attempting to appropriate the medium to create online communities or legitimate educational experiments. In ignoring or dismissing these aspects of computerization, they fall back into a more or less disguised determinism.

The posthumanist approach to the computer inspired by commentators in cultural studies suffers from related problems. This approach often leads to a singular focus on the most "dehumanizing" aspects of computerization, such as anonymous communication, online role playing, and cybersex (Turkle 1995). Paradoxically, these aspects of the online experience are interpreted in a positive light as the transcendence of the "centered" self of modernity (Stone 1995). Such posthumanism is ultimately complicit with the humanistic critique of computerization it pretends to transcend in that it accepts a similar definition of the limits of online interaction. Again, what is missing is any sense of the transformations the technology undergoes at the hands of users animated by more traditional visions than one would suspect from this choice of themes (Bakardjieva and Feenberg, forthcoming).

The effective synthesis of these various approaches would offer a more complete picture of computerization than any one of them alone. In my writings in this field I have tried to accomplish this. I did not set out from a hypothesis about the essence of the computer, for example, that it privileges control or communication, humanist or posthumanist values, but rather from an analysis of the way in which such hypotheses influence the actors themselves, shaping design and use.

The lifeworld of technology is the medium within which the actors engage with the computer. In this lifeworld, processes of interpretation are central. Technical resources are not simply pregiven but acquire their meaning through these processes.[12] In Latour's language, the "collective" is reformed around the contested constitution of the computer as this or that type of mediation responsive to this or that actor's program. However, under the influence of theorists like Latour, technology studies have become suspicious of the very terms of the actual debates surrounding computerization. Indeed, Latour's symmetry principle makes it difficult to recognize the uniquely significant role of the contests between control and communication, humanism and posthumanism, that I argue must be the focus of the study of innovations such as the Minitel and the Internet. As computers developed, communication functions were often introduced by users rather than being provided as normal affordances of the medium by their designers. To make sense of this history, the competing visions of designers and users must be introduced as a significant shaping force, not dismissed as irrelevant ideologies. How can one adopt the actors' perspective if it contradicts the premises of one's own method?

Consider the case of the current struggle over the future of online education (Feenberg 1999b,c). Over the past few years, corporate strategists, state legislators, top university administrators, and "futurologists" have lined up behind a vision of online education based on automation and deskilling. Their goal is to replace (at least for the masses) face-to-face teaching by professional faculty with an industrial product, infinitely reproducible at decreasing unit cost, such as compact disks, videos, or software. The overhead costs of education would decline sharply and the education "business" would finally become profitable. This is "modernization" with a vengeance.

In opposition to this vision, faculty have mobilized in defense of the human touch. This humanistic opposition to computerization takes two very different forms. There are those who are opposed in principle to any electronic mediation of education. This position has no effect on the quality of computerization, only on its pace. There are also numerous faculty who favor a model of online education that depends on human interaction on computer networks. On this side of the debate, a very different conception of modernity prevails. In this alternative conception, to be modern is to multiply opportunities for and modes of communication. The meaning of the computer shifts; instead of being viewed as a coldly rational information source, it becomes a communication medium, a support for human development and online community. This alternative can be traced down to the level of technical design; for example, the conception of educational software and the role of "asynchronous discussion forums."

These approaches to online education can be analyzed in terms of the model of deworlding and disclosure introduced earlier. Educational automation decontextualizes both the learner and the educational "product" by removing them from the existing world of the university. In this decontextualized world, the learner becomes a technical subject confronted by menus, exercises, and questionnaires rather than with other human beings engaged in a shared learning process.

The faculty's model of online education involves a much more complex secondary instrumentalization of the computer in the disclosure of a much richer world. The original positioning of the user is similar: the person facing a machine. However, the machine is not a window onto an information mall but rather opens up onto a social world. The user is involved as a person in a new kind of social activity and is not limited to the role of individual consumer by a set of canned menu options. The corresponding software opens the range of the subject's initiative far more widely than an automated design. This is a more democratic conception of networking that extends it across a wider range of human needs.

The analysis of the dispute over educational networking reveals patterns that appear throughout modern society. In the domain of communication media, these patterns involve playing off primary and secondary instrumentalizations in different combinations that produce either a technocratic model of control or a democratic model of communication. Characteristically, a technocratic notion of modernity requires a positioning of the user that sharply restricts potential initiative, while a democratic conception enlarges initiative in more complex virtual worlds. Parallel analyses of production technology or environmental problems would reveal similar patterns that could be clarified by reference to the actors' perspectives in similar ways.

Conclusion: Toward Synthesis

Let me conclude now by returning briefly to my starting point. I began by contrasting the theoretical revolutions of Marx and Kuhn and promising to bring them together with a method of analysis that would reconcile modernity theory and technology studies. Can a phenomenology of technical worlds do the job? Recall that Marx emphasized the discontinuity introduced into history by what has come to be called "rationalization," the emergence of modern societies based on markets, bureaucracies, and technologies. This view seemed to imply a universalism that erased all cultural difference. By contrast, Kuhn, or at least his followers, subverted the notion of progress implied in Marx's vision of an increasingly rational social process and offered us a history subordinate to culture.

I argue that rationalization describes the generalization of a particular type of deworlding involved in technical action. That such deworlding uproots nature and traditional ways is clear. In this account, rationalization no longer stands opposed to culture as such, but appears as a more or less creative expression of it, disclosing new worlds. In practice this means that there may be many paths of rationalization, each relative to a different cultural framework. Rationality is not an alternative to culture that can stand alone as the principle of a social order, for better or worse. Rather, rationality in its modern technical form mediates cultural expression in ways that can in principle realize a wide range of values in the design of artifacts. The poverty of the actual technoculture must be traced not to the essence of technology, but to other dimensions of our society, such as the economic forces that dominate technical development, design, and the media. This insight challenges us to engage in what Terry Winograd and Fernando Flores have called "ontological designing," the conscious construction of technological worlds that support a desirable conception of what it is to be human (Winograd and Flores 1987: p. 179).

We can fruitfully combine modernity theory and technology studies in an empirically informed, critical approach to important social problems. The triviality that threatens a strictly descriptive, empirical approach to such humanly significant technical phenomena as genetic manipulation, global warming, or online education, can be avoided without falling into the opposite error of a priori theorizing. There are ways of recovering some of the normative richness of the critique of modernity within a more concrete sociological framework that does allow entry to a few facts. Concepts such as "rationality," which technology studies have set out to demystify, can be employed in a new way, and the implicit emancipatory intent of that demystification can be brought to the surface as an explicit goal. Perhaps someday soon the disciples of Marx and Kuhn will be able to lie down together in the fields of the Lord.

Notes

1. Before I enter into my theme, I should add that I do not intend to survey all the activity in these two very active fields. An overview of the huge literature they have generated is a subject in itself, and not my subject here. In particular, I am leaving out of my account the many scholars who work on concrete problems with a range of tools drawn from both. My justification for this oversight is twofold: first, I have not yet found among these crossovers a satisfactory theoretical mediation between the two fields; and second, the most influential figures writing theory in these fields are not seeking such a mediation, but on the contrary ignore or exclude each others' contributions. Clearly, this situation deserves treatment on its own terms.

2. The notion of rationality as a cultural form is suggested by Weber's concept of rationalization. Lukács's theory of reification refined that concept by identifying the tensions between the type of rationality characteristic of capitalist society and the lifeworld it enframes (see Feenberg 1986: chap. 3).

3. For explorations of the relation between Marxism and modernity theory, see Berman (1982) and Frisby (1986).

4. There is an enormous literature on Kuhn. For an interesting recent critique, see Fuller (2000).

5. I have tried to reformulate Habermas's position to take technology into account (Feenberg 1999a: chap. 7).

6. The early Marxist Lukács already identified this plausible outcome of differentiation as a consequence of "reification." According to Lukács, capitalist society is characterized by the rationality of the "parts"—individual enterprises for example—and the irrationality of the whole, leading to recurrent crises (Feenberg 1986: pp. 69–70).

7. I have independently proposed something similar in Feenberg (1992) and Feenberg (1991: pp. 191–198). What I call "subversive" or "democratic rationalization" resembles Beck's "subpolitics," and his "code syntheses" is similar to the social interpretation of the theory of concretization I have developed. There seems nevertheless to be a difference in our relation to the field of technology studies, which should become clear to readers of Beck in what follows.

8. Richard Feynman defends the standard view of the accident, which he helped to shape. His observations are based not on constructivist methods but on common sense. Feynman's account is devastating for NASA management.

Consider, for example, the reaction of programmers to his praise for their very thorough testing programs: "One guy muttered something about higher-ups in NASA wanting to cut back on testing to save money: 'They keep saying we always pass the tests, so what's the use of having so many?' " (Feynman 1988: p. 194).

9. "[S]i je ne parle pas de 'culture', c'est parce que ce nom est réservé pour l'une seulement des unités découpés par les Occidentaux pour définir l'homme. Or, les forces ne peuvent être partagées en 'humaines' et 'non-humaines', sauf localement et pour renforcer certains réseaux."

10. In a review of *Questioning Technology*, Douglas Kellner (2001) objects that the term "instrumentalization theory" biases the analysis of technology toward modern instrumentalist interpretations of technical practice. This was not my intent. I do believe that peoples in all societies are capable of talking intelligently about their own technical practice in ways *we* would consider "instrumental" even if *they* do not routinely distinguish technique from other activities as we do. Analyzing this aspect of their culture in an "instrumentalization theory" does not necessarily imply that they share our conception of technique. I discuss this problem in Feenberg (1995: pp. 225ff.) I hope that the introduction here of the terminology of world-making helps to cancel the unfortunate connotation of my earlier choice of terms.

11. In Feenberg (1999a) I break instrumentalization down into eight correlated operations, including the primary instrumentalization of the subject, which I call "positioning," and the corresponding secondary instrumentalization, "initiative." Positioning is the general term for occupying the specific locus from which technical action is possible: the "driver's seat." So located, the subject finds itself before a "world" of affordances that invite initiatives of one sort or another.

12. I have developed this argument in relation to computerization in a detailed analysis of the Minitel (Feenberg 1995: chap. 7).

4

Critical Theory, Feminist Theory, and Technology Studies

Barbara L. Marshall

All social life is essentially practical. All mysteries which mislead theory into mysticism find their rational solution in human practice and the comprehension of this practice.—Karl Marx, *Theses on Feuerbach*

The questions raised by the terms of reference of this volume reconstruct some of the most enduring lines of debate in sociology. The counterposing of "modernity" and "technology"—more specifically of theories of modernity and studies of technology—calls forth a series of dualisms that are only too familiar to sociologists. Among these are the oppositions of abstract and concrete, theory and fact, modern and postmodern, universal and particular, and structure and agency. Any attempt to develop methodological strategies for grasping the co-construction of modernity and technology is thus thrust head on into a long-standing sociological conversation that is far from over. This is a conversation that has been enriched, in the past several decades, by a number of feminist interventions.

The juxtaposition of feminist theory, technology studies, and theories of modernity cuts to the heart of some critical debates. One thing that feminist studies and technology studies share is the conviction that they have something to say, not just about women or technology, but about the "social" more generally.[1] However, as Judy Wajcman suggests in her recent state-of-the-art review of gender and technology studies, obstacles remain to creating a more productive dialogue. Not the least of these is that "despite the emphasis on the way innovations are socially shaped" it remains "incumbent on feminists to demonstrate that this 'social' is also a matter of gender relations" (Wajcman 2000: p. 451).

One of the problems that theories of modernity and studies of technology seem to share is an inability to think about gender as more than a categorical variable that describes some already manifest difference between men and women, and even then, only when women are visible subjects. What I hope to demonstrate in this essay is that the insistence on gender as a crucial analytical category—a task undertaken by feminist scholars analyzing both theories of modernity and studies of technology—introduces important disaggregative and normative considerations that hold potential for pointing a way out of the theoretical and methodological impasses that frame this volume. That is, I argue that the introduction of a third term—gender—into the modernity and technology nexus makes clear the need to draw on richly empirical approaches that disaggregate all three terms while retaining the ability to produce limited generalizations. At the same time, the normative character of feminism insists that we be able to envision alternatives, and such an exercise necessarily invokes a focus on practice. Because feminist sociology has developed in the context of a broad-based social movement, theoretical and methodological questions must always return to practical questions. As the epigram from Marx suggests, this is the only antidote for theoretical mysticism.

My overall aim in this essay is to clarify some methodological strategies for grasping the co-construction of modernity, gender, and technology that are often obscured by theoretical polemics. I first briefly review some of the contours of the debates about modernity and the human sciences from a feminist perspective. I then suggest how some methodological criteria might be drawn out of critical theory's engagement with modernity and how these might be redeemed with particular reference to the relationship between modernity, technology, and gender. Finally, I briefly outline my own research on sexual technologies as an illustration of how some of the theoretical and methodological entanglements of modernity, gender, and technology congeal in a concrete empirical context.

I begin with some assumptions for which I do not provide a detailed defense here. First, I assume, rather than make a detailed case for, the general legitimacy of feminist scholarship. My specific concern here is to argue that feminist theory has the potential to be a particularly robust form of critical theory with distinct methodological contributions. Second, I assume general agreement that we are now in a postpositivist era of social research, and do not need to rehearse the arguments against positivism. Third, I take as a point of departure that it is both possible and desirable to transcend the dualisms that structure theoretical and methodological debates in both technology studies and social science more generally. It is to this end that I suggest that modernity, technology, and gender are all concepts that may be understood as having currency at a variety of levels of abstraction, and that a distinctly sociological and feminist approach is instructive in grasping this.

Modernity in Crisis?

Modernity, as the central problematic of sociology, has metaphysical, institutional, and normative dimensions. Against the backdrop of the Enlightenment, modernity is associated with the release of the individual from the bonds of tradition, with the progressive differentiation of society, with the emergence of civil society, with innovation and change. As a metaphysical attitude, modernity invokes the self-conscious modern subject, the triumph of reason over passion, and mastery of nature. Its institutional reconfigurations include the growth of capitalism, industrialization, urbanization, secularization, and separation of the public and the private. The normative content of modernity has taken the sovereign individual as its icon, confidently located within a discourse of material progress and political potency.

Postmodernity too, is both a descriptor of social configuration and intellectual orientation. What has been termed the "postmodern problematic" (Felski 1989) in the humanities and social sciences includes a nexus of social and philosophical-theoretical shifts. These include attention to the increasingly media-saturated and information-based nature of postindustrial societies; assertion of *de*differentiation rather than differentiation as a social logic; an analytical focus on language, discourse, and representation; a skepticism toward grand narratives; and a critique of essentialism and foundationalism.[2]

While these thumbnail sketches provide concise stories to tell our students, and have considerable rhetorical force, neither holds up well under close scrutiny. Modernity has never existed in pure form, and could be said to have been in crisis since its inception. Just as the language of modernity has obscured the extent to which tradition has persisted, so too does the language of postmodernity obscure the extent to which the modern persists. The unhelpfulness of the language of modernity versus postmodernity is particularly revealed by feminist work. Re-examination of the sociological canon has revealed a skewed story, which effaces the extent to which the changes associated with modernity were profoundly gendered processes. The social differentiation so central to the sociological account of modernity was a distinctly masculine account, and several decades of feminist scholarship have unequivocally put trousers on the "individual" who stars in the story (see, e.g., Bologh 1990; Kandal 1988; Marshall 1994, 2000; Sydie 1987; Witz 2000). Historians have demonstrated that some periods of supposedly progressive change have been not so progressive for women, and feminist critiques of science have contributed greatly to the more general crisis of faith in scientific rationality. At the same time, feminist claims draw freely on the idiom of the Enlightenment, speaking of rights, equality, and autonomy of the person.

Only beginning to be elaborated is the extent to which the distinction between sex and gender is itself both a product of, and shaper of, modernity. It is also a distinction that has always been, and continues to be, technologically configured (Hausman 1995). The philosophical and institutional transformations of modernity, including its technological dimensions, were not just mapped onto already existing gender differences, but actively constructed and invoked difference. Technoscientific reconfigurings of sex and gender are also taken as a harbinger of the postmodern—of a new world of cyborgs (Haraway 1991), where bodies themselves are only temporarily stabilized, "posthuman" technological accomplishments. These divergent accounts signal an ambivalent relationship of feminism to both modernity and postmodernity that suggests that we reject both as grand narratives.

In sum, the ongoing desire to tell a story of modernity (and I include here those postmodern accounts whose sequel narrates "its" dissolution) perpetuates an inability to come to terms with particularity and complexity. To adequately capture the varied and complex ways that men and women have been constructed in and through modernity requires a relinquishing of any simple narrative—whether that be a story of the modern that sees it as a progressive march toward a common humanity, or as a totalizing logic of patriarchal and technological domination that can only be transcended by embracing the postmodern. Modernity is not something to be categorically for or against; casting the issues in this way only shores up some one-sided stories that accord modernity far more coherence than is warranted.

Social Science in Crisis?

The debates about modernity versus postmodernity have, however, crystallized a number of metatheoretical issues related to the forms and possibility of the social sciences and their "knowledge projects." As Gill and Grint (1995: pp. 2–3), among others, suggest, the "questions about modernism and postmodernism" are integral to the debates about how to get at the relationship between gender and technology.

Contemporary feminism has not been immune to the generalized crisis of knowledge embodied in the modernity-postmodernity debates, and feminists have contributed some of the most interesting interventions. Yet some are struck with an overwhelming sense of déjà vu. As one critique suggests, some of the supposedly cutting-edge insights read like Europeans discovering the New World, in that they seem to have perceived "a new and uninhabited space where, in fact, feminists have long been at work" (Mascia-Lees et al. 1989: p. 14).[3] After all, feminist rejections of identity as an unproblematic reflection of some natural essence long predate postmodern and poststructuralist deconstructions of the subject.

A long history of feminist engagements with traditional disciplines exemplifies a profound critique of disciplinary categories and the power-knowledge nexus they embody.[4] Within sociology, at least 30 years worth of feminist work has illuminated the socially constructed and partial nature of mainstream academic discourse through textual analysis. Many of the recent debates seem to be treading well-worn feminist ground, and have in the process tended to set up abstract oppositions— modern and postmodern, material and discursive, essentialist and constructivist—that do not accurately reflect the complexity of the logics in use in feminist or sociological analyses. Abstractions such as gender and technology (however precarious and unstable those categories may be) have been muddied, not so much by the theoretical interventions of postmodernism, but by their practical investigation. There is an analytical dynamic here that belies any simple narrative of a theoretical shift from the material to the discursive, or from modern to postmodern. The empirical always strikes back at the abstractions—whether modern or postmodern—that we construct to tame it. Abstractions are simply that—abstractions—and not ossified laws for which history or empirical research can only provide illustration. Capitalism, patriarchy, scientific-technical rationality—name your poison—exist only in and through particular and concrete manifestations.

If the gap between theory and empirical research is taken to be indicative of a crisis, then the social sciences have been in crisis for a long time. More than 40 years ago, C. Wright Mills (1959) launched a stinging critique of sociology's retreat from the classical tradition, calling for a renewal of its heritage as a critical public discourse. He advocated a "sociological imagination" that eschewed both the fetishism of method and the fetishism of concepts in favor of a deeply social and historical approach, where neither system nor practice had an independent existence. It is a call that remains deeply relevant today, and that may find its best chance for fulfillment in contemporary critical theory. Of all the varieties of western Marxism, it is the tradition of critical theory that has most explicitly recognized the political character of science and technology in modernity, even if this has not informed a strong program of empirical research.

Critical Theory as an Orientation

While the term "critical theory" is traditionally associated with the Frankfurt Institute for Social Research and with Jürgen Habermas, the best-known heir of that tradition, it may be more broadly conceptualized to include a variety of postpositivist and reconstructivist (as opposed to deconstructivist) approaches that retain a commitment to social science as "the critical consciousness of modernity" (Delanty

1997: p. 4). While critical theory today cannot properly be called a "consensus," there are enough family resemblances to at least justify calling it an orientation. As Gerard Delanty suggests: "Perhaps convergence is too strong a term, but we can detect in all of these intellectual movements a recognition of the need to theorize new forms of mediation between agency and structure, culture and power, lifeworld and system, experience and rationalization" (Delanty 1999: p. 180).[5]

Similarly, Craig Calhoun (1995: p. 11) suggests that it is not a school, but an a "interpenetrating body of work which demands and produces critique."[6] In elaborating what "critique" entails, Calhoun stresses the significance of understanding historically the social conditions that permit specific forms of practice, including intellectual practice.

Following Morrow (1994: p. 269) I define critical theory in an ecumenical sense, as a research program "with many rooms," which includes metatheoretical reflection, a substantive and historical theory of society, and normative critique. Rather than engage in detailed explication and comparison of the substantive theories of those I am including under this rubric, I summarize what I see as some of the shared points of departure:

1. There is a continuing concern with modernity as a meaningful concept, although with variation in how this is conceptualized (as late modernity, a post-traditional order, radicalized modernity, and so on).[7] A defining characteristic of modernity thus understood is some sort of differentiation of spheres, most commonly formulated via a distinction between system integration and social integration. This distinction is identified as opening potentials for reflexive or communicative action, and for new forms of identity formation, albeit in ways that are only selectively institutionalized.

2. There is a commitment to understanding the duality of structure and action, which is culturally mediated.

3. There is an incorporation of insights from a range of theoretical traditions (including, but not limited to, those identified with Marx, Weber, Freud, Mead, linguistic philosophy, and hermeneutics)

4. They have a shared rejection of positivism and empiricism, while retaining some commitment to the Enlightenment belief in the possibility of social scientific knowledge.

5. There is an insistence that relativism is not the only alternative to objectivism.

Framing all of these dimensions of critical theory as I understand it is a desire to recover "the promise" of the "sociological imagination" as formulated by Mills. Making good on this promise requires a more clearly elaborated methodological program.

One of the most frequently invoked characteristics of critical theory, at least among its advocates, is its potential to ground politically progressive, empirical research. Almost as frequently invoked by its critics—both those who are sympathetic to the overall project, and those who are not—is its failure in this respect. Despite a long list of promissory notes,[8] the relationship between the theoretical and the empirical in this tradition has remained contentious. Critical theory continues to be received, not as an orientation to research, but as a grand theory—rich in abstraction, but with little application to empirical research or political practice. Thus, while critical theory provides a useful point of departure, it is a tradition that has on the whole failed to live up to its own methodological criteria. However, I argue that an examination of feminist theory in use shows it to be a useful application of the spirit of critical theory, albeit one that has incorporated methodological insights from a broader range of theoretical traditions.

Elements of a Critical Methodology

Let me first clarify what I mean by "methodological" in this context. I do not propose to specify any particular technique or form of "data" collection as the definitive critical or feminist approach. The conflation of methodology with technique, as represented by many mainstream sociological texts on methods, is symptomatic of an impoverished conception of the relationship between the theoretical and the empirical. This "conventional methods discourse," as Morrow (1994: p. 24) terms it, results in "a methodology that is not atheoretical, but theoretical in undeclared ways." Methodology, as I use the term here, refers to what Gareth Morgan has termed "forms of engagement," including the ways in which we constitute and render subjects amenable to study (Morgan

1983: p. 19), not just the techniques that we use to study them. Thus, methodological questions are never, strictly speaking, just about methods. The familiar oppositions between inductive and deductive strategies, and particularistic narratives and general theories are boundaries determined by deeper assumptions about just what and why we study what we do. Neither gender studies nor technology studies are unique in crystallizing these issues, but they are illustrative of some of the substantive contexts in which the methodological crisis outlined earlier has been played out. In the remainder of this essay, I attempt to summarize some of the ways we might move forward, loosely organized around the themes of methodological strategies, levels of analysis, and contexts of mediation.

Methodological Strategies

As Andrew Feenberg outlines in chapter 3 in this volume, the tension between theories of modernity and studies of technology concerns basic analytical categories and research methods. If modernity theories err on the side of abstraction and overgeneralization of processes such as rationalization and progress, overly particularistic technology studies risk being unable to rise above the level of case histories.

Within critical realist philosophies of social science, this problem has been addressed through the distinction and relationship between intensive and extensive research designs (Harre 1979; Sayer 1984; Morrow 1994). Extensive research designs, oriented toward nomothetic explanation, are concerned with generalizable patterns. Intensive research designs, by contrast, are oriented toward ideographic modes of explanation and are more concerned with individual cases, taking into consideration their particularity and uniqueness. As Morrow (1994: p. 252) notes, intensive research designs are usually more appropriate to the questions posed by critical theory, but he usefully stresses the "mutual necessity" of "individual explication (an ostensibly ideographic exercise) and comparative generalization (a weakly nomothetic activity)." In other words, "one cannot even begin to describe a 'case' without a sense of 'types' of cases and their shared properties." As he summarizes it: "Comparative generalization is a logic complementary to intensive explication. Here the strategy is one of comparing the patterns disclosed through intensive explication across a finite set of historically comparable cases" (Morrow 1994: p. 212). It is the mutually necessary character of these approaches that tends to get lost in theoretical polemics that seek to dissolve, or just ignore, the tension between the particular and the general, rather than recognize it as intrinsic to critical inquiry.

Kathy Ferguson's (1993) interrogation of strategies of interpretation and genealogy as metanarratives in feminist theory provides another, more politically grounded manifestation of this tension. Interpretation searches for commonalities from which general categories or conceptual unities can be constructed. Genealogy, on the other hand, wants to challenge those generalizations, drawing attention to idiosyncrasies and differences. Interpretation rests on an "ontology of discovery"—that there is some order and meaning to the world that can be uncovered—while genealogy rests on a counter-ontology—that there is no order to be discovered.[9] Each strategy reveals some (incomplete) understanding, and each represents a challenge to positivist conceptions of knowledge. As Ferguson notes, it is possible to become "enframed" in either framework, "seeing only the battles each practice names as worthy and missing the ways in which contending interpretations or rival deconstructions cooperate on a metatheoretical level to articulate some possibilities and silence others" (Ferguson 1993: p. 7). In methodological practice, we can see a fluidity between interpretation and genealogy that belies this opposition. Interpretation, as a strategy, always employs genealogical strategies to ground its critique of dominant interpretations. By the same token, genealogy depends on interpretation "to provide something to deconstruct." As she summarizes it:

genealogy and interpretation can . . . be seen as postures toward power and knowledge that need one another . . . Interpretation produces the stories we tell about ourselves and genealogy insists on interrogating those stories, on producing stories about the stories. This interrogation could go on forever; "stories about the stories about stories about . . . ", the infinite regress of metatheory. But . . . one can insist on (unstable) bridges between interpretation and genealogy, with a commitment to continue, combined with a recognition of limits. (Ferguson 1993: p. 29).

The holding in tension of the interpretive and genealogical moments in feminist theory has its corollary in the multivocal and coalitional politics that have increasingly come to define feminism. Coalitional strategies have been forged through recognition that our categories are never absolute, never unproblematically linked to stable axes of power, but shift temporally and in relation to materialized interests.[10]

The first methodological principle that I want to emphasize, then, is that strategies of narrative and analysis, or intensive explication and comparative generalization, are not only complementary, but continually invoke one another. To forswear macrolevel and comparative analyses in the name of recognizing contingency is as mistaken as assuming that historicity and contingency immediately confer unintelligibility. To put it bluntly, just as one need not become a full-fledged determinist to identify certain objective conditions (however temporarily stabilized) of the gender-technology relationship, neither need one embrace an unbridled indeterminacy to recognize the insights of constructivist approaches.[11]

Perhaps it is the fact that many feminist analysts of technology have come to their research, not from science studies, but from more practical interests[12]—in the labor process, the domestic sphere, health and illness, and so on—that grounds their pragmatic insistence on this mutuality. Engaging science and technology for feminists has been more than an interesting theoretical puzzle. Inquiry is rooted in the political project of identifying and analyzing sites of the production and reproduction of sex and gender divisions and inequalities. This practical grounding underlies the insistence on moving beyond the particulars of a given instance of technological innovation to some form of (however limited) comparative generalization, and recognition that both moments are integral to an adequate analysis. For example, Adele Clarke's (1998) recent study of the development of the reproductive sciences originated in her interest in the politics of conception. While she draws on the strategies of intensive explication characteristic of constructivist approaches to technology,[13] the "social worlds" approach she elaborates owes much to the traditions of symbolic interactionism and "grounded theory." Thus, she is able to relate the reproductive sciences as "communities of practice and discourse" (Clarke 1998: p. 15) to broader contexts, including institutional shifts in the organization of scientific disciplines; philanthropic and birth control movements; and at the most general level, within modernity—industrialization, rationalization, professional-ization, and specialization in terms of market value and effectiveness (Clarke 1998: p. 259).

Instructive as well is the well-known study of the microwave oven by Cynthia Cockburn and Susan Ormrod (1993). They, too, find ap-proaches developed in social constructivist and actor-network studies useful, but limited given their interest in generating insights beyond the specific case. In seeking to understand gendering as integral to technol-ogy—from invention through design, production, marketing, distribu-tion, and use—their analysis required them to "maintain a distinction between local action and individual agency on one hand, and on the other, the longer-lived and more widely spread social structures, partic-ularly those of class and gender, that shape probabilities and incline or dispose our individual and collective choices of behavior and thought" (Cockburn and Ormrod 1993: p. 10). Without some sense of those "longer-lived and more widely spread social structures" which are the stuff of feminist theories of modernity, technology studies are left with a particularly anemic conception of power, have no way of opening to critical analysis how the relevant social groups are constituted, who has been left out of the analysis and why, and how case studies might gener-ate insights that extend beyond that particular case.

Levels of Analysis

Clearly required is a willingness to work with constructs such as tech-nology, rationality, and gender at different levels of abstraction, always understanding that we are freezing complex and fluid social relation-ships as we do so (compare Edwards, chapter 7, this volume). This ques-tion of levels, however, leads us back to the more general problem of the duality of subject and structure, or structure and agency, that critical theory takes as a central problematic. This is also a problem that I think has been muddied, rather than clarified, by some cul-de-sacs in the post-modern turn.

The distinction between system and lifeworld that is integral to modernity, as developed by theorists such as Giddens and Habermas, is not identical to the micro-macro distinction as reiterated in mainstream sociology. It is the duality, or their mutual constitution, that is central.

While sympathetic critics such as Feenberg (from the perspective of technology) and Fraser (from the perspective of feminism) are correct to problematize the manner in which Habermas has developed the distinction between system and lifeworld,[14] the distinction itself is important in holding in productive tension the relationship between system integration and social integration. It is also significant methodologically. As Calhoun (1998: pp. 868–869) suggests, part of the value in the distinction is the ability to "distinguish dimensions of social life that could be understood well in agent-centered accounts . . . and those that would be missed or systematically distorted if only understood in such a way." Clearly both gender and technology are such dimensions of social life.

One of the analytical shifts identified as part of the postmodern turn is a move from material to discursive modes of analysis, or from things to words (Barrett 1992). This has been particularly contentious within feminism (see, for example, Jackson 1999) because it is seen to block reflection on both material inequality and the systemic nature of inequalities. The "turn to culture," with its emphasis on discursive construction and interpretation, also has a distinctive presence in constructivist approaches to technology. While feminists have certainly found this tack useful, there are problems in fully extending its premises. By way of illustration, we might look at a debate on constructivism and feminism conducted in *Science, Technology and Human Values* (Grint and Woolgar 1995; Gill 1996; Woolgar and Grint 1996). Grint and Woolgar criticize purportedly antiessentialist (and especially feminist) approaches to technology for perpetuating essentialist assumptions.[15] They challenge antiessentialists to eradicate all conceptions of nondiscursive elements. Technologies are not merely shaped or affected by social process, but are fully constituted by them. Methodologically, textual interpretation is *all* that counts. Gill (1996) objects to what she perceives as their assumption of "semiotic democracy" in the interpretation of technology and wants instead to permit a form of analysis that will admit to imbalances of power and that will admit ethical considerations. This insistence is the reason so few feminists are willing to do an antiessentialist "full monty."[16] More important for our purposes here is the manner in which Grint and Woolgar phrase what they see as the important question: "To ask whether an artifact is male or female or neutral is to miss the point; not only are these properties themselves socially constructed and therefore flexible, but the important question is how certain artifacts come to be interpreted (and this may well be disputed) as male or female or as neutral" (Grint and Woolgar 1995: p. 292).

If "how artifacts come to be interpreted as male, female or neutral" is what we are after, then analysis that takes as its focus the encoding and decoding of technology as text might suffice.[17] I would suggest, however, that this is not really the important question for feminists. Of primary concern is *why* artifacts come to be interpreted as male or female or neutral, and this admits questions of larger social structures and processes. I introduce the "why" here, not in a functionalist sense, but in the sense of what makes this process—the gendering of artifacts—intelligible in the first place. Then we might ask how technology is implicated in configuring gender as a social relation, and how the social relations of gender configure technologies, not just how masculinity and femininity are encoded or decoded in particular texts (whether those be technological or anything else). A particular concern that Grint and Woolgar have is debunking the determinism (here of a social character) they see implied in asserting that technology has effects. A different approach might reframe this to see both technology and gender as having effects, not in the causal sense in which they seem to use the term, but in the sense of constraining or enabling forms of practice. This would include seeing gender, understood as a social relation, as constraining or enabling certain forms of scientific or technical practices. Thus, the sort of analysis that Grint and Woolgar advocate can best be seen as contributing to, but not exhausting, a mediational level of analysis that admits that we do not (to poach one of Marx's most enduring insights) encode, decode, or otherwise construct things entirely under conditions of our own choosing. How else are we to meet the demand to "not efface real differences of power, access and control in relation to technology along gender, class, 'racial', and other lines"? (Gill and Grint 1995: pp. 25–26)

Morrow (1994) has identified three "moments" of inquiry: systemic, actional, and mediational. The first includes macro-structural analysis of open, historical social formations, and focuses on the level of system integration. The second is located at the level of social integration, focusing on skilled actors constructing reality through praxis. The third moment of sociocultural mediation seeks to identify those points of dereification where ruptures are possible between systemic structure and social action. As he suggests, "there are practical methodological grounds for research designs that focus on one or other of these dimensions, at least as long as their ultimate unity in mutual constitution is never completely forgotten" (Morrow 1994: p. 269). That is, if we are to preserve the potentials of the system and lifeworld distinction in identifying emancipatory potentials, then all three moments must be enacted in research. Indeed, from a critical perspective, each of these modes of analysis presupposes the other even though polemical priority disputes often obscure this complementarity" (Morrow 1994: p. 277). This is the second methodological principle that I want to suggest: that system, action, and mediational levels of analysis be recognized not as discrete, but as mutually constitutive moments of critical research.

Andrew Feenberg's development of a critical theory of technology (1995, 1999a) suggests the outlines of such an approach. Central to it are his concepts of "implementation bias," "technical code," and the related framework he develops of different dimensions of "instrumentalization." The concept of implementation bias builds on the notion of interpretive flexibility as developed in the social construction of technology literature, which suggests that there are alternative ways that technologies might be taken up. There is, however, a technical code, reflecting the "cultural horizon" of a given social formation (such as capitalist modernity) that mediates design and implementation. Feenberg argues that technology itself is a form of mediation of social activities (thus conceptualizing technology as analytically analogous to markets or bureaucracies as forms of mediation). As he summarizes it: "markets, administrations, technical devices are biased and embody specific value choices" (Feenberg 1999a: p. 174).

Feenberg offers a two-level theory of instrumentalization. Primary instrumentalization, or deworlding as he has more recently referred to this process, is reifying and abstracting, and is similar to what earlier dystopian analyses of the modernity-technology relation characterize as the embodiment of technological or instrumental rationality. Secondary instrumentalization, or reworlding, reembeds technology in social context and introduces possibilities for reflexivity, action, and dereification. Such a theory clearly lends itself to informing research along the lines of the critical methodology sketched out earlier. More important, Feenberg explicitly introduces the potential democratization of technology as a central component of the model, by attending to points of recontextualization and possibilities for change. This requires a more sustained attention to the question of mediation.

Contexts of Mediation

While I am in broad agreement with the theoretical framework that Feenberg develops, in several respects I find it requires further development to be up to the task of understanding the gender-technology-modernity nexus. Certainly, he is right to recognize the tension between system and lifeworld as integral to a critical theory of technology, and to identify the mediation of this tension as integral to opening up spaces for transformation. This pushes us to think about technological democratization as something much more than access to technology. However, I worry that posing the manner in which deworlding and reworlding get played out as producing "*either* a technocratic model of control *or* a democratic model of communication" (emphasis added) might close off some important questions (Feenberg, this volume). Feenberg appears to use a fairly weak conception of democratization as "participation,"[18] which lends itself to what might be seen from a feminist perspective as an overly optimistic prognosis for the development of a "technical public sphere."

The notion of secondary instrumentalization needs to more explicitly recognize that in the recontextualization of technology, the nontechnocratic participation of social interests and values may not always be progressive, in the sense of advancing the sort of technological democracy that he advocates. Let me provide a brief example from research in which one of my colleagues is involved, which has examined the development and use of the Sexual Assault Evidence Kit (SAEK) as a forensic technology (Parnis and Du Mont 1999, 2002; Du Mont and Parnis 2000).[19] The SAEK was developed in the late 1970s, in part as a result of pressure by feminists working politically and professionally in the area of sexual assault. It was designed as a means of obtaining more standardized and reliable evidence in sexual assault cases, with one of the desired ends being to help women attain more positive judicial outcomes (Parnis and Du Mont, 1999: 76).[20] So far, so good for participation. However, Parnis and Du Mont's analysis of both the design and utilization of the SAEK demonstrates multiple points of recontextualization (including through various professional cultures) at which social interests and values are inserted that perpetuate traditional biases rather than eliminating them or opening them to critical scrutiny. For example, the kit was not consistently administered in cases where the assault did not involve full penetration or in cases of acquaintance or spousal assault, reflecting long-standing myths about what constitutes "real" rape. As Parnis and DuMont conclude (2002, in press), "the kit may carry a legitimacy and a symbolic value which exceeds the capabilities of science to objectively determine defining the 'facts' of a case." This stands as a cautionary example of how participation in technological innovation and utilization may in fact contribute to the process of interpretation being removed from political contestation via that symbolic value. Thus, the third methodological point that I want to stress is that the level of mediational analysis must attend to the complex construction of contexts for agency, including an awareness of ways in which hegemonic codes may be both subverted and/or reproduced in the dynamic relationship of structure and agency. This requires more sustained attention to the construction of both agents and contexts, particularly if the intent is to identify and nurture emancipatory potentials.

While Feenberg acknowledges the feminist critique of abstract constructions of modernity (for example, via Nancy Fraser's [1989] critique of Habermas), he seems unwilling to more fully extend that acknowledgment to the recognition that gender, like technology, may be understood as a sort of code that has profound theoretical and methodological significance.[21] This reluctance, combined with the rather weak conception of democratization as participation, further complicates Feenberg's contention that democracy is something that can simply be extended to technology,[22] and this tends to weaken the potential of his framework for really grasping co-construction. Thus, we see an imaginary concept at work, which seems to assume that individuals enter into the democratic public already formed in their identities, with already existing (if, perhaps, latent) interests.[23] What a good deal of feminist work has demonstrated, however, is that the very formation of those identities and interests is at the heart of any political process. As Nancy Fraser (1989: p. 172) notes, "groups of women have politicized and reinterpreted various needs, have instituted new vocabularies and forms of address, and so, have become 'women in a different, though not uncontested or univocal, sense'."

Gender is not just taken as a political point of departure, but is actively constructed as an identity. It is only the explicit politicization of gender—which always reflects the practices that work to exclude or suppress it—that makes gender relevant, and this will not necessarily occur as an organic component of the democratic process, whether that be the democratization of technology or anything else.[24] In seeking to open spaces for democratic transformation, the real challenge is undermine the "grammar of liberalism" (Young 1997) that risks letting the system versus lifeworld distinction be conflated with the public versus private distinction, with identity formation occurring in the latter.[25] It should also be clear that if we need to retain a sense of active and ongoing formation in the construction of democratic agents, then we cannot accept less in our conception of the contexts in which they act. As Slater (chapter 5, this volume) persuasively argues, we cannot just simply put "things" (such as particular technologies, or specific forms of gender relations) in context, because the latter is "produced by the very 'thing' one is trying to put into it." The burden on the analyst is to grasp the concrete situational dimensions of this process while keeping one's eye on the systemic ball.

Let me attempt to briefly summarize my argument to this point. I have suggested that an adequate methodological approach to disentangling the co-constructions of gender-technology-modernity requires a more explicit grappling with the tension—conceived here as a productive tension—between system integration and social integration that is central to theories of modernity. I have also suggested that in order to engage in this sort of critical inquiry, we need to recognize that what theoretical polemics tend to set up as oppositions are more fruitfully conceived of as mutually constitutive. The interesting ground, I contend, is at the mediational level—the level of practice—which recognizes how both abstract systemic logics and concrete situational factors join to constrain or facilitate the development of particular identities, agency, and contexts, including the fostering or blocking of democratic tendencies. Furthermore, implicit in the framework I have outlined is that modernity, technology, and gender are all concepts that can be instantiated at multiple levels in this process. The next section briefly outlines some of my current research on sexual technologies as a means of illustrating some of these ideas.

Sexual Technology and Heterogendered Bodies: Deworlding the Genitals?

My current research is investigating biotechnical remedies for sexual dysfunction, and has its roots in the media frenzy over Viagra. Introduced to the American market in 1998, Viagra (sildenafil citrate) was lauded as the first effective treatment for erectile dysfunction. My interest in this specific technology was piqued when, while researching antifeminist interpretations of the concept of gender, I came across the following assessment of Viagra's potential from Bob Guccione (publisher of *Penthouse* magazine) in a *Time* magazine cover story: "Feminism has emasculated the American male, and that emasculation has led to physical problems. This pill will take the pressure off men. It will lead to new relationships and undercut the feminist agenda" (cited in Handy 1998: p. 44).

I was fascinated by the manner in which a pharmaceutical product was being granted causal agency to both counter a political movement and to ground masculine identity. It seemed an extreme example of technological determinism, and a manifestation of the tendency to locate gender ever more deeply within the body, more resolutely presocial and hence less open to contestation and reconstruction. Along with other feminists, I had found myself problematizing rather than assuming the lingering substrate of gender: sexed bodies. Their very construction through sexual medicine, which both invents and seeks to remedy pathologies of sex, seemed fertile ground indeed for pursuing this problematization. My research to date[26] suggests that there is no point at which technology and modernity are not joined in some way in the production of sexual bodies. It is, in fact, only through attempting to disentangle the relationship between modernity and biotechnology in this concrete context that the deep ontology of sexual difference that underpins both is illuminated. In other words, while the clinical and market success of Viagra has prompted both the scientific and popular literatures to speak of a "new age" in human sexual relations, assigning it a causal role in social change (and in particular, in affecting gender relations), only a deep ontology of gender and sexual difference made possible both the scientific research and the technological development behind Viagra and other sexual technologies.

The story behind Viagra is a complex history of the manner in which sexual dysfunction has been constructed and reconstructed in relation to a range of distinctly modern phenomena—including the rationalization and medicalization of sexuality (Jackson and Scott 1997; Tiefer 1996), the increased importance of expert systems and knowledge in managing everyday life (Giddens 1991; Rose 1996), and the expansion of consumer culture (Slater 1997). A historical analysis shows numerous junctures where shifts in scientific and medical conceptions of the sexual body have occurred, disease models of sexual dysfunction have been constructed and revised, and users of (and markets for) sexual technologies have been configured. By examining the implicit social claims embodied in this history, the extent to which biomedical anxieties over sexual function reflect broader social anxieties about gender and sexuality becomes apparent. I can only briefly allude to some of these themes here,[27] but I hope it will be sufficient to illustrate some of the analytical themes suggested in earlier sections of this chapter.

Without recounting a detailed history of sexual science, two significant shifts should be noted: first, the rise of science as the authenticating voice on what constitutes the normal and the abnormal, and second, a reframing of the abnormal to emphasize dysfunction rather than moral danger (Hawkes 1996). What is the function that sexual dysfunction threatens? Quite simply, it is penile–vaginal intercourse in the marital (or at least stable heterosexual) unit. The function is successful intercourse, which is functional for the couple, which is functional for society. It is not that this understanding of sexual function is overtly repressive of other forms of sexual expression or behavior, but that it operates through an increasing valorization of, and eroticization of, marital intercourse. This has a long history, from classical sociology's emphasis on the function of marriage in regulating passion (Sydie 1994), through the proliferation of "marriage guidance" that eroticized marital sex in the first half of the twentieth century (McLaren 1999). The increasingly scientific turn of sexology did not divest it of this normative framing. Key contributors to the modern science of sex framed their work as science in the service of the greater social good—as facilitating successful "marital coitus" (Kinsey et al. 1953) and curing "sexual inadequacy in the marital unit" (Masters and Johnson 1966). As part of its concern to constitute itself as an authoritative voice on such matters, sexual science has increasingly asserted a physiological basis for sexual problems within a medical paradigm of diagnosis and treatment. The medicalization of sexuality has rendered it amenable to intervention and management according to a biomedical model. That biomedical model accepts scientific rationality as a basic premise, which seeks universal truths about the body as a biochemical machine (Gordon 1988).

Nowhere is this more apparent than in the technologization of the penis in the construction of erectile dysfunction. As one of the scientists puts it:

Few fields in medicine can match the rapid progress that has been made in our understanding of male erectile function. These changes have been profound, and fundamental. Baseless speculation about the essential vascular mechanisms of erection and the belief in a predominantly emotional etiology have given way to the identification of the molecular events resulting in an erection and to effective pharmacological treatment of their alterations. The current state of the art is a pre-eminent example of what is achievable by systematic and conscientious application of basic research and clinical observation. (Morales 1998: p. xv)

This neatly encapsulates the story told by the scientists—it is a narrative of progressive discovery, assisted by new techniques of visualization, which has allowed them to get at the truth about "the molecular events resulting in an erection." Erectile dysfunction becomes a simple mechanical problem. As another scientist puts it, "The man needs a sufficient axial rigidity so his penis can penetrate through labia, and he has to sustain that in order to have sex. This is a mechanical structure, and mechanical structures follow scientific principles" (Dr. Irwin Goldstein, cited in Hitt 2000: p. 36).

The penis, however, is only partially deworlded here; on the one hand it is conceptualized as a fairly simple hydraulic mechanism, but on the other hand it is never entirely decontextualized from the act of heterosexual intercourse. This is what makes sexual technologies such a rich site for exploring the issues I have tried to raise in this paper—the abstract systemic logic of technical rationality as it is refracted through the construction of functionally sexual bodies both depends on and shapes the lifeworld, that locus of the concrete experience of intimate relationships. While research and production related to sexual dysfunctions and their biotechnical remedies occur within an international network of scientists, not to mention a global biotechnology and pharmaceutical industry, distress or dissatisfaction with sexual experience and the search for and/or consumption of technological expertise occur in very specific contexts. Particularly interesting is the manner in which these different worlds—of the scientist, the pharmaceutical company, the clinic, the sufferer of erectile dysfunction and his sexual partner—are articulated.

The task of decoding these articulations has been facilitated by the very public presence of the penis and its discontents in mass-media coverage of advances in sexual medicine and the proliferation of mass-market paperbacks on erectile dysfunction and its remedies. A critical reading of these irreducibly cultural interpretations of technology demonstrates the extent to which heterogendered sexual bodies are a crucial link between abstract systems and concrete lifeworlds. As these cultural products—television programs, newspaper and magazine articles, self-help books, advertising—consciously seek to act as translators between the worlds of science and technology and intimate relations, they not only mediate the relationship between them, but reveal much about how actors in each understand the world of the other. For example, in a segment introducing the topic of erectile dysfunction in an episode of a popular Canadian science program devoted to the penis,[28] we are introduced to an older man, his wife by his side, who tells us that when "the erections just weren't what they used to be" they decided to "do something about it." They proceeded to see what technologies were out there to help, and a vacuum pump now "lives" on their bedside table. Another couple is thrilled when, after unsuccessful results with previous treatments, the husband is asked by his doctor to be part of an

open-label trial of Viagra. He leaps at the opportunity, and they glee-
fully tell us how she just *knows* when he's taken a pill because he gets a
special sparkle in his eye. Whatever the situational specifics of their relationships, and however the technologies were reembedded in those relationships, neither seriously questioned that there was a technological fix
either available or immanent, and should that one prove unsatisfactory,
another would come along in due course. Their faith in expert systems
and scientific and technological progress was clear. This theme is consistently reiterated in case study after case study, as they are recounted in
both the popular and clinical literatures.

It is also clear that scientific and technological advances in the treatment of sexual dysfunction do not proceed strictly on the basis of abstract
logics of rationalization or technological progress. The development of
sexual technologies is premised on socially rich conceptions—albeit
overly universalized and objectified conceptions—of who the potential
users of these technologies are and what their motivations are. Nothing,
we are told, "not even cancer or heart disease" (Melchiode and Sloan
1999: p. 17), can be as devastating to a man's self-confidence or as dam-
aging to a relationship as a faulty erectile mechanism. Again, case study
after case study recounts the very tangible anxieties involved—worries
about aging, about the ability to satisfy one's partner, about the consequences to the relationship if they don't.

As in earlier manifestations of sexual science, technologically oriented
sexual medicine has a clear sense of its social mission, which reaches far
beyond the amelioration of personal troubles. Erectile dysfunction is (especially given anxieties over aging populations in western societies) of
potentially epidemic proportions, and poses a serious public health concern (Aytac et al. 1999; Hatzichristou 1998). It is not just a medically
manageable disease, but is increasingly framed as a progressive condition, with phases and early warning signs (Lamm and Couzens 1998).
Remedies such as Viagra are seen as part of a broader regime of bodily risk
management and "penile fitness" (Drew 1998; Seiden 1998; Whitehead
and Malloy 1999). The Human Genome Project is hailed as holding out
the hope of prevention through gene therapy (Christ 1998).

Science has more recently turned its attention to female sexual
arousal disorder—the corollary of erectile dysfunction in men is vaginal

engorgement and clitoral erectile insufficiency syndromes in women (Goldstein and Berman 1998). While the initial clinical trials of vasoactive drugs such as Viagra for women have been disappointing, the U.S. Food and Drug Administration has recently approved the first mechanical therapy,[29] and clinical trials with various pharmaceutical products, including hormonal therapies, continue. In reading the scientific and clinical literature, one cannot help but be struck by the limitless horizon envisioned.

One way of reading the bodily configurations being produced here is suggested by postmodern analyses, whereby we might see technologically enhanced genitals as an illustration of "cyborg bodies" (Haraway 1991), or "hybrids" (Latour 1993). I find a different reading more compelling. While academics may see the proliferation of sexual technologies as a harbinger of a postmodern age, where bodies have no limits, and the nature-culture division is irreparably blurred, there is no reason to suppose that those developing or availing themselves of these technologies share that interpretation. In the case of technologizing the genitals, it is mastery over, not playful transformation of, the body that is at stake. What is being sold in these technologies is not flexible, malleable bodies—it is reliability, predictability, and calculability, all within the context of rather rigidly heterogendered performance expectations. What makes the whole enterprise intelligible is the distinctively modernist framing shared by the scientists, pharmaceutical companies, physicians, and consumers: that the diligent application of scientific rationality will result in discoveries that will lead to technological innovations that will solve objectively defined problems, that those with problems will seek out the appropriate expertise to advise them and products to help them, and that scientific progress will result in even better solutions in the future.

We may, as Latour (1993) asserts, have never *been* modern, but we certainly think and act as if we were. Certainly the effects of technologies produced and marketed within that frame are never foreclosable in advance, and may, in fact, reconfigure bodies, sexuality, and gender relations in ways that are unpredictable, and which belie the distinction between nature and culture. They may even contribute to delightful erotic experiences and satisfying relationships. They may also retraditionalize and renaturalize, rather than radicalize, phallocentric and gendered heterosex, and close off as many possibilities as they open.[30] The very possibility of such technologies presupposes certain assumptions about both the bodily and cultural parameters of sexuality. As Menser and Aronowitz (1996: p. 12) suggest, "pushing the boundaries" is not the same as eliminating the materiality of sexuality.

The sexual body is pivotal to processes of both system and social integration and to the tension between them.[31] It is only at the mediational level that we are able to unravel the entanglements of system and lifeworld, and open space for a critique (and transformation) of the manner in which an instrumentally rational logic is refracted through even our most intimate experiences. This is not to suggest a straightforward colonization of lifeworlds by strategic technical systems. This sort of one-way thinking lingers, as Judy Wajcman has pointed out, in the residual technological determinism that continues to shape empirical research on gender and technology. As she suggests (Wajcman 2000: p. 460): "while at the theoretical level, we all take for granted that gender and technology are mutually constitutive, I would still argue that the weight of empirical research is on how technology shapes gender relations, rather than on how gender relations are shaping the design of technologies."

To really get at co-construction, we need to look for the relationship between system and lifeworld (and for potential points of transformation) by grasping its instantiation through grids of power that situate bodies and subjectivities in particular (and analytically comprehensible) ways. In the case of biotechnologies of sex, this instantiation is, to put it bluntly, an extraordinarily profitable anatomization and renaturalization of cultural heterogender. This is not to suggest, however, that things could not be otherwise. It does suggest that opening possibilities for transformation requires further inquiry into the ways in which discursive closure on deep-rooted assumptions about sexuality and gender is enacted through technology, and how these might be opened up to foster new forms of, and sites for, sexual agency.

If critical research on the gender-technology-modernity nexus is to be able to envision alternative futures, then we need to be able to distinguish what is specific and contingent about the concrete entanglements that we study, and what can be abstracted as more general and enduring. If we cannot do this, then we end up in either the postmodern fantasy world of unbridled fluidity and flux, or in what Andrew Sayer (2000: p. 722) has termed the fatalistic "nothing-can-change-until-everything-has-changed dilemma." Is it possible to envision technological futures—including sexual technologies—that neither construct nor reproduce deep ontologies of sex and asymmetries of gender? Only, I think, if we take the technological present as an instantiation of both historically and culturally produced lifeworlds and more abstract and systemic forms of rationality that can only be analytically separated from their concrete manifestations. It is in that analytical separation that the critical space for understanding and potential transformation obtains.

Conclusions

The methodological principles that I have emphasized (the mutual necessity of intensive explication and comparative generalization; the mutually constitutive moments of system, action, and mediational levels of analysis; and attention to the complexity of both reproduction and disruption of hegemonic codes at the mediational level of recontextualization) are in fact already the defining characteristics of good, critically inclined research, including feminist research.[32] Thus, my intent in systematically drawing them out here is not so much prescriptive as it is to argue for their more general applicability in critical research on technology. In suggesting this, I think that there is already much to agree on in the various communities of critical theory, feminist theory, and technology studies. That the technical is social, and the social is technical is now widely accepted. That there is a discernible relationship between gender and technology is not in dispute. To emphasize practice as the point at which the social shaping of technology occurs seems noncontroversial. However, as MacKenzie and Wajcman (1999: p. xvi) suggest: "If the idea of the social shaping of technology has intellectual or political merit, this lies in the details: in the particular ways technology is socially shaped, in the light these throw on the nature both of 'society' and of 'technology'; in the particular outcomes that result; and in the opportunities for action to improve those outcomes."

In other words, the framework that I have outlined in this essay cannot in itself generate knowledge about modernity, technology, gender, or their mutual constitution. This remains the task of detailed empirical and historical investigation that is able to enact the various moments of critical research suggested here. Theory—whether of modernity, technology, gender, or anything else—is best advanced through engagement with concrete problems, and these rarely present themselves in tidy packages.

Notes

1. For an example of this argument from the perspective of science and technology studies, see Woolgar (1996: p. 235), who suggests that "STS is no longer merely concerned to convey substantive findings about science and about technology, but instead finds itself involved in attempts to 'respecify' key notions such as 'social,' 'society' and 'agency'." As I have argued in an earlier work, feminist theory is also centrally concerned with a rethinking of "the basic analytical categories of social theory" (Marshall 1994: p. 2).

2. I will not attempt to sort through the troublesome distinction between postmodernism and poststructuralism. Maintaining a careful distinction is made difficult by their continual conflation in the literature. Briefly, postmodernism is a theory of society and social change. It rejects the possibility or desirability of resuscitating the Enlightenment project of normatively grounding an emancipatory practice, seeing these aspirations as historically passé. Poststructuralism is an analytical stance grounded in a refusal of the coherent subject, and concerned with language, discourse, and representation. I believe that poststructuralist insights and methods of analysis may be usefully appropriated without accepting the premises of postmodernism.

3. Stevi Jackson makes a similar argument on behalf of sociology in general: "Sociologists have long been aware, for example, that there is no essential pre-social self, that language is not a transparent medium of communication, that meanings shift as they are contested and re-negotiated, that knowledge is a social construct rather than a revelation of absolute truth" (Jackson 1999, section 2.2).

4. For one of the most influential texts in this respect, see Smith (1974).

5. Here he is referring to the collective work of Jürgen Habermas, Alaine Touraine, Zygmunt Bauman, Ulrich Beck, and Anthony Giddens.

6. Calhoun's (1995: p. 34) list includes, in addition to Habermas and "more direct heirs of Horkheimer and Adorno," Pierre Bourdieu, Michel Foucault, Donna Haraway, Dorothy Smith, and Charles Taylor.

7. There is justification here for including in this definition some sociological conceptions of postmodernity which argue that the present social formation is

significantly unique to justify a distinction from modernity, but do so in a way that does not overexaggerate a radical break. In this category I would include those of Zygmunt Bauman and David Lyon.

8. Here we might include Horkheimer's distinction between traditional and critical theory, Adorno's intervention in the "positivist dispute in German sociology" (Adorno 1976), Habermas's statement of the tasks of critical theory, and Giddens' confidence in the empirical applicability of structuration theory. For assessments (mostly negative) of their success in informing research in different substantive contexts, see Blaug (1997), Dryzek (1995), and Jahn (1998).

9. While this ontological opposition is certainly reflected in tensions within feminist theory between the articulation of women's experience on one hand, and efforts to deconstruct the category of woman altogether on the other, it is not a simple restatement of this tension. That it is frequently reduced to this, and the related debates about voice and the authority to speak for women, reflects a slippage between ontological, epistemological, and political questions. For elaboration, see Ferguson (1993), Hennessy (1993), and Marshall (2000).

10. I discuss this more fully in Marshall (2000). See also Young's (1994) use of Sartre's notion of seriality, and Nicholson's (1994) use of Wittgenstein's concepts of games and family resemblances.

11. Such a tack also avoids the tendency, displayed ironically by many of those who reject such modernist arrogance, of positing a linear conception of theoretical and methodological progress, with postmodernism and deconstruction as its apogee.

12. Hapnes and Sørensen (1995) make a similar observation, although for a somewhat different purpose.

13. Specifically, she locates her work in relation to the constructivist approaches represented by both actor-network theory and social construction of technology (SCOT), citing those such as Bijker and Law (1992), Bijker et al. (1987), Law (1991), and Latour and Woolgar (1986).

14. For a particularly valuable collection of feminist engagements with Habermas that constructively criticize the system-lifeworld distinction, see Meehan (1995).

15. The charge of essentialism has also become the *j'accuse* of many contemporary feminist debates. While it is beyond the scope of this essay to more fully discuss the debates around essentialism, a few words are in order regarding the manner in which this has been taken up in the gender-technology literature. Many of the theoretical objections to feminist essentialism in technology studies are directed at the rather functionalist and ahistorical conceptions of patriarchy and rigidly categorical conceptions of gender which have now been the subject of at least two decades of sustained criticism within feminism. I have discussed this at length in other publications (Marshall 1994, 2000). In short, rather than "black-boxing" gender, as these critiques suggest, the unpacking of that box has been one of the primary themes in feminist work for quite some time.

16. This is reflected in the admission of, in even the most deconstructive feminisms, concepts such as "strategic essentialism" or "contingent foundations."

17. A somewhat analogous set of debates in media studies has highlighted the extent to which the economic and political contexts in which texts are encoded and decoded is obscured through purely textual readings. Stuart Hall's work is a useful point of reference here (Hall 1980).

18. I owe this phrasing to Andrew Light (2000), who has discussed this in relation to environmental politics. I thank him for sharing this paper with me prior to publication. In contrast to Light, however, I maintain that a stronger conception of democratization is required to sustain Feenberg's argument for alternative modernities.

19. Their research has analyzed the design of the kit, the medical and police records of over 200 cases of women presenting for sexual assault treatment, surveys of a range of professionals involved in the use of the Sexual Assault Evidence Kit (nurses, nurse examiners, physicians, police officers), and interviews with forensic scientists who analyze and interpret the physical findings. In addition to numerous instances of subjective bias that permeate the kit's path through the medical, scientific, and legal cultures, little or no relationship between medical forensic evidence collected through this technology and positive legal outcomes has been found.

20. At least from the perspective of its feminist proponents, this was one of the objectives. As Deborah Parnis has suggested to me in a personal communication, from the point of view of many (including Crown Attorneys and many medical personnel), its purpose was viewed as getting at "the truth" about what "really" happened.

21. This has been developed in Cohen and Arato's critique of Habermas and Fraser, especially chapter 10. As they suggest: "That power operates through gender codes, reducing the free selectivity of some and expanding that of others, is the most important and paradigmatic core of any theory that might be labeled feminist" (Cohen and Arato 1992: p. 542).

22. While Feenberg implies this in several publications, the notion of democracy expanding and advancing clearly frames his book, *Questioning Technology*, in which he begins with the assertion that "technology is now about to enter the expanding democratic circle" and concludes that technology can be "swept into the democratic movement of history" (Feenberg 1999a, p. 225). As I have argued elsewhere (1994: p. 2000), the democratic public sphere is not an already realized ideal, and thus cannot simply be extended to now include technology.

23. To be sure, this is not a problem unique to Feenberg. It is also haunts Habermas's (1996) work on deliberative democracy.

24. Analyses of the "triumph" of democracy in post-Soviet eastern Europe are instructive here. It was widely implied in both academic (including some feminist accounts) and nonacademic commentary that the observed decline in rights for women was a legacy of the association of gender equality with discredited

policies of Communism, and would eventually be ironed out with the extension of a democratic political culture. However, as Peggy Watson (1997: p. 156) has argued, it was democratization itself that permitted gender to become a social characteristic of political significance, which "engaged and mobilized" difference in the construction of political identities. Her analysis of Poland traces how deep differences of sex had to be constructed and invoked to legitimate forms of masculinist politics in the newly democratized public.

25. It is on this point that I see the greatest contribution of standpoint epistemologies, as developed, for example, by writers such as Harding (1996) and Hartsock (1987) with reference to gender, and West (1988) with reference to race. As Rosemary Hennessy (1993: pp. 95–99) argues in her excellent discussion of standpoint theories, their import lies in "pushing on the boundaries of Western individualism" by situating historical constructions of subjects in a systematic analysis of the social production of difference.

26. I am still in the relatively early stages of a long-term project on sexual medicine and sexual technologies. The material reported here is based on a critical reading of approximately 100 articles in medical and scientific journals, 9 mass-market paperbacks, advertising materials from pharmaceutical manufacturers, television programs, and numerous articles in newspapers and magazines. For an overview of the research, see Marshall (2002).

27. A number of papers are currently in preparation that expand on this material and develop lines of analysis that I cannot enter into here. These include (1) explorations of the new focus of interest in female sexual dysfunction which has emerged in urology, (2) a case study of resistance to the medicalization and "technologization" of women's sexuality, (3) how a distinction between technologies for the treatment of sexual dysfunctions and sex toys is constructed and marketed, (4) historical and contemporary conceptions of sexual fitness and the aging male body (with Stephen Katz), and (5) a historical account of the construction of the heterogendered body in the late nineteenth-century development of sexual medicine.

28. "Phallacies" was the 1999 season opener of "The Nature of Things," a popular documentary series on science hosted by David Suzuki for the Canadian Broadcasting Corporation. It originally aired on 4 October 1999, and has been repeated several times since.

29. The EROS-CTD (clitoral therapy device) is a small, battery-powered suction pump designed to stimulate blood flow to the clitoris. While one can imagine that providing "gentle suction directly to the clitoris" (Urometrics 2000) may indeed be a good thing from the perspective of women's sexual pleasure, conceptualizing this as clitoral therapy could only occur once female sexual arousal disorder was constructed as a vasculogenic deficiency. See also Maines (1999) on the history of the vibrator.

30. As Potts (2000: p. 99) suggests, "very few men might ever actually experiment with the sensations of the non-erect penis due to the prioritization of the erection in notions of healthy and satisfying male sex."

31. Crossley (1997) provides an interesting overview of embodiment and the subject-body as providing a vantage point for understanding system, lifeworld, and the relationship between them.

32. Thus, as Morrow (1994: p. 268) notes, "most social scientific definitions of feminist methodology are clearly a species of critical methodology whose identity stems from its focus on gender/power issues as the object of inquiry."

II
Technologies of Modernity

5

Modernity under Construction: Building the Internet in Trinidad

Don Slater

The perils of connecting modernity and technology include the danger of replicating on the side of modernity the very same false objectivities that have been so roundly deconstructed on the side of technology. That is to say, if scholars are now comfortable with the social construction of technology and the co-construction of social and technical relationships, then they should be profoundly *un*comfortable with presuming a global and abstract notion of modernity. In particular, there would be painful ironies if at the micro level we dissolve objects and contexts into a dynamic dialectic only to reassert at the macro level a pregiven and assumed context, as if modernity could be conceptually established—using quotes from Giddens or Habermas—as an overarching structure into which technologies are inserted and in terms of which they are then to be understood. Surely modernity is itself a complex object which has to be disaggregated, not only in the interest of revealing its heterogeneity, but also of establishing its nature ethnographically. Can we do an ethnography of both technology and modernity, of the two together, which grounds them in lived social relationships? And can we use ethnography as a basis for moving from the particularities of technology and modernity to more general theoretical formulations?

The dangers of objectifying and totalizing the terms "modernity" and "technology" encompass issues of both presumption and homogenization: how do we presume to know what either of these terms mean in advance of a fine-grained engagement with a particular social configuration? And why should we assume that either represents a uniform phenomenon that can be easily generalized across cultures? This is not to say that there are not essential levels of generalization—in the end these are what we aim at—but rather that ethnographic attention to particularity and an ethnographic basis for comparison of different encounters with modernity and technology might provide a sounder basis for arriving at them. If we should no longer be trying to ground an account of Internet uses in a global definition of technology as such, why would we presume to place a technology in a context of modernity, as globally defined by a social theorist? Yet we find authors seeking to produce a general theory of the Internet on the basis of such a general theory of modernity (e.g., Slevin 2000).

The issue posed here clearly parallels the contradiction mapped out by Feenberg (chapter 3 in this volume): modernity theories of various brands point us toward an implacable modernist logic, whereas technology studies point us, ethnographically, toward the socially complex, local, and contingent construction of technical objects. In agreement with Feenberg, the intention of this essay is not an ethnographic reduction of abstract modernity to particularistic contingency and relativism: we certainly need an ability to generalize outward from the particular and to identify globally shaping forces. But I want to suggest that we need to arrive at a notion of modernity by a different route, and not by assuming a particular logic for modernity. Above all—and especially when wearing the hat of an empirical researcher—I am concerned that concepts of modernity should be to some extent the outcomes of investigation rather than of methodological presuppositions.

Modernity cannot be presumed as a ready-made context into which we then fit the various phenomena, but must be a conceptualization that is at the very least responsive to the different kinds of modern experiences that we find empirically. This is partly for the obvious reason that modernization and modernity take different forms and are experienced differently in different social places. Modernity is not simply a mode of existence that originated in the West and was then exported (practically and intellectually) everywhere else. The quite different Trinidadian experience that is discussed in this chapter is just as much modernity as the European one.

However, there is something much more important at stake than mere difference: these different modernities are very definitely not contingent, but are connected by histories of colonialism, emigration, industrialization, political revolutions, and so on. An ethnographic sense of modernity is not simply a relativistic acceptance of heterogeneity. It is first a recognition of the role that Trinidadians and all the other Others have played in constructing "our" (European) modernity, a modernity to which colonial and other global histories are intrinsic and constitutive. Modernity contains within itself European modernity's encounters with its "others," including their resistances to and incorporations of modernity. I need only cite the work of the Trinidadian C. L. R. James (2001), whose *Black Jacobins* is foundational to this argument, and contemporary authors such as Gilroy (1993). Modernity was (and is) produced "there" as much as "here," and in the relation between here and there.

It is second a recognition that more general accounts of modernity, insofar as they actually emerge from these diverse experiences, can be offered as a dialogic framework, as an account of shared disruptions and powers, experienced by many, but very differently. In brief, it is intellectually and politically fundamental that the modernity we recognize in our research is a clearly heterogeneous and emergent phenomenon, something that can be traced to the diverse social contexts in which it is actually produced. What we need is a methodological orientation that studies modernity under construction.

An incisive example is provided by Daniel Miller's (1994) *Modernity: An Ethnographic Approach*, a book that in important respects forms the backdrop to the study presented here. On the one hand, contemporary accounts of modernity (and even more so, postmodernity) conventionally assume uniform experiences of dislocation, individualization, rationalization, and post-traditional culture and uniform strategies for contending with them, such as Giddens' (1991) "reflexive project of the self." These are theoretically projected onto the wider world as already globalized structures, or as imminently universal. Miller's ethnographic encounter with Trinidad indicated quite other experiences and strategies, most notably ontologies of the self and projects of individualization that involved relationships to time, meaning, and morality that were in stark contrast to those by which Giddens characterized modernity as such.

Trinidadians, as it were, responded to the dislocations we define as modern through their particular circumstances and cultural filters. Above all, close attention to mass consumption and the ways in which the appropriation of material culture mediated and articulated core contradictions gave Miller access to the specific style and substance of Trinidadian modernity and its relation to global modernity. On the other hand, Miller was clear that while his account of Trinidad made nonsense of those accounts of modernity that superimpose European intellectual paths on a society of the periphery, there were nonetheless higher-order accounts of modernity that could be brought into productive tension with Trinidadian experience.

At the most abstract level, Trinidadians and Europeans live in the same modern world in the sense that they inhabit a time of such rupture in material circumstances that modernity can be characterized by "a new temporal sense [that] has undermined the conventional grounds for moral life" (Miller 1994: p. 76). In this account, modernity is not a homogenizing process—on the contrary, the term is used to bring to light a world of different experiences and responses—but it can be specified at a level that brings "us" and "them" within a single framework of historical intelligibility. Indeed, Miller goes further to argue that while most people assume that modernity is best exemplified by the metropolitanism of Paris or New York, it is equally likely that the rupture from traditional life brought about by Trinidadian experiences of slavery, indentured labor, and forced migrations made this a place where one might actually find a more exemplary or even "vanguard" form of modernity.

In this chapter, I simply want to flesh out what this means for a very particular encounter with a modernizing technology: the assimilation of new Internet technologies within a range of Trinidadian social relations. I am reporting on a project (Miller and Slater 2000) in which we attempted to situate the Internet in relation to Trinidad and Trinidad in relation to the Internet. This work was conducted over approximately 18 months. It involved first, interviews with diasporic Trinidadians in London and New York, analysis of hundreds of Trinidadian websites, and participation in Trinidadian online chat and email. Second, it included fieldwork in Trinidad that consisted of a house-to-house survey of four residential districts, followed by selective in-depth interviews; participant observation in several cybercafes; and interviews with personal, commercial, and governmental users of Internet media. The goal was to accomplish the traditionally holistic aims of ethnography in which phenomena (Internet uses) and contexts (the personal, the commercial, political economy, global modernity) made sense of each other.

Disaggregating the Internet

Let us first start on familiar ground: technology. What seemed apparent in starting this study was not only that much of the academic Internet literature avoided a proper ethnographic engagement with the particularities of Internet use, but also that this stemmed from a highly specific relationship between that literature and the very question of modernity. Specifically, *because* that literature was so obsessively engaged with global notions of modernity, it projected onto the Internet, as its essential properties, equally global images of antimodernism. These were above all poststructuralist and postmodern concepts of subjects, sociality, and globalization. The point here is not so much to contest the usefulness of these terms in themselves as to question the methodological status they have been accorded. It appeared that the first generation of Internet literature largely ignored all the lessons of the social construction of science and technology by starting research from the presumption of inherent technical properties. This can be most readily grasped through two of the buzzwords it persistently employed: "virtuality" and "disembedding." Both terms started from the presupposition that Internet media inherently constitute forms of online sociality, with a convincing reality status, that transcend or replace their offline contexts. Discussion then focused on whether this virtuality was a good or a bad thing; on what kind of ontological status could be ascribed to its participants or their relationships (is a virtual relationship anything serious?); and on describing the structures of the interactions that were carried out within the already methodologically bracketed space of the "virtual" (e.g., accounts of netiquette, interaction orders, and conversational rules).

Virtuality had its correlate in the notion of disembedding. While the former emphasized the separation between online and offline realities by focusing on the self-constituting realities of Internet social interaction, disembedding also assumed radical separations of Internet users from their immediate spatiotemporal location because of the technology's ability to connect dispersed people with a casual immediacy. Even more insistently than virtuality, this term pointed to a continuous modernist theme: the capacity of technologies to produce global connections that marginalized both local and national identities and social orders.

Three issues emerge here: The first is that these presumptions of virtuality and disembedding give a false unity to the new media themselves. Examples could be given from anywhere in the world, but the Trinidad study immediately indicated that it would be wrong to treat the Internet as a single technology. Consider two cases. On the one hand is a Trinidadian school pupil who might typically use an online chatting and paging system like ICQ to continue conversations started in the schoolyard that will persist into a later meeting at a "liming spot" (place to hang out), and who uses the World Wide Web (WWW) to research school projects and locate scholarship sources for further education in North America. On the other hand is a Trinidadian business for whom the Internet ideally means intranet integration of departments and work processes combined with web-based integration of consumers and suppliers into the corporate operation through online ordering, order tracking, and customer relations. In what sense can both of these uses be deemed "*the* Internet," let alone be characterized in terms of such global features as virtuality and disembedding? Does it not make more sense to establish ethnographically how people assemble their own Internets from the use of different technological potentials and features within different projects, institutions, and constraints?

Second, virtuality and disembedding cannot be assumed at the level of methodology without doing violence to the diversity of constructions of the media themselves. The virtuality of the Internet media is a matter of ethnographic investigation, not of starting principles. For example, my own first Internet ethnography—of "sexpics trading on IRC" (Rival et al. 1998; Slater 1998)—conducted entirely online, indicated that virtuality was not only a central issue but also was one topicalized by participants. They invested a great deal of time in interrogating the reality status of what they were doing online. Some of them were very concerned to separate their online from their offline activities, others to absorb one into the other. Moreover, the interaction order online could be highly variable, depending on whether participants were trying to constitute their activities as virtual or not. That is to say, I had to investigate virtuality as a potential social accomplishment of participants, one they might or might not be seeking to achieve, rather than as a methodological presumption about the nature of the technology.

In Trinidad, on the other hand, virtuality could only be raised as an issue by imposing it upon what we encountered: it simply did not feature in our observations and interviews, and made little sense of what we did observe. To the contrary, in almost every Internet use we dealt with, the participants stressed continuities between online and offline projects, generally understanding Internet media in terms of their potential to further activities that were firmly grounded in a concrete sense of the everyday (Slater 2000a,b; Miller and Slater forthcoming). This applied, for example, to religious organizations which, while extremely articulate about the revolutionary properties of online communication for such encounters as confession and religious study, went to great lengths to debate the appropriateness of these media in terms of long traditions of theological debate. It also applied to young people who formed romantic attachments, through online chat, with people in other continents, but who went to great lengths to integrate these encounters into their relationships with those physically around them. Virtuality, in the sense of construing Internet activities and relationships as self-constituting and sealed off from other spheres of life, was neither valued nor pursued. This does not mean that virtuality is never a feature of Internet use, but simply that it was not in Trinidad and that it would have meant doing violence to our material if we had assumed that it was and hence a thing to be investigated.

Third, the first two issues involving the centrality of virtuality as a unifying concept of the Internet seemed clearly to have emerged from the very particular relationship that many Internet researchers (and the Euro-American netizens they studied) bore to western modernity. The Internet was received as virtual within self-consciously antimodern intellectual currents—postmodern, poststructural, postfeminist, postcolonial—in which the desire to deconstruct modernist presumptions about rationality, the subject, and the body into textual or performative accomplishments seemed to find their apotheosis and objective correlate in a new social space that was disembodied, textual, and fluid. Again the issue is not whether these theoretical trajectories are valid or fruitful (I could not be a responsible western intellectual today without engaging them dialogically and hence accepting some common premises and lines of thought). The issue is whether it is valid to project this problem, a priori, onto the social actors and practices one is investigating so that we *start* by looking for deconstructive or performative practices as constitutive of Internet sociality. This was the case with much of the early cyberutopian and cyberfeminist literature in which the Internet appeared as the space in which the revolutionary subject of antimodernist theory (cyborgs, hybrids, rhizomes, posthumans) could socially emerge, constituting a model and stage for avant-garde subjectivity. This is simply to impose, in the guise of a methodological first principle, an agenda arising from the encounter of a rather slim stratum of the western intelligentsia with their own modernity.

Trinidadian use of the Internet clearly could not be clarified by looking through the lens of this relationship to modernity. In almost every area we studied, as will be elaborated later, Trinidadians appropriated the Internet within identity projects that emerged from a quite different experience of and relationship to modernity. It was a relationship that very strongly embraced a sense of themselves as modern in a proud and positive sense, while at the same time centered on the ways in which they were disrupted and marginalized by modernity. The Internet was largely perceived as a modern means to be *more* modern, rather than a means to deconstruct modernist identities. This could be no more clear than in the centrality of national identity to Trinidadians' use of the Internet. It would be ludicrous to imagine an American or European teenager decorating their personal website with the Northern equivalents of Trinidadian flags, national anthem, core symbols of local culture (Carnival, beaches, beer), prominently displaying links to the Trinidadian national industrial development and tourist agency, or actually constructing their entire websites as encyclopedic guidebooks to Trinidad (including history, dictionaries of dialectic, recipes, music, and so on). Yet this is what Trinidadian teenagers routinely did.

It would therefore have been utterly misplaced to approach the Trinidadian Internet through a relationship to modernity that emphasized the marginalizing of place and indeed nation and national culture, or the transcendence of local identity through globalized neotribes (e.g., Trinidadian teenagers seemed to identify very little with a concept like "HipHop Nation"). But it would have been positively outrageous to treat such a relationship to modernity (disembedding) as if it were an inherent property of the Internet as such, as is regularly assumed in much of the literature.

Finally, cyber versions of the Internet have piggybacked on the voices of various social actors who are also theorists of (anti-)modernity, such as hacker culture, postpunks, riot grrrls and cyberfeminism, new economy pundits (such as *Wired* magazine) and many others. In a very real sense, we too had to attend to a range of Trinidadian social actors who were highly articulate theorists of modernity, and who located and appropriated the Internet within their own discourses of modernity. These were as evident at the governmental level where we listened to modernization discourses that linked new media to national and regional development plans as it was in the conversations of young people thinking through the art of gaining employment in a globalized labor market in relation to which the new media seemed to reposition them. At the same time, the same Trinidadian youth might be well aware of cyber discourses. Their version of modernity might well include a response to these themes (without sharing or adopting them).

Indeed, terms such as "Internet" and "modernity"—and above all, the rhetorical connections that could be drawn between them—had in the first instance to be studied as terms used strategically and politically by participants, and as potent signifiers over which they contested or mobilized ownership. Paradoxically, although I have stressed that the Internet must be ethnographically disaggregated into specific assemblages of technology and use, nonetheless Trinidadians clearly did talk a lot about "*the* Internet" as a single phenomenon: they experienced it as a massive global development in terms of which they have to rethink their own future development, and which plays a crucial role today in redefining what modernization means at the level of the individual, household, community, nation, region, and so on. As noted earlier, they were aware of much of the global discourse on the Internet, including all the hype. At this level of engagement they certainly encountered and used "*the* Internet" as a preeminent signifier of global modernity and as a means of renegotiating their relationship to it. Indeed, it was virtually a synonym for "the future," and an active term in every discussion of the future (personal, corporate, or national). Hence, the emergent and heterogeneous character of both Internet and modernity must include the theoretical and strategic practices of those who have to deal with such terms in the first place.

Disaggregating Modernity

This is not the place for a full treatment of Trinidadian modernity. However, I would like to sketch out broadly an account of why the Internet might have particular uses and meanings in relation to their experience and construction of modernity. As a generalization we might say that Trinidadians see the Internet as a modern means of overcoming the dislocations and indeed catastrophes that modernity has inflicted upon them. The country was largely assembled through a history of slavery and indentured labor within the colonial projects of a modernizing Europe. The experience of disembedding is a historical one for Trinidadians—to put it rather mildly—and hardly *post*modern. It is also an experience perpetuated through a twentieth-century history of diaspora and mass emigration, in which Trinidadians in great numbers have been further dispersed by modern economic and political forces, following the tug of labor markets and educational opportunities to London, Toronto, New York, and elsewhere.

Hence, some of the most notable Internet uses in Trinidad fit into a far longer-term relationship to modernity and were experienced less as a ruptural development than one by which earlier ruptures of modernity might be repaired. For example, Trinidadian families—the great majority of which are diasporic at the nuclear level—have long had to deal with their modern problem of how to be a family at a distance. The Internet, specifically email, was taken up with alacrity simply as a cheap and convenient alternative to traditional means of communication, such as travel, letters, and telephone. What seemed quite specifically new and valued in this new medium was its mundane character rather than its virtual possibilities. For example, using email, a mother in Trinidad could nag her daughter in New York on a daily basis about what she was wearing and how late she stayed out, demanding from her the frequent replies that make for a sense of everyday contact and therefore replicate a face-to-face family (we interviewed the beleaguered daughter rather than the surveilling mother). Email was a modern means by which the mother felt she could reconstitute a normative Trinidadian parental role that had been disrupted by modernity. This relationship to modernity permeated Trinidadian society and mediated people's sense of what the Internet was good for. Similar examples might include the ways in which religious groups such as Hindus, Moslems, and Pentecostals could *re*connect with the global dispersed cultures from which they had been sundered.

At the same time, the very notion of "being Trinidadian" was itself defined, often very proudly, as a profoundly modern identity, indeed one that had long been construed in cosmopolitan and global terms (by which they also frequently distinguish themselves from other Caribbean people, and certainly from any sense of being Third World). Trinidad is not an agricultural or tourist-based island. Its relative wealth is based on petrochemical industries arising from the oil fields it shares with Venezuela. It therefore has a history of rapid, thorough, and often leading-edge industrialization going back to the 1930s. This combines with and supports near-universal literacy based on a very effective and highly valued education system. On this basis, most Trinidadians feel themselves to be both able and willing to be highly entrepreneurial in business and highly accomplished in education and culture. That is to say, they feel they should rightfully be considered full participants in global modernity, in terms of their capabilities, orientations, and achievements. They readily point to such indicators as their economic and political leadership in their region, their disproportionate representation in top American universities, their Nobel Prize winners, and so on.

The contradiction articulated here is between their sense of being almost *more* modern than anyone else and their experience of being marginalized and kept off the international stage of global capitalism and culture, an experience that is again historically continuous from colonial to neoimperial and contemporary eras. The crystallizing example we encountered in fieldwork was that of a teenage Trinidadian who was asked during an online chat with a North American, not "Where is Trinidad?", but "What is Trinidad?" His bemusement was doubled by his certainty that by virtue of his education and cosmopolitan modernity, he knew far more about North America than the North American he was chatting with. Again, this vignette is repeated in countless areas of social life: the business people who feel they are capable of leading-edge enterprise, but have been relegated to the margins; the musicians who know that their highly accomplished musical traditions can absorb all global influences, yet are ignored throughout the globe.

The Internet yet again does not appear as a rupture with the past or a conduit to some new global identity as a modern or postmodern. Quite the contrary: it was largely understood as offering a range of opportunities for Trinidadians finally to be as modern as they always thought they were, and for this to be recognized by global modernity. Again, there are deep continuities here. As Miller (1994) noted, if "being Trinidadian" was partly defined around being modern, being cosmopolitan, and being where things are happening (a core identity replicated as much through dress and music styles as through technology), then it has not been uncommon for Trinidadians to feel that they could only be properly Trinidadian if they were living in New York or Toronto rather than back home. The Internet appeared to many as providing, at long last, an opportunity to be properly modern—and therefore Trinidadian—without having to leave Trinidad.

A central example of this arose during our fieldwork: the Miss Universe 1999 competition and website. A beauty contest was certainly a politically paradoxical event with which to demonstrate modernity, but that is precisely what it did for many different sectors of Trinidadian society (although with some ambivalence). Trinidad, which also boasted a disproportionate number of winners of such contests, was host to the event and felt (probably wrongly) that the eyes of all the world were on it. Moreover, being the home of Carnival, their national identity hinged somewhat on their ability to throw world-class parties and festivals.

A local and highly ambitious new media and technology company won the contract for the website from the international governing body and did indeed produce far and away the most sophisticated use of Internet media yet seen in the region. The site was large, graphically complex, and striking; it included chat facilities, considerable backoffice integration, and multimedia. Moreover, the local company achieved this by working in close collaboration with major North American corporations, by which they not only accomplished considerable technology and skill transfer but also, more important, demonstrated themselves to be a world-class business organization. They had placed themselves on a world stage on which they felt they had a rightful place, but for once they could do it from Trinidad. Because of Internet media, they could be in close everyday contact with North American partners while carrying out high-tech, high value-added, and leading-edge technological and corporate projects that marshaled and developed the potentials of Trinidadian labor and entrepreneurship. They did this both practically and symbolically: the fact that the local company logo was on the same web pages as Microsoft, Oracle, Real, and so on literally placed them in the same global space.

The point here is not to promote undue optimism as to whether this is a durable model for national development, and certainly not to argue that the Internet will solve Trinidad's marginality with respect to global capitalism. The point is rather that this encounter shows how the Internet could be widely seen as a technology that far from imposing a dominant global modernity onto a peripheral region, might actually be the vehicle by which Trinidadians could assert and accomplish the modernity from which they have felt wrongfully excluded. They might well be wrong in their hopes. They were certainly facing global structuring agents and forces seeking to impose their own versions of modernization, and these indeed have to be part of the picture. For example, there were huge multinational interests taking such forms as the World Trade Organization and huge pressures from telecommunications companies attempting to open up and dominate the strategic Caribbean market. Nonetheless, the Trinidadian framing of this situation and their response to it also has to be an essential part of the story, and one of the most common metaphors in which the Trinidadian Internet came wrapped was that of "the level playing field," a sense that the Internet allowed them a direct access to global markets and forums by which they could finally compete on equal terms and show their worth. The Internet spoke to a kind of populist capitalism in which the vested privileges of the international giants might now count for much less against the abilities and energies of these other moderns.

Finally, we need to take up this complex relationship with modernity at the level of nation and national modernization projects. The assumed conjunction of the Internet, globalization, and disembedding in so much contemporary academic literature, plus proper distrust of the entire history of nationalism, left us as researchers wholly unprepared for the extent to which the Internet in Trinidad was so embedded in a sense of national project. We have argued elsewhere (Miller and Slater 2000: chap. 4) that when Trinidadians, both home based and diasporic, went online, their behavior had to be at least partly understood in terms of both "being Trini" and "representing Trinidad." That is to say, first, their online activities were very largely mediated through a constant awareness of a broad cultural identity (e.g., using Trinidadian terms to describe Internet activities, engaging online in consciously Trinidadian forms of sociality such as "liming" and "ole talk"); and second that they persistently felt in their online presence the responsibility to be a good representative of Trinidad, and to represent the country well through their websites and chat. The Internet, again, was not understood as a conduit to new identities that stood in contrast to their local one. On the contrary, it offered modern possibilities to ensure Trinidad's representation in a global space.

Similarly, at an overtly political level, the Internet was widely understood as the key to Trinidadian modernization in relation to which many older battles between state and civil society were revived. For example, the central public issue during the research period directly tied the national development of Internet infrastructure to a whole history of modernization. The local telecommunications monopoly, Telecommunication Services of Trinidad and Tobago (TSTT), jointly owned by the Trinidadian government and by the multinational but also excolonial company, Cable and Wireless, was construed as the preeminent bottleneck in establishing both infrastructure and cheap and abundant access. TSTT was talked of as an almost treasonous force, as well as legacy of a British colonial past, which would keep Trinidad from taking advantage of the small window of opportunity by which the country could achieve its rightful place on a North American axis of modernity.

Ethnography, Generalization, and Comparison

It should be clear from the foregoing discussion that doing an "ethnography of the Internet" is not a simple matter of putting the use of a specific technology in context. Above all, the idea of ethnographically relating technology to modernity cannot be glossed as putting the Internet in a social context. This is the case for at least two reasons: First, context cannot be treated as a methodological given: the level at which one defines it arises partly from the kinds of things that one is putting into it. As the technology disaggregates into uses, so do the contexts. We certainly did not expect, when we started the project as good European netizens, that Trinidad as a nationalist project would turn out to be the crucial context of Internet use, or that its relation to global modernity would be so highly mediated by such contexts as the international Hindu diaspora. In fact we did not recognize them as contexts at all until the research was under way. Second, context is not substantively given: the context of a technology is also partly a consequence of that technology; it is produced by the very "thing" one is trying to put into it. Ideas and manifestations of Trinidad as a modern society are not only a context of Internet use, they are also precisely what people are often trying to produce by using the Internet. It would be unwarranted (and rather missing the interesting points) for us to define modern Trinidad in terms of global notions imported from Euro-American social sciences, instead of looking at what is being defined and produced through the local Internet practices. Modern Trinidad—the context of Internet use—is changed, extended, redefined, and indeed reinvigorated through the ways in which Trinidadians make sense of Internet use.

However, the point of advocating an ethnographic approach to both the Internet and modernity is definitely not to retreat into an extreme particularism, relativism, or nominalism. The argument is not that modernity can be reduced to what particular constituencies make of it, to local definitions (whether they are Trinidadians or northern academics), but rather that modernity needs to be treated as a complex and heterogeneous accomplishment, not as a fact or assumption. The ways in which Trinidadians respond to constructions of modernity (their own and others), and the ways in which they pursue projects that aim to construct themselves as modern, have to be part of the story of modernity under construction. The issue of generalization remains, but it is no longer posed as a contradiction between (an abstract, overwhelming logic of) modernity and (a grounded, socially diverse) technology. Both are under construction, and at least partly in relation to each other.

To be clear, I am not arguing against higher-order theories and models. The issue is simply the status accorded to such generalizations, their responsiveness to the particular, and their ability to retain a sense of the provisional and constructed character of modernity and technology. In this spirit, my concern in this chapter is not to revise any particular notion of modernity or to offer an alternative. Rather, let me conclude by pointing toward a further mediating level of analysis and methodology: the comparative. Between ethnographic particularity and higher-order generalization, we can think about modernity as a global phenomenon that nonetheless emerges from particular and divergent conditions. A comparative approach asks us to uncover and account for different modernities, in the sense of different constructions of what it is to be modern; different strategies for dealing with a modernity that is construed by people as "out there" (in the West, the North, the East); different understandings of how technologies map out potential historical paths, power relations, spatial imbalances. It is, moreover, through the comparative mapping of differences that we might get a more grounded sense of whether different localities are confronting the same global forces, disruptions, and reconfigurations, so that they may be labeled as "modernity" in a more encompassing way.

The task of comparative ethnography in relation to the Internet and modernity is to identify some common themes, dilemmas, or dynamics that would sensitize us to what is most important in making sense of both terms in a wide range of different places. It should also allow us to ask intelligent questions about the similarities and differences in peoples' responses to new communication possibilities. In the course of the Trinidad fieldwork, we came up with four dynamics (Miller and Slater 2000: pp. 19–21), which retrospectively brought out what we considered to be the central features of the construction of Internet and Trinidad. Each one seeks to focus on general social issues and transformations in a peoples' encounter with the Internet, but without assuming any particular outcome. Our suggestion is that these dynamics can be applied across a wide range of cultures, but they are expressly designed to sensitize researchers to the very different ways in which each dynamic will work out in different social settings. Their claim to generality stands or falls on the grounds of their usefulness in making sense of Internet use in places other than Trinidad. If they do bring out the important features of Internet use that emerge in ethnographies conducted elsewhere, then it is because we have identified common dilemmas and tasks that affect people who are occupying a transforming communication environment anywhere on the globe. In the context of this chapter, I am also suggesting that the task of identifying this commonality has a close relationship, substantively and methodologically, to the task of characterizing modernity from an ethnographic perspective: can we build a sense of modernity under construction from the ground up rather than from the top down?

Dynamics of Objectification   We might start by thinking about Internet media as part of the material culture of particular places. People can use the Internet to objectify different senses of themselves, or find particular versions of themselves reflected back from these media. In the case of Trinidad, as argued earlier, people largely viewed the Internet as affording possibilities for realizing and projecting a sense of themselves as modern and even in the vanguard of global modernity, an identity from which they felt themselves to have previously been excluded. In Miller and Slater (2000) we discuss this as an example of "expansive realization." Internet presences and performances are objectifications of an identity, so that people seem to "find themselves" or realize a sense of who they think they are, though on an extended, indeed global, scale. Hence the ability to use email to maintain mundane family relations at a distance or to use chat facilities to enact Trinidadian forms of sociality were experienced as means to be or to perform "Trini-ness." This might contrast with Euro-American cyberculture, which aims at the production of potential identities rather than the realization of more continuous ones, though even in this case it is interesting how the Internet can objectify a tension between a discourse of transgression and a normatively conventional performance of identity (Rival et al. 1998; Slater 2000b).

Dynamics of Mediation   Technologies are reconfigured in use (very literally in the case of information technologies that can be materially reconfigured by users reprogramming them). We therefore need to investigate how people engage with new media *as media*. How do they come to understand, frame, and make use of features, potentialities, dangers, and metaphors that they perceive in these new media? This is a dynamic that requires disaggregating the Internet into the processes of making sense of media potentials within specific practices: the mother to whom email represents a practical and mundane social connectedness versus the local entrepreneur who sees, in the Miss Universe website, the potential to link Trinidad to the core of capitalist modernity.

This also requires us to think about the overall media mix in people's practices. For example, many Trinidadians, in their family uses, seemed to understand email more in relation to phones, mail, and personal travel than in relation to the Internet as a totalized medium; similarly, ICQ and other chat facilities (by the far the most popular use of the Internet among teenagers) seemed to have very little to do with the WWW, and everything to do with continuing specifically Trinidadian modes of socializing, from face-to-face interaction through phone calls to online media and back to physical meetings. From a completely different direction, we talked to religious groups who were attempting to puzzle out the nature of various kinds of online communication in relation to older genres of communication (for example, the confessional) and in relation to different modes of social organization. (For example, how does networked communication affect older forms of hierarchy and authoritative communications such as orthodoxy passed on in texts and pronouncements?)

Once again we would expect fairly localized and divergent understandings and practices, although possibly in relation to similar issues. For example, even within Trinidad itself, different religious groups exhibited different understandings and different tensions in relation to the issue of hierarchy. A young Moslem couple went online to find out what marriage rituals could be authoritatively regarded as orthodox (only to find that Islam was far more geographically diverse than they had realized). A Pentecostal group believed that the diversity of Internet connections could provide for a more dynamic and decentralized spreading of the word (and indeed that the networked structure of the Internet indicated God's will in this regard).

Dynamics of Normative Freedom   New communication technologies and forms always involve new normative structures. What kinds of sociality, ethics, and regulation are appropriate to these forms of communication? What kinds of sociality, ethics, and regulation are the new media appropriate for? This kind of consideration often gets stalled at the level of describing netiquette and other rules of interaction, or the different values placed on offline and online relationships. And yet new modes of communication point to profound questions about social relationships (as evidenced in continual moral panics).

A central issue in the emergence of the Internet has been "freedom." In Europe and America it certainly came wrapped in various libertarian discourses, often radically contradictory ones. For example, the Internet was embraced equally by far right libertarians, left-leaning anarchists, free marketers, and free speech advocates as an exemplary space in which to be "free." And yet, freedom only makes sense in particular contexts; it always takes particular normative forms that have to be interpreted in their contexts. In Trinidad, for example, concepts of freedom often took a highly entrepreneurial, antistate, and even neoliberal form, which was clearly articulated in the way people understood the Internet. It was commonly seen in terms of the populist capitalism described earlier, offering the aspiring individual leverage against vested and conservative interests of states and large corporations, both at home and abroad. It was very difficult to disentangle the Internet in Trinidad from the rhetoric of the free market as the path to modernization.

These issues may play out rather differently in other places. Aside from the obvious issue of how various states are dealing with issues of privacy, government regulation, property, and so on (for example, by attempting tight control over access), we can point to interesting and characteristic contradictions within Europe and America. On the one hand there is an exuberant discourse of freedom as definitive of the modern citizen, but which in practice is restricted to commercial, consumerist "freedom of choice." On the other hand there is a profoundly antimodern social conservatism that seeks to "protect" subjects from such effects of freedom as pornography, marginalization of nation or family, fragmentation of traditional identities, and so on.

Dynamics of Positioning How do people engage with the ways in which Internet media position them within networks that transcend their immediate location and that comprise the mingled flows of cultural, political, financial, and economic resources? This means looking at the ways in which people understand the new "place" in which the Internet puts them in relation to global modernity and global capitalism, and the strategies they adopt to exploit or defend themselves in this new context. As already indicated, it is misguided to reduce this entire theme to a presumed disembedding effect of the technology, which is then traced according to impacts predicted by various modernity theories (decline of nation, neotribalism, glocalization, and so on).

In a more ethnographic approach, one starts from the way in which people conceptualize not only how the Internet might reposition them but also the kinds of strategies and opportunities for repositioning that the Internet opens up for them. For example, it is a fairly common experience around the world that the introduction of the Internet threatens the very idea of a "local price": market boundaries, market information, distribution relations, and so on change so that the entire economic geography is potentially reconfigured and experienced in different terms. One could easily imagine a comparative analysis of the construction of new economic geographies in different places, often taking the form of different strategic responses to very similar social forces.

Cultural repositioning is equally important. For example, in terms of music culture, many Trinidadians saw the Internet as a new (intensified) means to develop the local "soca" music according to an older pattern: it is the tradition of Trinidadian music to proceed by confidently assimilating global flows of musical development (in taxis you could find yourself listening to ear-splitting fusions of rap and soca, techno and soca, ska, reggae and soca, or virtuoso arrangements of the European classical music canon for steel band orchestras of over a hundred parts). The most common response to the Internet among Trinidadian youth was excitement at immediate (and free) access to every manner of world music. This excitement certainly included participation in global cultures, but it always also involved an excitement at being able to incorporate that music and finally to be able to present the results to wider audiences.

The point of these examples, again, is not to argue that what we have learned from Trinidad can be easily generalized to anywhere else in the world. On the contrary, it is to suggest that if these dynamics are applied to other contexts, they are likely to bring into view the significant features through which a local culture must confront what is indeed a global phenomenon or potential. It may be the case, for example, that people in Indonesia, in stark contrast to the Trinidadian case, encounter the Internet as a means to deconstruct the repressive modernist national identity of Indonesia into recovered ethnicities, that they focus on the media potentials for highly local communication, and that they construe the liberating potentials of the Internet in terms of state programs rather than commercial enterprises and as a means to position themselves in relation to Pacific markets rather than a North–South axis (this is all speculation). If so, by pointing up these contrasts, we might well develop a solid foundation for theorizing relations to both technology and modernity as a diverse, heterogeneous, and open-ended sociohistorical accomplishment.

Conclusion

The Trinidad research started from an intuition that the Internet literature of the 1990s—the literature of virtuality, cyberspace, and new economy—was premised on a particularly Euro-American version of (anti-)modernism, and therefore on a desire to find in these new media the vehicle to deconstruct and escape an implacable logic of modernity. Ironically, this way of framing the new media came to be generalized in a rather characteristically modern way: the Internet was totalized, positivistically, as a "thing with properties," a technology that promised (or threatened) certain identities and experiences. The first step in moving toward an ethnographic account was to disaggregate this technology into something that emerged, in unpredictable and often unstable forms, from complex local practices and appropriations. However, this is not enough if it leaves an abstract notion of modernity as a context into which the technology is placed. The crucial move was to look at how

Trinidadians used the Internet (both practically and discursively) to mediate their relationship to modernity. Neither Trinidad nor modernity were given contexts; they emerged from developments such as new media through which Trinidadians are redefining themselves and their relationship to other places.

6
Surveillance Technology and Surveillance Society

David Lyon

Surveillance is a distinctive product of the modern world. Indeed, surveillance helps to constitute the world as modern. Detailed personal information—pursued by many organizations through requests to fill out forms, to produce identification, or to undergo ordeals such as fingerprinting or urine tests—was never routinely and systematically demanded of people before modern bureaucracies came into being. Today, in the so-called advanced societies, people's everyday lives are circumscribed by the record keeping, monitoring, and supervising of multifarious agencies and organizations. Moreover, the ways that personal details are collected, stored, processed, and retrieved are also very modern in the sense that they depend on rational techniques and, increasingly, on relatively new technologies.

Personal experiences of life are shaped, among other things, by relationships with organized social life, and this includes how organizations try to influence, manage, and control us through surveillance. Sociology concerns itself with trying to understand these relationships as they develop historically. Those relationships are "continuously constructed in time" (Abrams 1982: p. 16), which is why any sociology worth the name is historical. However, these relationships are also increasingly mediated by new technologies. We cannot pretend, for instance, that the television and the telephone have been somehow neutral in their effects. Today, in addition to those older technologies, computers mediate more and more between what was once thought of as "individuals" and "society" and, as I argue, those computers do not have merely "social impacts." They are also, even in their personal data processing functions, socially shaped as well as socially influential.

Curiously enough, some of the ways in which those apparently very rational machines—computers—have played a part in mediating between people in everyday life and large-scale social processes have had less than predictable outcomes. The searchable databases that make bureaucratic administration more manageable have also found a place in corporations, for managing not only workers but also consumers. Database marketing is a multi-billion dollar industry that seeks personal data on consumers' spending habits, preferences, and lifestyles in order to profile and track current and potential customers in many distinct realms of life. The result is not a neat pyramidlike structure of control, such as the classic bureaucracy, but something much more like a creeping plant that sends out shoots here and there, growing rhizomically. This may be thought of as aspects of a "surveillant assemblage" (Haggerty and Ericson 2000), a sort of postmodern mutation of earlier practices that is decentralized, polycentric, and only very partially predictable.

However, that is only part of the story. In order to explain things, sociology, like history, has to invoke theory. But theory, as Barb Marshall reminds us (chapter 4, this volume) is best advanced through engagement with concrete problems, and such problems do not come in tidy packages. If anyone in North America was under any illusion that social life involves "tidy packages," that illusion was decisively shattered on September 11, 2001, when the World Trade Center in New York, and the Pentagon in Washington, D.C., were devastatingly attacked by suicide pilots flying hijacked commercial aircraft. Even if it turns out that the course of life in the twenty-first century is not shaped irreversibly by those dramatic events, they certainly made a radical difference to the ways in which ordinary people and politicians were prepared to think about surveillance technologies. In the wake of those attacks, internal domestic security measures were tightened as if for war.

Many proposals have been made—as I write in October 2001—for the use of national identification cards and immigration cards, often containing biometric devices such as digitally stored fingerprints; for closed-circuit television surveillance at airports and in other public places; and for the extension of wiretapping from telephones through email to Internet clickstream monitoring. These techniques have already been tested in either criminal or consumer contexts; what is new is their proposed use by law-enforcement agencies of the nation-state for curtailing or preventing terrorist activity. This "big event," in other words, will play a part in deciding which technologies are adopted and in the ways that they are technically shaped. In turn, the use of these technologies will likely alter the relationships between individual citizens and those nation-states, making it easier for authoritarian control to appear and reviving talk of "Big Brother." They will also influence relationships between consumers and large corporations because personal data gleaned from more and more "innocent" commercial and entertainment activities are now sought by the state.

In what follows, I start by looking a little more closely at the sets of relationships that obtain among routine surveillance, technologies, and everyday life, and I hint at the sorts of issues, both theoretical and ethical, that these relationships raise. In the central sections of the chapter I look first at modernity and surveillance, and show how new technologies were required for surveillance but also how they were influential in shaping social development as they mediated social relationships. I then argue that surveillance techniques help to constitute a postmodern condition, just as earlier surveillance techniques helped to constitute modernity (which is not, of course, to suggest that modernity has itself disappeared!). Next, my focus is on the co-construction of surveillance practices as a case in point of technologies both mediating social relationships and themselves being socially malleable. Finally, I turn to the issues of ethics and politics which, it seems to me, cannot responsibly be evaded in any discussion of surveillance and society.

Routine Surveillance, Technology, and Everyday Life

Technological developments and social processes mutually influence and shape, or co-construct, each other. The case discussed here is the rapid expansion of surveillance technologies in the later twentieth century, which, I argue, illustrates some central aspects of modernity and a putative shift toward more postmodern conditions. Whereas the nation-state and the capitalist workplace are the primary sites of surveillance in modern times, computerization has not only augmented surveillance in those sectors but also moved decisively into the consumer sphere.

Boundaries between surveillance sectors have blurred, even as a shift is under way from past records and present activities to the anticipation and management of future behaviors through simulation. Small- and medium-scale alterations in surveillance practices both reflect the broader changes described as postmodern and contribute to them.

Surveillance is so much an intrinsic part of daily life today that it is sometimes hard to explain exactly what it is and how much it has changed in less than a generation. We unblinkingly produce passports for scanners to read at airports, feed plastic cards with personal identifiers into bank machines, fill out warranty forms when we buy appliances, key confidential data into telephones or online transactions, drive through automated toll sensors, make cell-phone calls, or use bar-coded keys to enter offices and laboratories. How inefficient and inconvenient it would be if we were obliged to pay cash for everything, or to be interviewed by officials each time we crossed a border! Nonetheless, at each encounter we leave a trail of personal data that is tracked and processed in ways that influence our activities and our life chances. Surveillance is always Janus faced.

Some surveillance practices have been a feature of modern life for a long time, of course. Medical records, voting lists, housing registries, tax files, and employee numbers are part of what living in the twentieth century was all about, at least in urban industrial societies. Indeed, they facilitate modern life by providing evidence for eligibility and entitlement to benefits and privileges, while placing power in the hands of those who handle that information. The potentially totalitarian threat that accompanies the spread of surveillance was given sinister reality in various state socialist and fascist regimes in the mid-twentieth century, sparking rhetorical warnings, the best-known of which came from the novelist George Orwell. Discourses of privacy have been a means of limiting, and sometimes of resisting, surveillance power within legal and political frameworks.

The major change that took place in the latter part of the twentieth century was the wholesale computerization of surveillance. At first, this simply made the processes of bureaucratic administration easier, thus reinforcing government and workplace surveillance as classic hallmarks of modernity. You could say that Max Weber's rational institutional ordering of the office was upgraded from an iron cage to an electronic cage, or that Frederick Taylor's detailed work-task monitoring system shifted from scientific to technological management. The top-down style of management and administration, based on a rigid hierarchy of officials, was reinforced by computerization. This process made possible a holding operation at just the moment when bureaucratic structures were crumbling under their own unwieldy weight. It also facilitated a fresh attention to the minutiae of workplace activities at a time when managers had an increasing number of processes on their minds that were harder to keep together.

However, as personal databases proliferated within government departments, the very idea of centralized control became less plausible in many sectors. And as capitalist enterprises turned their attention toward managing consumption in addition to organizing workers, surveillance spilled over into numerous other areas, further diffusing its patterns within the social fabric. Numerous research studies have documented the ways in which computerization actually permitted new surveillance practices in the office and on the shop floor, thus adding the potential for qualitative as well as quantitative change to such settings (Gill 1985).

At the same time, questions were raised whether "modernity" adequately describes contemporary conditions. Leaving aside the debates over aesthetics and architectures that often appear under the rubric of postmodern*ism*, and the discussions of *anti*modern tendencies that sometimes enter the same arena, it can be argued sociologically that postmodern*ity* poses questions about novel social formations. In particular, postmodernity may be used to designate situations where some aspects of modernity have been inflated to such an extent that modernity becomes less recognizable as such. The sociological debate over postmodernity has leaned toward examining either the social aspects of new technologies or the rise of consumerism, but a good case can be made for combining these two forms of analysis to consider postmodernity as an emergent social formation in its own right.

It has taken some time to appreciate that surveillance technologies are vitally implicated in the processes of postmodernity. Analysts of consumerism have tended to underestimate the extent to which surveillance is used for managing consumers, while analysts of technologies are just now exploring the imperatives of consumer capitalism (see Strasser et al. 1998; Kline 2000; Blaszcyzk 2000). None of this means, of course, that "postmodernity" should necessarily be preferred to "network society" or "globalization." Each of these concepts points up significant aspects of contemporary social formations. However, postmodernity, understood as the complementary development of communications technologies and consumer capitalism, does raise some important questions about surveillance.

The question of surveillance systems is central to the tilt toward postmodernity. There are pressing questions, not only of the role of surveillance in constituting postmodernity, but also of how surveillance should be conceived in ethical and political terms. While discourses of privacy have become crucial to legislative and political efforts to deal with the darker face of surveillance, they frequently fail to reveal the extent to which surveillance is the site of larger social contests. If the following argument is correct, then surveillance practices and technologies are becoming a key means of marking and reinforcing social divisions, and thus are an appropriate locus of political activity at several levels.

Modernity and Surveillance

Modernity is in part constituted by surveillance practices and surveillance technologies. In order to establish the administrative web with which all moderns are thoroughly familiar, personal details are collected, stored on file, and retrieved to check credentials and eligibility. This is a means of creating a clearly defined hierarchical management order, of rational organization, in a variety of contexts. It lies behind what has come to be called the information society (see Lyon 1988 and Webster 1995) or, more recently, the network society (Castells 1996). But it also lies behind their extension and alteration as electronic technologies have been adopted to enhance their capacities. As we will see, in a technological environment the social practices of bureaucracy cannot be separated from their technological mediation.

Theoretically, then, an approach deriving from the work of Max Weber is entirely appropriate for discussing modern surveillance

(Dandeker 1990). In this approach, a rational bureaucracy is seen as an effective and enduring mode of surveillance. It connects the daily and ubiquitous processes of tax collection, defense, policing, welfare, and the production and distribution of goods and services in all modern societies with issues of state and class power in a rapidly globalizing world. Such an approach also characterized the first systematic study of the shift from paper files to computer records, James Rule's *Private Lives, Public Surveillance* (1973). Indeed, one important finding from Rule's ongoing work is that the introduction of computers, particularly in the workplace, has had unintended consequences for surveillance, even though those consequences make sense in the context of capitalist enterprise.

Rule et al. (1983: p. 223) suggested that surveillance be thought of as "systematic attention to a person's life aimed at exerting influence over it," and that this has become a standard feature of all modern societies. The nation-state, with its bureaucratic apparatus, and the capitalistic workplace, with its increasingly detailed modes of management, exemplify surveillance of this routine kind. Surveillance as intelligence gathering on specific individuals to protect national security, or by police to trace persons engaged in criminal activities also expanded, but it is the routine, generalized surveillance of everyday life that became a peculiarly modern aspect of social relations. As Anthony Giddens (1985) rightly observes, modern societies were in this sense information societies from their inception. One might equally say that modern societies had a tendency from the start to become surveillance societies (Lyon 2001).

That tendency became increasingly marked as surveillance practices and processes intensified from the 1960s onward, enabled by large-scale computerization. Computerization was used at first primarily to add efficiency and manageability to existing systems, organized as they were in discrete sectors—administrative, policing, productive, and so on. Cumbersome bureaucracies acquired the means to handle vastly larger volumes of data at much greater speed. The routine processing of personal data was increasingly automated, whether for welfare benefits, insurance claims, payroll management, or tax calculation. Such generalized surveillance, using computing machinery for the calculating and processing of data, was described as dataveillance.

For Roger Clarke (1988: p. 2) "dataveillance" referred to the "systematic monitoring of people's actions or communications through the application of information technology." He concluded that the rapid burgeoning of such dataveillance, given that it tends to feed on itself, demanded urgent political and policy attention. In the same year, Gary Marx (1988) released his study of undercover police work in the United States, which also warned of some broader social implications that he dubbed the "new surveillance." Among other things, he showed how computer-based surveillance was increasingly powerful yet decreasingly visible. He also noted the trend toward preemptive surveillance and "categorical suspicion." This refers to the ways that the computer matching of name lists generates categories of persons likely to violate some rule. One's data image could thus be tarnished without a basis in fact.

Other kinds of consequences of computerized surveillance became evident during the 1980s and 1990s, including particularly the increasing tendency of personal data to flow across formerly less-porous boundaries. Data matching between government departments permitted undreamed-of cross-checking, and the ineffective limits on such practices permitted leakage even under routine conditions. The outsourcing of services by more market-oriented government regimes and the growing interplay between commercial and administrative sectors—in health care, for instance—mean that the flow of personal data has grown to a flood. And as Roger Clarke and others have remarked, a centralized surveillance system—the archetypal modern fear—is unnecessary when databases are networked. Any one of a number of identifiers will suffice to trace your location or activities.

The "big event" of September 11, 2001, demonstrated that the U.S. government's surveillance at all levels was surprisingly loose and ill coordinated. This can be seen dramatically in the mutual recriminations of the Central Intelligence Agency and the Federal Bureau of Investigation, which each argue that if only the other agency had cooperated in data sharing the terrorists could have been apprehended beforehand. The intensified quest for new security arrangements has led to deepened surveillance in specific areas (airports, public sporting events, tall office buildings), as well as more general antiterrorism legislation enacted in the countries of Europe, the United States, and Canada. The pervasive enthusiasm for high-tech solutions, given the absence of clear evidence that they actually had the capacity to prevent acts of terror, suggests something about their cultural meanings. Surveillance technologies tend to be trusted implicitly by government agencies as well as by their corporate promoters. At the same time, populations vary widely in their responses to the new measures, with many seeing them as unnecessary and irreversible applications of intrusive techniques. Despite its best efforts, modernity does not always overcome ambivalence, but often creates it.

The primary questioning and criticism of surveillance has been carried out very much along modern lines. From the start, "Orwellian" became the preferred adjective used to condemn computerized administration perceived as overstepping democratically established limits to government power. Orwellian concerns are still the ones addressed most frequently. Thus Orwellian arguments defeated the proposal for a national electronic identification card in Australia in 1986, and similar initiatives have met a similar fate elsewhere, such as in Britain, the United States, and South Korea. More mundanely, "Big Brother" was found in the "telescreens" of factory supervision or, by the 1980s, in stores and in street-level video surveillance systems.

Until the 1990s—consistently with Orwell's *Nineteen Eighty-Four*—the greatest dangers of computerized surveillance appeared to be in the augmented power of the nation-state, with the capitalist corporation as a decidedly secondary source of risk. Thus the strategies for resisting and limiting surveillance power were similarly modern in style. Legislation relating to data protection (in Europe) or privacy (North America) muzzled the more threatening aspects or methods of surveillance. Privacy advocates were understandably slow to recognize the peculiar traits that computerization had added to modern surveillance. Only in the late 1990s, for instance, did Canadians (outside Quebec) start to take seriously the privacy issues raised by the personal data-gathering activities of private corporations. Orwell had no inkling of these!

Another notable feature of political life in the 1990s was the raised profile of "risk." This is relevant to surveillance in at least two ways. On the one hand, as more and more organizations turned their attention to the future, to capture market niches, to prevent crime, and so on, they adopted the language of risk management. The events of September 11,

2001, raised awareness of risk, including technologically generated risks such as biological terrorism, and the risks that civilian populations face from enhanced surveillance. Surveillance is increasingly seen as the means of obtaining knowledge that would assist in risk management, with the models and strategies of insurance companies taking the lead. The growth of private security systems, noted later, is one example of how risk management comes to the fore in stimulating the proliferation of surveillance knowledge.

On the other hand, surveillance itself presents risks, an aspect of what Ulrich Beck calls the "risk society" (see Brey and Mol, chapters 2 and 11, this volume). Beck has in mind the ways that the modern industrial production of "goods" seems to carry with it a less obvious production of "bads" in unforeseen side effects and in environmental despoliation. His description of how the risk society arises in "autonomised modernization processes which are blind and deaf to their own effects and threats" (Beck et al. 1994: p. 6) certainly resonates with the expansion of computerized surveillance since the 1960s, even if the risks appear in Orwellian terms. In his perspective, the two faces of surveillance may be thought of as securing against, and unintentionally generating, risk.

At the same time, the scope of insurable or securable risks seems constantly to expand. Video technologies, mentioned earlier, could be used to monitor public as well as private spaces, especially through the application of closed circuit television (CCTV). Most if not all of the world's wealthy societies today use surveillance cameras to guard against theft, vandalism, or violence in shopping malls, streets, and sports stadiums. Biometric methods such as thumbprints or retinal scans may be used to check identities, or genetic tests could be introduced to exclude the potentially diseased or disabled from the labor force. Risks may be managed by a panoply of technological means, each of which represents a fresh surveillance technique for collecting and communicating knowledge of risk. What is particularly striking about each of these, however, is their dependence on information and communication technologies.

In each case of technological scrutiny that goes beyond personal checking and dataveillance, it is information technologies that provide the means of collating and comparing records. Any video surveillance that attempts to automatically identify persons will rely on digital methods to do so. Likewise, biometric and genetic surveillance methods depend for their data-processing power on information technologies. Computer power enables voice recognition to classify travelers on the Saskatchewan–Montana border between the United States and Canada; and computer power allows researchers to screen prospective employee or insurance applicants for telltale sighs that indicate illicit drug use or early pregnancy.

Information technologies are also at the heart of another surveillance shift. Not only does surveillance now extend beyond the administrative reach of the nation-state into corporate and especially consumer capital-ist spheres, it also extends geographically. Once restricted to the admin-istration of specific territories, surveillance is steadily experiencing globalization. Of course, part of this relates to the activities of nation-states acting in concert to protect their interests by enhancing their con-ventional intelligence capacities, as seen in the Echelon system that embraces the United Kingdom, the United States, Australia, and a num-ber of other countries around the world. However, the globalization of surveillance also relates to the stretching of social, and above all com-mercial relations enabled by information and communication technolo-gies. The partnerships between major world airlines, for instance, stimulates the global circulation of personal data. Similarly, the advent of electronic commerce entails huge surveillance consequences. One major Internet company, Double-click, collects surfing data from 6400 locations on the web; and a rival, Engage, has detailed surfing profiles on more than 30 million individuals in its database (Ellis 1999).

By the 1990s, then, surveillance had become both more intensive and more extensive. Using biometric and genetic methods, it promises to by-pass the communicating subject in the quest for identificatory and diag-nostic data obtained directly from the body. Through video and CCTV, the optical gaze is reinserted into surveillance practices, which for a while seemed to rely mainly on the metaphor of "watching" to maintain their power. So what is new about these developments? From one point of view, they return us to classic sites of surveillance—the body and the city—reminding us of some long-term continuities in the surveillance practices of modernity. Even the projection of surveillance onto global terrain could be viewed merely as a quantitative expansion, a logical and predictable extension of the quest for control that gave birth to modern surveillance in the first place. But at what point do quantitative changes cross the threshold to become qualitative alterations in social formations and social experiences?

The situation at the turn of the twenty-first century resembles in some respects the surveillance situations of the earlier twentieth century. The surveillance technologies that helped constitute modernity are still present, as is modernity itself. Persons find themselves subject to scrutiny by agencies and organizations interested in influencing, guiding, or even manipulating their daily lives. However, the widespread adoption of new technologies for surveillance purposes has rendered that scrutiny ever less direct. Physical presence has become less necessary to the maintenance of control or to keeping individuals within fields of influence. Not only are many relationships of a tertiary nature, where interactions occur between persons who never meet in the flesh; many are even of a quaternary character, between persons and machines (see Calhoun 1994 and Lyon 1997a).

Moreover, those relationships occur increasingly on the basis of a *consumer* identity rather than a *citizen* identity. The most rapidly growing sphere of surveillance is commercial, outstripping the surveillance capacities of most nation-states. And even within nation-states, administrative surveillance is guided as much by the canons of consumption as those of citizenship, classically construed. At the same time, administrative records sought by the state increasingly include those gleaned from commercial sources—telephone call data, credit card transactions, and so on. Thus is formed the "surveillant assemblage" (Haggerty and Ericson 2000).

Enabled by new technologies, surveillance at the start of the new century is networked, polycentric, and multidimensional, including biometric and video techniques as well as more conventional dataveillance. These same information and communication technologies are the central means of time-space compression, in which relationships are stretched in fresh ways involving remoteness and speed, but are still sustained for particular purposes, including those of influence and control. In some respects, those influences and controls resemble modern conditions. However, in the consumer dimensions of surveillance, the influences involved are at once less coercive and more comprehensive. All of which leads one to ask: are we witnessing the postmodern power of data-processing?

Postmodernity and Surveillance

Postmodernity, as understood here, refers to a substantive social transformation in which some key features of modernity are amplified to such an extent that modernity itself becomes less recognizable as such. Postmodernity serves as a tentative, interim descriptor for an emergent social formation. The key features in question include a widespread and deepening reliance on computers and telecommunications as enabling technologies, and an intensification of consumer enterprises and consumer cultures. Both technological dependence and consumerism characterize modernity, of course, as does surveillance. What is new is that surveillance increasingly depends on information and communication technologies and is driven by consumerism (Lyon 1999).

Postmodern surveillance raises questions of meaning and political issues. Sociological accounts of the postmodern, with few exceptions, pay scant attention to technological development as such (Lyon 1997b). Thus while there is a robust literature on postmodernity, the technological shifts that I argue are central to it, while often mentioned, are infrequently investigated in sufficient empirical detail. An examination of the co-construction of these emergent social and technological formations, as seen through the case of surveillance, promises to throw light on both of them. In a significant sense, the postmodern modes of surveillance are constitutive of postmodernity. This may be better understood by examining in turn surveillance networks, surveillance data, and surveillance practices.

Surveillance networks operate in so many parts of daily life today that they are practically impossible to evade, should one wish to do so. The establishment of information infrastructures and of so-called information superhighways means that many of our social encounters and most of our economic transactions are subject to electronic recording, checking, and authorization. From the electronic point-of-sale machine for paying the supermarket bill or the request to show a bar-coded driver's license, to the cell-phone call or the Internet search, numerous everyday tasks trigger some surveillance device. In these cases, it would be a check to determine that sufficient funds are available, a verification of car ownership and past record, a timed locator for the phone call, and data on sites visited drawn from the parasitical "cookie" on the hard drive that targets advertising on your computer screen.

No one agency is behind this attention focused on our daily lives. Centralized panoptic control is less an issue than polycentric networks of surveillance, within which personal data flow fairly freely (Boyne 2000). In most countries the flows are more carefully channeled when they are found in government systems, where "fair information principles" are practiced to varying degrees. By contrast, commercial data move with less inhibition, as personal data gleaned from many sources are collected, sold, and resold within the vast repositories of database marketing. These polycentric surveillance flows are as much a part of the so-called network society as the flows of finance capital or of mass media signals that are taken to herald the information age or postmodernity. Zygmunt Bauman's stimulating analyses of postmodernity (1992, 1993), which highlight its consumerist aspects, are remarkably silent about the mushrooming surveillance technologies used to manage consumer behavior.

The fact that a single agency does not direct the flows of surveillance does not mean that the data gathered are random. The opportunities for cross-checking and for indirect verification through third-party agencies are increased when networks act as conduits for diverse data. Such networks make it easier for a prospective employer to learn about traits, proclivities, and past records not included in a résumé; for taxation departments to know about personal credit ratings; or for Internet marketers to target advertising to each user's screen. These surveillance flows erode the dikes between different sectors and institutional areas, leading to traffic between them that might not have been anticipated by the subject of the data. Enhanced efficiency of administrative and commercial operations goes hand-in-hand with the greater transparency of individual persons. These surveillance flows also undermine any sense of certainty that data disclosed for one purpose, within one agency, will not end up being used for other purposes in far-removed agencies. We are just beginning to think through the social and political implications of a world in which personal data may be retrieved and collected through networked systems even though no central agency may be involved.

The second area worth examining is that of *surveillance data*. Who or what is included within the scrutiny of surveillance? Within modern surveillance systems, some sort of symmetry exists between the record and the individual person; the one represents the other for administrative purposes. But with the proliferation of surveillance at all levels, enabled by new technologies, the very notion of a fixed identity, to which records correspond, has become more dubious. At one end of the social spectrum, the carceral net has been spread more and more widely, although not necessarily as coercively as analysts such as Stanley Cohen (1985) or Gary Marx (1988) have argued. While prisons are extending their surveillance capacities within their walls, their use of less overt means of segregation and exclusion, from parole to electronic tagging, diffuses surveillance systems throughout society. At the other end of the spectrum, the rising crescendo of calls for credentials, and other—usually plastic—tokens of trustworthiness, means that another range of discretionary and screening powers has grown up, largely distinct from those of government.

The vast and growing array of means of identification and classification that circulates within electronic databases has given rise to questions about how far the data image or the digital persona may be said to correspond to the "real world" person. This is, you notice, a modernist construal of the situation. Beyond this, however, Mark Poster (1996: n. 18) argues that in a Foucaldian sense, individuals are in a sense "made up" by their digital classification (or more properly, "interpellation"). Poster sees the world of electronic surveillance as a "superpanopticon" in which the principles of Bentham's original prison plan are expressed within a virtual realm. Subjects are now reconstituted by computer language, refuting the centered, rational, autonomous subject of modernity. Now "individuals are plugged into the circuits of their own panoptic control, making a mockery of theories of social actions, like Weber's, that privilege consciousness as the basis of self-interpretation, and liberals generally, who locate meaning in the intimate, subjective recesses behind the shield of the skin" (Poster 1996: p. 184).

The idolatrous dream of omniperception embodied in the panopticon, however, may be connected with a yet more ambitious goal of perfect knowledge, in which simulation steadily takes over from knowledge of past records. This more Baudrillardian vision of simulated surveillance is explored by William Bogard (1996) as "hypersurveillance." Bogard prophesies surveillance without limits, which aspires not only to see everything, but to do so in advance. He connects this with the desire for control as the long-term goal of many technologies, insisting that simulation's seductive claim is that "any image is observable, that any event is programmable, and thus, in a sense, foreseeable" (Bogard 1996: p. 16). Simulation is the "panoptic imaginary" that animates and impels constant upgrading and extension of surveillance. This explains, for Bogard, what Marx, Giddens, and others grope toward, but ultimately misunderstand as "new"—the technical refining of surveillance strategies. Rather, says Bogard (1996: p. 24), this is Baudrillard's "control by the code" in which the order of simulation transcends all previous references and signs, becoming entirely self-referential. Supervision and monitoring still exist, but are also "paradoxically inflated" to surpass and complete them.

Much may be learned from accounts of the superpanopticon and simulated surveillance, but it does sometimes seem that theorists themselves succumb to seduction, the allure of the metaphor. Bogard's description of surveillant simulation is essentialist to say the least. The fact that individuals may be made up by their digital image, or that self-referential simulation seems to have intoxicated surveillance systems, should not be taken to mean that the digital personae are somehow randomly constructed, or that all references have been removed from categories that sift and sort one group from another.

Third, an examination of *surveillance practices* underscores the ways in which modes of control and influence are connected with deeper determinants of life chances and of social ordering. Such practices are the nexus of co-construction, where technological potential meets social pressure. While Mark Poster (1989), in particular, is sensitive to the quest for a mutual dialogue between poststructural and critical theory approaches, one wonders if others have not forsaken what they see perhaps as an overly modern project, of "penetrating the visible forms of the present in search of the concealed mechanisms that organize contemporary life" (Boyne and Rattansi 1990: p. 8).

High-technology surveillance systems, however self-referential, are not self-financing. They are set up by those with specific kinds of interests in control and influence. As mentioned earlier, varieties of risk management lie behind much expansion of surveillance facilitated by electronic technologies. Behind them, in both commercial and government sectors, one frequently finds market values, so-called. Richard Ericson and Kevin Haggerty (1997) show how contemporary policing practices are increasingly preemptive and geared to risk communication. Police act as brokers of knowledge—personal data—used to satisfy the demands of institutions, especially those of insurance. Thus the making up of individuals according to certain categories useful to those institutions produces databases used in the effort to eliminate or at least to minimize criminal behavior.

Police and private security services today are concerned less to apprehend criminals after the fact than to anticipate criminal behaviors, classify them on a risk calculus, and contain or preempt them (Marquis 2000). Despite the "failure" of any surveillance techniques to predict terrorist attacks in 2001, faith in those techniques seems undaunted. Here in high profile may be seen just those processes of control by codes, and of self-referentiality, discussed by Bogard, but now placed within specific settings in which the interests of actors are more effectively laid bare. In the setting of police work, the augmentation of surveillance has meant the establishment of data-gathering and processing systems cut loose from previous rationales and goals of policing. Moral wrongdoing seems pushed to the edge of the picture as has, at least for the majority of routine police work, the discovery and bringing to justice of lawbreakers. Yet this is not a function of the application of surveillance technologies per se. The relentless drive for efficiency noted in earlier studies, and now the quest for speed and simulation, is an attitude, an obsession, perhaps even, as I noted above, an idolatry. That such attitudes become embedded in technology-dependent surveillance practices, thus giving the impression that the system produces these amoral, self-referential effects, is not in question.

Similar traits—control by codes, self-referentiality, and so on—may be noted within other surveillance sectors, and similar critiques may be mounted. Urban geographers, for instance Boyer (1996), have noted the ways that Foucault-type disciplines are overlaid by pervasive webs of electronic systems that assert control by distributing bodies and uses in space, using road transport informatics, home communication and information technologies, or downtown CCTV systems. But as Stephen Graham (1998: p. 486) effectively argues, essentialist accounts such as Bogard's lack "the degree of close empirical detail necessary to unravel clearly the complex social practices and political economies through which surveillance and simulation become interlinked in the production of new material geographies."

In specific ways, then, electronics-based surveillance, using databases to enable and support a whole panoply of practices, may be complicit within the peculiar emergent formations of postmodernity, hinting at new conduits of power and new modes of control. Dependence on new technologies, a basic trait of postmodernity, may contribute to what Foucault calls "governmentality," based on biopower, or intimate, everyday knowledge of populations. Such governmentality is exercised increasingly by agencies other than the welfare state, in which at first it was most widely practiced. As we saw in the policing example, even agencies that once were directly aligned with the legal power of such states are now as answerable, if not more so, to commercial organizations such as insurance companies. This governmentality is thus inherently connected with the second major trait of postmodernity as understood here; namely, consumer enterprises and consumer experiences.

Co-construction and Surveillance Practices

The case of surveillance illustrates well the mutual shaping and influence of technological developments and social processes. It also shows how that mutual shaping and influence may be imbricated within larger sociocultural shifts such as that described between modernity and postmodernity. Modernity is characterized by an increasing reliance on bureaucratic apparatuses for social administration and control, but once those systems are augmented by computerization, certain other features appear and are amplified through greater technological capacity. Surveillance practices that once relied on a generalized knowledge of populations contained in paper documentation and classified within files now are capable of simulating future situations and behaviors. This suits nicely the burgeoning enterprises of consumer capitalism, which now have the means to more directly manage their markets rather than merely the flows of raw materials or the activities of productive workers. With enhanced simulation capabilities, market criteria become more significant within decision making, not only within commercial agencies, but also within government organizations and departments such as the police. Thus, subtly and imperceptibly, the shape of modernity morphs into the postmodern. This may be seen in several related contexts.

The large-scale social transformations referred to here under the rubric of postmodernity may be considered in terms of the growing social centrality of surveillance. The accelerating speed of social transactions and exchanges, which makes them all the more fleeting, generates means of trying to keep track of those interactions. The increased geographical and electronically enabled mobility that characterizes the postmodern requires more sophisticated surveillance practices in order to ensure that rules are kept. It is not just on a macro level that the postmodern is in a mutually augmenting relationship with surveillance technologies. The world of consumption and of information and communication technologies not only generates surveillance but is itself under scrutiny.

Take for example the case of Internet-based commerce. The new networks are themselves a means for discovering consumption practices and may be used even when consumers are not actually shopping online. "zBubbles," for instance, is a program that works for Alexa, a subsidiary of Amazon.com. Programs like this offer shopping advice (the ostensible task of zBubbles) and simultaneously collect data about the computer-user's files and surfing habits to send back to profiling and marketing companies. Even some games, such as the popular Everquest, may include means of searching for hacker software on users' hard drives (Cohen 2000). Other kinds of data—including personal medical data—may also be sought by such web-based systems. Pharmatrak Inc., for example, places identifying codes on computers that visit its web sites in order to follow similar transactions or just site hits relating to, say, HIV, or prescription drugs (O'Harrow 2000).

This process by which data have become crucially valuable also stimulates the development of private security in order to protect online communications. Indeed, one could argue that private security systems, once used to protect spaces or goods, are now used above all to protect information itself. So-called business intelligence systems are rapidly burgeoning to counteract the dangers of interception of data, especially within digital and biotechnology companies (Marquis 2000: p. 18). In this case, then, the technology-enabled transmission of sensitive consumer and industrial information—itself often the product of surveillance practices—spurs security-oriented activities, which in turn require yet more surveillance. While the data may not in every case be identifiable, systematic attention is nonetheless paid to those personal data that ultimately affect the subject's life chances, either directly or indirectly.

At the more local level, however, surveillance systems are still set up with a view to making places and property secure. So-called information society or network society features are simply superimposed on already existing social conditions, so that while they may well alter them, they do not necessarily supplant them. At the same time, the alterations may contribute to an accenting of certain aspects of social situations that render them more postmodern than modern in character. So when it comes to making property secure, for instance, the trend seems to be toward "protecting profit" rather than "preventing crime" (Beck and Willis 1995: pp. 40–41).

Unlike the concerns of modern criminal justice systems, which might still stress the moral wrongdoing of, say, shoplifting, security and surveillance interests focus more mundanely on containing behaviors on the basis of a profit-and-loss calculation. Surveillance cameras may appear in the workplace and in retailing premises because petty crime raises insurance costs, thus threatening profitability (McCahill and Norris 1999: pp. 209–210). This is how insurance categories come to feature so prominently in surveillance communication, but it is also how what might be called actuarial justice becomes the norm. And actuarial justice, lacking reference points in moral codes, also helps to propel societies more fully into postmodern modes.

In a final example, that of call centers, we may see again the ways that surveillance technologies are woven into both local social arrangements and struggles, and into major societal shifts. Call centers are themselves the product of postmodern times, as understood here. They utilize the new technologies to reduce overheads associated with commerce, and may be located wherever sources of labor are relatively cheap. They illustrate well the shift toward consumer capitalism and the flexible labor force. They may be established for a variety of reasons, including handling customer call-in orders, service queries, data input, and account verification and payment. However, call centers rely on a rapid turnaround of incoming calls and workers are monitored to try to improve performance and to punish dilatory styles.

In their study of computer-based monitoring systems in England, Ball and Wilson (2000) found that how well surveillance worked depended on a number of social factors and not just the supposedly inherent properties of the technologies themselves. In a debt collection center, social relations in the workplace differed with the gender of the workers and their (supposed) knowledge of the performance monitoring technologies. Those who were "in the know"—often younger, male workers—fared better than older, female workers, and their relations with management were more cooperative. By contrast, in a data input center, where workers were fired for not meeting performance targets, relations were much more strained and antagonistic. Impersonal management styles, using the computer monitoring equipment, led to misunderstanding and resentment and to different levels of compliance with the surveillance technologies.

Kirstie Ball (2000), one of the researchers, suggests that actor-network theory allows the subject to return to a mediating role between the social and the technological and that it also permits an understanding of how information categories both produce surveillance and are produced by surveillance. Her comments facilitate a view of surveillance that once more distances it from modern accounts, where "new technologies" play a misleadingly determining role. Whether the results of this kind of research lead in more postmodern directions remains to be seen, but they certainly open the analytical door to more nuanced studies of surveillance in which outcomes are not "read-off" allegedly panoptic or Orwellian technologies.

Surveillance, Ethics, and Politics

The co-construction of technologies and social relations does not occur in a social or an ethical vacuum. The ad hoc practices of organizations as well as the self-conscious political stances of those who question and resist encroaching surveillance are inextricable elements of that co-construction process. Thus mere recognition of the social reality of the co-constructive drama is inadequate. The role of valuing and of moral positions within theoretical explanatory frameworks itself plays a part in the sociotechnical outcomes that are the topic of analytical scrutiny. The theorist is inevitably informed by a normative stance or stances in addition to the empirical findings collated through systematic study. In the case of surveillance technologies, just such implicit stances are revealed in Weberian analyses of bureaucratic monitoring and supervision, along with a legal stress on privacy as having the potential to mitigate the dangers of surveillance.

The Weberian approach laments the loss of substantive values as bureaucratic efficiency takes over. In surveillance terms, the complex actions of self-conscious actors are stripped down to their basic behavioral components, as that an amoral approach is ascendant, or "adiaphorized" as Bauman (1993) would say. The modern response is classically consistent; namely, to argue that concern for privacy is an antidote to such technicized surveillance, so that the sacrosanct self can be sheltered behind legal limits on personal data collection. This neatly answers the problem as perceived by the subjects of the data collection and thus should be treated seriously. It also resonates with deep philosophical commitments, for example, to the self as communicative and as possessing inherent dignity. Moreover, the development of privacy policies has itself contributed at least tangentially to the shaping of surveillance systems, and this has to be recognized as a factor in co-construction.

Can this help us in postmodernizing contexts, where surveillance is an increasingly powerful means of reinforcing social divisions, as the super-panoptic sort relentlessly screens, monitors, and classifies to determine eligibility and access, to include and to exclude? Today the social fractures of modernity have not so much disappeared as softened, becoming fluid and malleable. Surveillance has become much more significant as an indirect but potent means of affecting life chances and social destinies. Technological developments and social processes interact to produce outcomes which, although not necessarily as stark as the rigid class divisions of early modernity, nevertheless raise analogous questions of fairness, mutuality, and appropriate resistance.

To say that electronic technologies and consumer cultures contribute to the rise of postmodernity is not for a moment to say that the novelty of the circumstance entirely displaces all earlier concerns for human dignity or for social justice. Indeed, the inscribing of panoptic categories and surveillant simulations on the practices and patterns of everyday life is a challenge to redouble analytical and political efforts to ensure that disadvantage is minimized, especially for the most vulnerable. At the end of the day, to explore co-construction as the mutual shaping and influence of technological development and social process is to explore possibilities as much as it is to discover patterns of determination. As Judy Wajcman (1991) reminds us, if it is true that technology is socially shaped—as, indeed, social situations are also technically shaped—then it may also be reshaped for appropriate purposes.

However, while that notion of reshaping is good political rhetoric, a reminder of human agency, and at least a partial antidote to technological determinism, much more work must be done before it can become a reality. That work involves exploring the grammar of contemporary sociotechnological development, which as I have hinted, is a moral as well as an analytically inscribed grammar (Barns 1999). Demonstrating how the sociocultural and the technological interact on several different levels to produce outcomes that both facilitate social life and foster caution concerning postmodern surveillance processes is perhaps the more modest task and one that social analysts should undertake.

7

Infrastructure and Modernity: Force, Time, and Social Organization in the History of Sociotechnical Systems

Paul N. Edwards

The most salient characteristic of technology in the modern (industrial and postindustrial) world is the degree to which most technology is *not* salient for most people, most of the time.

This is true despite modernity's constitutive babble/Babel of discourses about "technology." Technology talk rarely concerns the full suite of sociotechnical systems characteristic of modern societies. Instead, at any given moment most technology discourse is about high tech, i.e., new or rapidly changing technologies. Today these include hand-held computers, genetically modified foods, the Global Positioning System (GPS), and the World Wide Web (WWW). Television, indoor plumbing, and ordinary telephony—yesteryear's Next Big Things—draw little but yawns. Meanwhile, inventions of far larger historical significance, such as ceramics, screws, basketry, and paper, no longer even count as "technology." Emerging markets in high-tech goods probably account for a great deal of technodiscourse. Corporations, governments, and advertisers devote vast resources to maintaining these goods at the forefront of our awareness, frequently without our realizing that they are doing so. Unsurprisingly, they often succeed.

Nevertheless, the fact is that mature technological systems—cars, roads, municipal water supplies, sewers, telephones, railroads, weather forecasting, buildings, even computers in the majority of their uses[1]— reside in a naturalized background, as ordinary and unremarkable to us as trees, daylight, and dirt. Our civilizations fundamentally depend on them, yet we notice them mainly when they fail, which they rarely do. They are the connective tissues and the circulatory systems of modernity. In short, these systems have become infrastructures.

The argument of this essay is that infrastructures simultaneously shape and are shaped by—in other words, co-construct—the condition of modernity. By linking macro, meso, and micro scales of time, space, and social organization, they form the stable foundation of modern social worlds.

To be modern is to live within and by means of infrastructures, and therefore to inhabit, uneasily, the intersection of these multiple scales. However, empirical studies of infrastructures also reveal deep tensions surrounding what Latour recently named the "modernist settlement": the social contract to hold nature, society, and technology separate, as if they were ontologically independent of each other (Latour 1999b). Close study of these multiscalar linkages reveals not only co-*con*struction, but also co-*de*construction of supposedly dominant modernist ideologies.

To develop these arguments, I begin this chapter by exploring how infrastructures function for us, both conceptually and practically, as environment, as social setting, and as the invisible, unremarked basis of modernity itself. Next I turn to a methodological issue that affects all historiography: the question of scale. How do infrastructures look when examined on different scales of force, time, and social organization? As Phillip Brey notes in chapter 2, "the major obstacle to a synthesis of modernity theory and technology studies is that technology studies mostly operate at the micro (and meso) level, whereas modernity theory operates at the macro level." I argue that infrastructure, as both concept and practice, not only bridges these scales but also offers a way of comprehending their relations. In the last part of the essay, I apply these methods and arguments to several examples from the history of infrastructures, including the Internet and the SAGE (Semi-Automatic Ground Environment) air defense system. Ultimately, these reflections lead me to conclude (with Brey) that social constructivism, as a core concept of technology studies, and the notion of "modernity" as used in modernity theory, are strongly conditioned by choices of analytical scale. A multiscalar approach based on the idea of infrastructure might offer an antidote to blindness on both sides.

What Is Infrastructure?

The word "infrastructure" originated in military parlance, referring to fixed facilities such as air bases. Today it has become a slippery term, often used to mean essentially any important, widely shared, human-constructed resource. The *American Heritage Dictionary* defines the term as (1) "an underlying base or foundation, especially for an organization or a system," and (2) "the basic facilities, services, and installations needed for the functioning of a community or society, such as transportation and communications systems, water and power lines, and public institutions including schools, post offices, and prisons."

In 1996–97 the U.S. President's Commission on Critical Infrastructure Protection (PCCIP) chose the following functions and services as fundamental to its own definition of infrastructure:

- transportation
- oil and gas production and storage
- water supply
- emergency services
- government services
- banking and finance
- electrical power
- information and communications

The Commission went on to explain: "By *infrastructure* . . . we mean a network of independent, mostly privately-owned, man-made systems and processes that function collaboratively and synergistically to produce and distribute a continuous flow of essential goods and services" (President's Commission on Critical Infrastructure Protection 1997: p. 3).

The free-marketeering sloganism of this definition should not distract our attention from its key concept: *flow.* Manuel Castells, one of the few scholars to succeed in fully characterizing the close interplay among sociotechnical infrastructures and the grand patterns of twentieth-century cultural, economic, psychological, and historical change, calls this relation the "space of flows" (Castells 1996). Given the heterogeneous character of systems and institutions referenced by the term, perhaps "infrastructure" is best defined negatively, as those systems without which contemporary societies cannot function.

It is interesting that although "infrastructure" is often used as if it were synonymous with "hardware," none of the definitions given here center on hardware characteristics. As historians, sociologists, and anthropologists of technology increasingly recognize, all infrastructures (indeed, all "technologies") are in fact *socio*technical in nature. Not only hardware but organizations, socially communicated background knowledge, general acceptance and reliance, and near-ubiquitous accessibility are required for a system to be an infrastructure in the sense I am using here.

An important caveat is in order here. This notion of infrastructure as an invisible, smooth-functioning background "works" only in the developed world. In the global South (for lack of a better term), norms for infrastructure can be considerably different. Electric power and telephone services routinely fail, often on a daily basis; highways may be clogged beyond utility or may not exist; computer networks operate (when they do) at a crawl. I will not attempt to integrate this very different—but equally "modern"—set of infrastructural norms into this chapter, which thus suffers from a form of idealism that might also be characterized as a western bias. Instead, I simply indicate this bias where it occurs and note that any adequate theory of modernity and technology would have to come to grips with this additional level of complexity. Other chapters in this volume, notably those by Slater and Khan, begin to move in this direction.

Infrastructure and/as Environment

As I noted earlier, infrastructures are largely responsible for the sense of stability of life in the developed world, the feeling that things work, and will go on working, without the need for thought or action on the part of users beyond paying the monthly bills. This stability has many dimensions, most of them directly related to the specific nature of modernity.

Among these is *systemic, societywide control* over the variability inherent in the natural environment. Infrastructures make it possible to (for example) regulate indoor temperatures, have light whenever and wherever we want it, draw unlimited clean water from the tap, and buy fresh fruits and vegetables in the middle of winter. They allow us to control time and space: to work, play, and sleep on schedules we design, to communicate instantaneously with others almost regardless of their physical location, and to go wherever we want at speeds far beyond the human body's walking pace. These capacities permit us, and perhaps compel us, to approach nature as a consumable good, something to be experienced (or not), as and when we wish (Nye 1997).

Infrastructures constitute an artificial environment, channeling and/or reproducing those properties of the natural environment that we find most useful and comfortable; providing others that the natural environment cannot; and eliminating features we find dangerous, uncomfortable, or merely inconvenient. In doing so, they simultaneously constitute our experience of the natural environment, as commodity, object of romantic or pastoralist emotions and aesthetic sensibilities, or occasional impediment. They also structure nature as resource, fuel, or "raw material," which must be shaped and processed by technological means to satisfy human ends.

Thus to construct infrastructures is simultaneously to construct a particular kind of nature, a Nature as Other to society and technology. This fundamental separation is one key aspect of Latour's "modernist settlement."

Infrastructure and/as Society

In the same way, infrastructures can be said to co-construct society and technology while holding them ontologically separate. As Leigh Star and Karen Ruhleder (1996) observe, knowledge of infrastructures is "learned as part of membership" in communities (quoted in Bowker and Star 1999: p. 35). By extension, such knowledge is in fact a prerequisite to membership. In the case of the major infrastructures listed earlier, these communities include almost all residents of societies in the developed world. The degree to which such knowledge is shared accounts in large part for the spectrum between familiarity and exoticism experienced in travel. Societies whose infrastructures differ greatly from our own seem more exotic than those whose infrastructures are similar to ours. Belonging to a given culture means, in part, having fluency in its infrastructures. This is almost exactly like having fluency in a language: a pragmatic knowing-how, rather than an intellectual knowing-that, so that the bewildered questions of an outsider might strike one as not only hilarious but also unanswerable. Infrastructural knowledge is a Wittgen-steinian "form of life," a condition of contextuality in which under-standing any part requires a grasp of the whole that comes only through experience (Edwards 1996; Wittgenstein 1958). In this sense, infrastructures constitute society.

At the same time, we treat infrastructures and society as ontologically separate. For example, the causes of infrastructural breakdowns such as power blackouts or telephone outages are nearly always reported either as "human error," which codes the problem as individual and allows the assignment of blame, or as technological failure. Although most breakdowns would in fact be better explained by complex relationships among operators, systems, natural conditions, and social expectations (Vaughan 1996), social causes are rarely invoked. Power outages or traffic jams cause most of us to think of downed power lines or inadequate roads, rather than to question our society's construction around them and our dependence on them. As for those few people in the developed world who choose to live without electricity or automobiles, we generally regard them as eccentrics who have "moved backward" or "live in another era"; they have chosen, as it were, not to be moderns (Kraybill and Olshan 1994).

Similarly, the notion of technological failure codes infrastructure as hardware (Perrow 1984). But most such failures can be anticipated and prevented through design and/or maintenance, which in turn require highly organized social commitments (La Porte 1991; La Porte and Consolini 1991; Rochlin 1997; Sagan 1993). The remarkably low accident rates in commercial air transport, for example, reflect the success of vigilant organizations, legal apparatus, and social learning about accidents as much as they demonstrate the quality of aircraft design and maintenance (La Porte 1988). Nevertheless, for most travelers, the social components of safe air transport are even more transparent than the airplanes in which they fly; people worry much more about the airplane than about the ground crew, the Federal Aviation Administration, or air traffic controllers. Thus while infrastructure in fact functions by seamlessly binding hardware and internal social organization to wider social structures, our commonsense perspective on infrastructure creates a "black box" that enables the rhetorical separation of society from technology in the modernist settlement (Latour 1999b).

Infrastructure and/as Modernity

Thus infrastructure is the invisible background, the substrate or support, the technocultural/natural environment, of modernity. Therefore, the question of infrastructure seems to me better posed than Heidegger's rather ill-formed "question concerning technology," which he, like most others, understood chiefly as "artifact" (Heidegger 1977). To paraphrase Langdon Winner, infrastructures act like laws (Winner 1986). They create both opportunities and limits; they promote some interests at the expense of others. To live within the multiple, interlocking infrastructures of modern societies is to know one's place in gigantic systems that both enable and constrain us. The automobile/road infrastructure, for example, allows us to move around at great speed, but also defines where it is possible to go; only a few modern people travel far on foot to places where there are no roads. When they do, it is chiefly as recreation ("being in nature"). Telephones, electric power, television, and other basic infrastructures offer many services, but also ensnare subscribers in webs of corporate bureaucracy, government regulation, and the constant barrage of advertising. Control, regularity, order, system, technoculture as our nature: not only are all of these fundamental to modernism as *Weltanschauung*, ideology, aesthetic, and design practice, but they are also (I want to argue) basic to modernity as lived reality.

This combination of systemic, technologically supported social possibilities and lawlike constraints leads to my first answer to the questions that motivate this book: Building infrastructures has been constitutive of the modern condition, in almost every conceivable sense. At the same time, ideologies and discourses of modernism have helped define the purposes, goals, and characteristics of those infrastructures. In other words, the co-construction of technology and modernity can be seen with exceptional clarity in the case of infrastructure.

Scale as Method

In the rest of this essay I want to explore a method for studying infrastructures that may help to clarify their relation to modernity. At the same time, this method draws attention to difficulties, contradictions, and fault lines within those concepts; thus it may help us further untangle their complexity, question their utility, and perhaps lead to reformulation of the question itself. The method involves looking at infrastructures simultaneously from a variety of scales of force, time, and social organization.

This technique was initially sparked by Misa's ideas about the importance of scale in the history of technology (Misa 1988, 1994). It also has something in common with Bowker and Star's method of "infrastructural inversion," which involves close attention to the normally invisible "bottom" layers of infrastructure, the levels of basic standards, classification schemes, and material bases (Bowker and Star 1999).[2] The general discussion in the rest of this section is followed by application to some examples in my own field, information infrastructure studies.

Force

I begin by considering scales of force that run from the powers of the human body (at the low end) to the geophysical. For most of human history, transportation and production systems depended primarily on human and animal power. Many modern infrastructures, such as transportation systems and electric power, create what appear on the human scale as amplifications of natural energies, beyond what unaided human beings or animals could achieve. "Modern" societies are practically synonymous with those in which such amplification is generally available. So (some) infrastructures can be characterized as force amplifiers, and the modern condition as a Heideggerian ready-to-handness of these amplifying powers. The sense of empowerment we gain from these is great indeed.

Many energy-based infrastructures thus occupy a scale of force intermediate between the human body and the geophysical. They create reliable, invisible, socially useful capacities to contain and control energy. Preindustrial infrastructures, of course, often relied directly on harnessing natural forces, such as water and wind, which also occupy this intermediate scale. A less-noticed point is that many modern energy-based infrastructures also rely, at least in part, on natural forces. Hydroelectric dams and air travel's use of the high-altitude jet stream are only two of many possible examples. This much is relatively obvious.

However, another, larger scale of force is usually ignored in discussions of infrastructure. As the Dutch (for example) know only too well, infrastructures function only within a particular range of natural variability; the system of dikes and pumping stations that keeps the ocean from reclaiming much of the Netherlands is occasionally overcome by unusual natural events. Similarly, floodplain residents across the globe regularly see their homes destroyed, only to rebuild them again. Earthquakes, tornadoes, global climate change, and other natural events represent scales of force beyond the range for which most infrastructures are, or even can be, designed.

At least in the United States, these events are known as "natural disasters." Among their social effects is to bring infrastructure suddenly and painfully to our awareness. Hurricane Floyd ravaged North Carolina and other East Coast states in September 1999; headlines about its aftermath frequently focused on the hardship, suffering, and even death resulting from the failure of water and power supplies. Power failures in major U.S. cities during the summer of 1999, when demand for air conditioning soared because of "unusual" heat waves, were blamed for a number of deaths and near-deaths. California telephone books warn residents to stock a week's worth of water, food, and cooking fuel, in case earthquakes take out electric power, water supplies, and/or gas lines. The severe destruction wrought by earthquakes in Turkey and India, in which many thousands perished, brought hand wringing about building codes, an important politico-legal standard for infrastructure. This list could be expanded indefinitely.

In the developed world, probably the large majority of "natural disaster"-related injuries and deaths are actually caused not directly by the natural event itself, but indirectly by its effects on infrastructures. For example, damage to roads, bridges, rails, and tunnels leads to automobile and railroad accidents; or municipal water supplies contaminated by flood waters and broken sewer mains cause disease. Flooding can result as much from shattered dams and levees, or silt buildup actually caused by flood-control systems, as from heavy rainfall. Edward Tenner calls these "revenge effects" of technology (Tenner 1996). The effects of such failures can be magnified by interdependencies among infrastructures. For example, natural cataclysms can cripple one infrastructure, such as the emergency services system, by taking out others, such as the telephone system and the roadway network. Indeed, we depend so heavily on these infrastructures that the category of "natural disaster" really refers primarily to this relationship between natural events and infrastructures.

Increasingly, modern societies are confronted with the forgotten relationship between built infrastructures and the assumed background of natural forces and structures upon which the former rely. Long considered essentially static, this background is now regarded not only as naturally variable, but also as subject to alteration by human activity. Global climate change, for example, is altering the parameters within which built infrastructures must function, in ways ranging from changing agricultural conditions to an increase in the frequency of severe weather events. Because of its inherently forward-looking, long-term perspective, the insurance industry—a fundamental financial component of virtually all modern infrastructures—has begun to incorporate climate change in its analysis of vulnerabilities to "natural" disaster, especially in low-lying coastal regions. As a political issue, climate change represents the dawning awareness that geophysical scales of force must be included in any complete analysis of infrastructure. This recognition could be understood as a fundamental, and fundamentally new, feature of infrastructure in modernity.

Time

Another, related scalar dimension is time. I will discuss scales ranging from the *human* (hours, days, years) through the *historical* (decades, centuries) to the *geophysical* (millennia and beyond).

The specific character of human time is one reason infrastructures fade into invisibility between moments of breakdown. Human time scales are set by our natural (animal) characteristics: the horizon of death; the salience of extremes; the fading and distortion of memory; the slow, faltering process of learning; and our restless, present-centered, single-focus attention, among many others.[3] Outside rare moments of creation or major transitions, infrastructures change too slowly for most of us to notice; the stately pace of infrastructural change is part of their reassuring stability. They exist, as it were, chiefly in historical time.

Partly because of this, infrastructures possess the power to shape human time, shaping the preconditions under which we experience time's structure and its passage. For example, the telegraph created a sense of simultaneity across huge distances, prefiguring McLuhan's "global village," while electric power extended working hours into the night.[4] Transportation infrastructure fixes the relationship of time to space, transforming human experience of both. Societies build infrastructures, of course, but because of their endurance in time, infrastructures then become the more important force in structuring society. This point is similar to Giddens' concept of "structuration," which he once defined as "how it comes about that social activities become 'stretched' across wide spans of time-space" (Giddens 1984: p. xxi).

Yet on geophysical, or even long-term historical, time scales, infrastructures are fragile, ephemeral things. The Roman aqueducts still stand, but most have carried no water for many centuries. The global telegraph network, mainstay of world communications even into the 1960s, has been largely replaced by the telephone. On this long view, time shapes infrastructures, rather than the other way round. In geophysical time, cataclysms far larger than anyone now living has experienced have occurred with monotonous regularity, while even apparently gentle forces, such as continuously dripping water, exceed the capacities of technological control (for example, in the still-unsolved problem of long-term storage of nuclear waste).

Thus, returning to my point in the preceding section, the irregularity with which "natural disasters" occur can be seen (on human force and time scales) as one vehicle for constructing properties of a modernist "nature" (as dangerous, unpredictable, and/or inconvenient), thereby separating nature from infrastructure and framing technology as control. Yet in geophysical time, this same irregularity becomes a fundamental, predictable property of nature, deconstructing the separation between them by illustrating the permanent imbrication of infrastructure in nature.

In other words, we might say that infrastructures fail precisely because their developers approach nature as orderly, dependable, and separable from society and technology—an understanding that is in fact a chief characteristic of modern life-within-infrastructures. Yet nature recalcitrantly refuses to agree to this modernist settlement. Alternatively, we could say that on long historical and geophysical time scales, breakdown is a natural property of infrastructures, or instead is a property of nature *as* infrastructure (on which all human-built infrastructures ultimately depend). Thus modernity can also be depicted as a condition of *systemic vulnerability*.

Consciousness of this vulnerability runs deep in modern thought. It is no accident that modern apocalyptic fear stems chiefly from two sources: nuclear war on the one hand, and ecological catastrophe on the other. The former represents, in a sense, the ultimate scientific/technological force amplifier. At its height during the Cold War—an utterly modern conflict of two gigantic systems whose military infrastructures permeated entire societies—widespread (and well-justified) fear of accidental nuclear war brought home the normality of breakdown, even in an infrastructure built with essentially unlimited resources (Borning 1987; Bracken 1983).

More recently, fear of global warming represents the permanent imbrication of industrial infrastructures within the planetary carbon metabolism. This again drives home the falsity of the modernist settlement; technological systems consume carbon, but they rely on nature to cycle it out of the atmosphere and back into the soil (and to produce it in the first place). As a global infrastructure, the fossil-fuel economy is simply a part of this larger process. Nature is thus in some sense the ultimate infrastructure. Ecological awareness, especially in its planet-management variants, explicitly recognizes this inseparability. We might imagine Beck's "risk society" (Beck 1992) as a description of an emerging *post*modernist settlement, which functions by rendering the natural and the sociotechnical commensurate via the omnipresent category of risk.

Social Organization

To force and time, let me now add a third scalar dimension: social organization. In contrast to its relatively straightforward application to time and force, the notion of "scale" applies to social organization only as a heuristic; the size of organizations is only one of numerous, not necessarily related, variables governing their relative importance. Still, for my purposes here it works as a rough, intuitive guide. The "scale" of social organization runs from individual families and work groups to governments, economies, and multinational corporations. It is multiply and crucially crosscut by categories such as gender, ethnicity, and other identity-constituting social formations. Here I begin to introduce empirical studies (the purpose of this volume) directly.

As I noted earlier, infrastructures exist on historical time scales. Under my definition, they also exist on large social and economic scales. Most are built and maintained by very large organizations (e.g. telephone and power companies, national and international regulatory bodies, etc.). They may connect millions, even billions, of individual and corporate users, who may employ them on a daily basis for a lifetime or more. Yet from the perspective of these users, infrastructures also exist on much smaller temporal and social scales. In some sense, every house is an individually configured infrastructure for a family or small group, built primarily by selecting commercially available components whose connectibility is ensured by standardized interfaces (e.g. wall outlets, telephone jacks, and television cables). Small, ephemeral social groups, such as those constituted by email lists or neighborhood telephone directories, may function largely or entirely through large-scale infrastructures.

Scales of social organization require a different terminology than the ones I used to describe force and time, so I will adopt Misa's useful categories:

- micro: individuals, small groups; generally short-term[5]
- meso: institutions, e.g. corporations and standard-setting bodies, generally enduring over decades or longer
- macro: large systems and structures, such as political economies and some governments, enduring over many decades or centuries

Here as earlier, I argue that a micro-scale approach to infrastructures produces one view of their role in modernity, while a macro-scale approach produces a quite different one. Each scale tells us something about the condition of modernity, yet the tensions among scalar views simultaneously call into question the category of "modernity." They also suggest a serious problem with the currently popular social constructivist approach to science and technology studies.

Meso Scales: Large Technical Systems

Let me begin with a meso-scale view. A number of empirical studies have treated aspects of the history and sociology of individual infrastructures, including highways (Goddard 1994; Lewis 1997; Seely 1987), the telegraph (Blondheim 1994; Standage 1998), radio (Douglas 1987), air traffic control (La Porte 1988; La Porte and Consolini 1991), and more recently the Internet (Abbate 1999; Hauben and Hauben 1997; Segaller 1998). The best and most successful of these have examined railroads (Chandler 1977; Yates 1989), electric power (Hughes 1983), and telephone systems (Fischer 1992).[6]

However, only a few such studies seek to address issues of infrastructure formation and development per se. The most systematic attempts began in the mid-1980s under the aegis of a loosely organized "large technical systems" group of European and American sociologists and historians (La Porte 1991; Mayntz and Hughes 1988; Summerton 1994). Hughes, the dean of American historians of technology and a prominent figure in the large technical systems group, set the agenda by arguing that on historical time scales, large technical systems tend to follow a well-defined developmental path. Initially, an unorganized, diffuse set of inventors and tinkerers create new technological possibilities. At some point, "system builders" see a way to organize these possibilities into a complete system with an important function, as Edison conceived a lighting system from generator through cable to light bulb, or as Morse imagined a transatlantic network made from telegraph keys, cables, and code. The vision of system builders must be simultaneously social and technical, since commercial success depends on understanding not only how a system might be built, but also what it might be good for and what might make it attractive to customers or clients (who usually already have some way of carrying out the function in question). In the terms I am using here, system builders imagine an infrastructure.

Following a diffusion stage, when variations on the original concept emerge, networks begin to acquire "technological momentum," characterized by analogues to mass, velocity, and direction (Hughes 1987). In this phase, some particular version of the system acquires a critical mass of users. The latter's collective financial and cognitive investment gradually acts to inhibit radical change in fundamental system properties.

At this point, standards emerge that limit the possible configurations. This is a critical stage, at which chaotic competition becomes organized around a relatively stable system concept. Eventually, competing networks must convert to these standards, find ad hoc ways to connect nonstandard equipment with them, or else die out. Standards reduce the risk to manufacturers and the cost to consumers, thus increasing the dominant system's overall momentum. In a consolidation phase, any remaining independents convert to the established standard. This creates a unified infrastructure, sometimes in the form of a public or quasi-public monopoly ("public utility"). More recently, some major infrastructures in the United States and Europe (especially Great Britain) have entered another phase: deregulation, in which government reduces or removes monopoly protection, recreating a (limited) free market for infrastructural services such as telephone and electric power.

Hughes also demonstrated that national infrastructures developed according to different "technological styles." Comparing the history of electric power systems in the United States, Germany, and England, he explained technical variations among systems through the influence of particular histories and political economies, and sometimes through more intangible factors, such as the desire to assert national identity through a unique technological style (Hecht 1998).

The large technical systems group convincingly showed that these and similar patterns can be found in the history of many major infrastructures. The lessons of these studies are twofold. First, individual infrastructures follow a life cycle, a developmental pattern *visible only on historical time scales*. Second, infrastructures consist not only of hardware, but also of legal, corporate, and political-economic elements. For example, the developmental pattern of the U.S. national railroad system had as much to do with federal land grants, the regulatory activities of the Interstate Commerce Commission, certain Supreme Court decisions, and corporate defenses against stock market speculation as with innovations in steam engines, railbed technology, and signaling systems. "Technology" is not only socially shaped; it is social through and through. Understanding *how* it is shaped demands appropriate choices of temporal and social scales of analysis. While individual system-builders like Thomas Edison, Thomas Watson Sr., or Bill Gates can matter greatly in the history of infrastructure, the real lesson of Hughes-inspired histories has been the crucial role of large social institutions.

Most of the patterns discerned by the large technical systems group apply directly to infrastructure development. However, the two concepts are not quite identical. The idea of "large technical systems" focuses attention on growth around a technological core. By contrast, infrastructures are not merely large systems, but sociotechnical *institutions*. Some infrastructures (such as school systems and constitutional legal systems) rely very little on technology, although I do not discuss this form of infrastructure here. Furthermore, some kinds of infrastructures—particularly digital information infrastructures—can be extended, interconnected, and "repurposed" almost infinitely, creating metalevel webworks that no longer fit the mold of a technology-centered system. A good example is contemporary "digital convergence," in which radio, television, recorded music, cellular telephony, and other media come together in new systems based on the Internet and World Wide Web (Edwards 1998a,b; Hanseth and Monteiro 1998). Clearly these emerging, interconnected systems do not fit the mold of electric power grids or telephone networks. As I use it, the notion of infrastructure invokes possibilities of extension in time, space, and technological linking that go beyond individual systems.

This description of infrastructure development clearly situates it as a modern phenomenon. Building regional- to world-scale infrastructures requires large institutions with long lifespans; enormous political, economic, and social power; and (on the private-sector side) great wealth. Individuals and small social groups do affect their course, but chiefly in earlier phases, before institutions have taken control. This understanding is generally compatible with the widespread view of modernity as the submergence of individuals and local communities beneath the imperatives of state and corporate power (Borgmann 1984, 1992; Foucault 1977; Vig 1988; Winner 1986). In this case, such imperatives operate by means of generalized and pragmatically unavoidable enrollment in the forms of life dictated by infrastructures.

Micro Scales: The User Heuristic

Yet views of infrastructure at other social scales offer different lessons. Under the rubric of the social construction of technology (SCOT), much recent scholarship in science and technology studies has concentrated on the micro scale of individuals and small social groups (Bijker and Law 1992; Bijker et al. 1987). Here I focus on Claude Fischer's social history of the telephone (Fischer 1992), which is among the most successful examples of this perspective.

Fischer studied telephone users in the years when the telephone was still acquiring its infrastructural status. He argued that *user* innovation shaped the social role of the telephone—more so than telephone company marketing. While early telephone companies thought of the phone by analogy to the telegraph, which was chiefly a business instrument, women (and others) rapidly adopted it for their own, nonbusiness-related purposes, such as what Fischer called sociability. This was initially seen by the telephone companies as "idle chatter" that wasted the system's value; only after decades of spontaneous, user-driven telephone sociability did telephone companies perceive the vast marketing opportunity this represented.

Working-class telephone users also innovated, creating sociotechnical networks within their communities that allowed them full use of telephone technology without subscribing to the system individually. For example, in working-class neighborhoods, young boys would monitor banks of public pay telephones, answering calls and then running off (literally) to find whomever the caller requested. This kind of system persisted for decades, even after the cost of telephone service made it possible for even the very poor to afford a home telephone. By using their own bodies and their existing community structures (neighborhoods, gathering places) as components, these users created an important variation on the infrastructure offered them by corporations and governments.[7]

Rather than assume that users are powerless pawns of dominating corporations or technological systems, Fischer argued, technology studies should adopt a "user heuristic." In other words, analysts should always *determine empirically* whether users are active agents of technological change. Fischer acknowledged important "system effects" on the micro scale (for example, the large disadvantages of not having a telephone once most people did). Nevertheless, he maintained, the empirical history of the telephone does not fit an a priori view of modernity as a condition of technological subjection and alienation. Instead, users appropriated telephone technology to their own ends, and they employed it for a decidedly pre-"modern" purpose: sociability.

Applying the User Heuristic to ARPANET/Internet history

Empirical studies of the ARPANET/Internet and the World Wide Web have brought to light stories quite similar to Fischer's account of the telephone network. In 1968–69, the ARPANET's designers imagined it as an official communication channel for research groups sponsored by the Defense Department's Advanced Research Projects Agency (ARPA) across the United States (but see the later discussion for another aspect of the ARPANET's design). The purpose was to allow ARPA computer science researchers to share programs and data quickly, cutting down on delays and inefficiencies in the existing channels, such as ordinary mail and telephone.

By 1972, however, ARPANET users had composed simple electronic mail programs that allowed them to use the system as an unofficial, general-purpose communications medium. Just three years after the ARPANET's creation, 75 percent of network traffic was email (Hafner and Lyon 1996: p. 194). This spontaneous user takeover of an official medium for unofficial purposes has many parallels in the history of information technology. For example, corporations using email for so-called group-work have sometimes felt it necessary to impose random surveillance to prevent employees from using the medium to "socialize." While their (modernist) power to do this has been upheld by U.S. courts (in-house email is legally considered official communication), the dampening effects of this strategy have often led corporations to remove surveillance later (Zuboff 1988).

Similarly, Usenet newsgroups were an unforeseen, entirely user-developed application of the ARPANET (Hauben 1996). Though initially many newsgroups were computer related, they, too, rapidly became a medium for general-purpose communication on a vast variety of topics. Today Usenet comprises tens of thousands of newsgroups, spanning subjects from scuba diving to Star Trek. This and similar phenomena have been widely discussed under the rubric of "virtual communities" (Rheingold 1993). Where telephone-supported sociability occurred primarily between people who already knew each other, and who continued to meet in person, these forms of Internet-supported sociability frequently involve strangers who never meet face to face.

The World Wide Web originated at CERN, the European high-energy physics laboratory at Geneva, in the late 1980s. Once again, its original purpose was narrow and official. The title of the document proposing what became the Web was simply "Information Management: A Proposal"; its author, Tim Berners-Lee, sought a way to cut down on the vast volume of CERN documents and data mailed around the world in support of the many physicists who collaborate on CERN experiments. Instead, he proposed a system by which such documents and data could be accessed easily, through a hypertext interface, via the Internet using a simple protocol (hypertext transfer protocol, or HTTP). Berners-Lee named this system the "World Wide Web" in 1990.

Yet what this really described at the time was the World Wide *High-Energy Physics* Web. Berners-Lee and Robert Cailliau wrote in their 1990 project proposal that

[A] universal hypertext system, once in place, will cover many areas such as document registration, on-line help, project documentation, news schemes and so on. It would be inappropriate for us (rather than those responsible) to suggest specific areas, but experiment online help, accelerator online help, assistance for computer center operators, and the dissemination of information by central services . . . are obvious candidates. WorldWideWeb (or W3) *intends to cater for these services across the high-energy physics community*. (Berners-Lee and Cailliau 1990, emphasis added)

Similar language characterized most of the early CERN project. Indeed, until mid-1993, virtually all the computer servers running HTTP were located at CERN and other high-energy physics laboratories around the world.

Here too, however, users very quickly began to add features and to use the system for general-purpose communication. Unlike the ARPANET's designers, however, Berners-Lee and his colleagues had intentionally built the system to allow users to add new material and expand the transfer protocol. With the 1993 release of a graphical browser (Mosaic) by the U.S. National Center for Supercomputing Applications (largely a support system for U.S. physics laboratories), the WWW began its explosive growth into the emerging infrastructure we know today.

These examples illustrate an important lesson of empirical studies for theories of modernity. Selective attention to the specifically "modern" aspects of infrastructures can produce blindness to other aspects that may in fact be "antimodern" (as Fischer called the sociability aspects of telephone systems). For example, modernity studies continually note the anonymity and geographically dislocated character of Internet virtual communities (Stratton 1997), but they tend to ignore, or to dismiss as utopian illusion, their well-documented qualities of spontaneity, self-organization, and sociability (Rheingold 1993, 1996; Sproull and Kiesler 1991). They point to the panoptic power of corporate surveillance in networked offices, but they fail to notice when employees find ways to work around surveillance systems (Zuboff 1988).

The key point here is that infrastructures (like all sociotechnical systems) have many and sometimes contradictory aspects. At the micro scale of social organization, "modernity"—as subjection, control, dominance of systems, panopticism—becomes slippery and difficult to locate.

Macro Scales: Functional Approaches to Infrastructural Change

Empirical studies at the macro scale—entire societies and economic systems—reveal yet another set of patterns, especially when they also employ a historical time scale. As Misa has noted, explanations on these scales tend to be functional and systemic, rather than constructivist, in character.

On societywide, historical time scales, infrastructures die. Gas lighting, the telegraph, the passenger railroad, and inner-city streetcars are all examples of once-major infrastructures that are dead or radically diminished in the United States. Any complete explanation of why they vanished requires a functional view of the reasons they came to exist in the first place. If we look at function rather than at the particular technology or infrastructure that fulfills it, we see not disappearance but growth. Gas lighting may be dead, but artificial light illuminates the world; the telegraph is gone, but far more intricate and capable long-distance communication systems have replaced it.

On this scale, we see that new infrastructures at first supplement, then sometimes replace existing ones. For example, the (expensive) telegraph supplemented (inexpensive) postal services. The telephone at first supplemented the telegraph, then replaced it.[8] At present, email supplements the telephone and is rapidly replacing postal services for personal, letter-length messages. The infrastructures delivering these services changed, but the fundamental functions they performed did not.[9] This perspective draws attention away from particular technologies, and it is scale-dependent. *On macro scales of time and social organization*, function matters more than form.

Beniger, for example, developed a theory of industrial capitalism centered around the problem of control, a functional issue linking technological, social, institutional, and informational dimensions. He argued that a generalized "crisis of control" resulted from the Industrial Revolution. Mass production techniques created control problems at the micro level of individual machines; such technologies as the steam engine governor and the Jacquard loom represented solutions at this level. But mass production also created a control crisis at the macro level of the entire production-distribution-consumption system. It rapidly produced more goods than local markets could possibly absorb. Therefore, finding new markets for this dramatically increased output soon became an urgent imperative. Faster, higher-capacity transportation systems could increase the flow rate of mass-produced goods to new, more distant markets (recalling the PCCIP definition of infrastructure). Therefore, transportation became a Hughesian "reverse salient" in the distribution system, overcome by technological innovations such as railroads, trucking, and air freight.

In order to handle the new, higher flow rates, manufacturers and distributors required better, faster control mechanisms of a different type. Information requirements—for inventories, orders, accounts, commissions, clients, and so on—grew enormously with the increasing scale of the production and distribution system. Solutions to information-processing and communication needs were both technological and social. Beniger argued that the rise of bureaucracies in the nineteenth century was a direct response to information-handling demands. Like Chandler, Yates, and others (Chandler 1977; Yates 1989), Beniger noted that railroads—the nineteenth century's largest and most complex infrastructures—deployed innovations in both human organizations and information technology to administer and coordinate their far-flung networks. Problems of scheduling, optimizing loads, transferring shipments from one railroad to another, technological standardization, and accounting were severe in the rapidly expanding national and even continental networks. Railroads resolved these control problems through both social innovation (complex administrative organizations, with multilayered managerial hierarchies and a high degree of functional specialization) *and* technological change (vertical files, standard reporting and accounting forms, etc.). These sociotechnical systems later became models for the administration (control) of other emerging infrastructures, such as the telephone network, which adopted and adapted them (Friedlander 1995a).

Control through information and communication was driven by two additional imperatives deriving from the production-distribution problem described earlier. First, efficient distribution across expanding, widely distributed sales networks required feedback; as flow rates increased, speed became more critical. Communication innovations such as the telegraph and telephone provided the possibility of near-instantaneous feedback, vastly increasing the control capacity of the overall production-distribution system. Second, Beniger argued, the problem eventually became one of creating new markets, as even distant markets became saturated with mass-produced goods. Advertising, a way of generating demand, often by creating "needs" from thin air, and market research, another form of feedback that acts to increase the efficiency of sales and distribution, constituted responses to this new reverse salient.

The macro scale of this functional view offers several unique advantages. First, it focuses attention not on "technology" but on *socio*technical solutions to large problems. Paradoxically, while many read Beniger as a technological or economic determinist, his functional view could also be seen as the ultimate in social constructivism, since it is fundamentally indifferent to whether solutions come in the form of hardware, organizations, micro-scale user innovations, or some combination of these.

Beniger's work is deeply problematic in many respects. In particular, some scholars have challenged the idea that an inherent functional logic drives industrial capitalism regardless of location or past history. The existence of widely different production techniques and structures across industrial sectors, nations, and time periods has been used to argue that the macro view fails to account for (or even correctly to describe) the historical realities (Sabel and Zeitlin 1985, 1997).

This debate is far from closed, and I will not presume to resolve it here. Yet something like Beniger's macro-scale, evolutionary perspective on industrial capitalism has been widely shared, most notably by Marxist scholars and world systems theorists. Whether or not it is correct in every detail, the macro-scale view is radically underappreciated. The "control revolution" concept allows us to understand, not only the genesis and growth of the many large infrastructures that characterize modernity, but also the process of linking these infrastructures to each other, beginning (perhaps) with the nineteenth-century co-development of the telegraph and railway systems.

Explaining Information Infrastructure: A Macro Perspective

The macro scale perspective has important implications for understanding the origins, evolution, and importance of modern information infrastructures and has relevance for modernity studies as well. Among these implications is that notions of a "computer revolution" or (more recently) an "information revolution" crucially miss the continuity of information infrastructures over time. Seen as infrastructure, information systems are ways to handle the functional problems of information storage, transfer, access, and retrieval; books and libraries remain our most important information infrastructures even today.

Ever since the vertical files, typewriters, and punch card tabulating equipment of the late nineteenth century, information-processing techniques and technologies have received enormous attention from innovators (Campbell-Kelly and Aspray 1996; Cortada 1993, 1996). Beniger's analysis explains why this should be so. The increasingly global markets of the post-World War II world presented renewed control challenges, as the speed and efficiency of transportation rose with air travel, intermodal freight, and other infrastructural innovations. Control requires information; the increasing speeds and/or sizes of the systems to be controlled required, in turn, faster and more powerful information-processing technologies. Better information processing is not a mere convenience but a sine qua non of the increasing speeds and scales at which the global material economy now operates.

Similarly, Manuel Castells' monumental three-volume study, *The Information Age: Economy, Society and Culture*, explored the functional role of computers and telecommunications in a new "informational mode of development," defined as "the technological arrangements through which labor acts upon matter to generate a product" (Castells 1989: p. 10). In the informational mode of development, information itself is both a raw material and a product. This feature generates an ever-faster development cycle; since each new process or product consists largely of information, it can instantly become input to a new round of innovation (Castells 1996: pp. 32–65). Information infrastructure thus plays a double, and doubly important, role as the fundamental basis, not only of information products and processes, but also of the global organization of *material* production and distribution. The informational mode of development takes different forms in different world regions, with material production concentrated in some areas and information production focused elsewhere. But information technology, he argues, creates everywhere a "networking logic" that integrates specific technologies into larger systems. I return to this point later.

The point here is not to make information technology the centerpiece in an ideology of progress. Instead, I simply want to acknowledge that for better or for worse, *on macro scales of time and social organization*, the co-evolution of industrial capitalism and its infrastructures displays a powerful, if never entirely determining, functional logic. As Hughes observed, this logic accounts for such historical phenomena as simultaneous invention; to those who understand a system's overall characteristics and potentials, reverse salients can become quite obvious and can command extraordinary theoretical, practical (engineering), and economic interest. The solutions adopted are not necessarily the "best" ones, if such a term is even coherent; they are simply those that endure in the market. The principles of technological change are frequently not "survival of the fittest," but merely "survival of the surviving." Neither Beniger nor Castells can explain why particular innovations occur, or why one is ultimately successful while another is not; for this one needs micro- and meso-scale views. Yet the macro perspective points to the centrality of technologies of information and control and to the ways in which overall system problems of industrial and postindustrial capitalism generate technological solutions which create, in turn, new system problems requiring further sociotechnical innovation.

At the largest scales, principles of increasing speed, volume, and efficiency drive the entire economy, with each increase in one area (e.g. production capacity) creating a reverse salient in another (e.g. market "development"). The overall system can be fruitfully described as posing a linked series of sociotechnical problems; the informational dimensions of many of these fall under Beniger's rubric of control. Just as Hughes used reverse salients to explain the phenomenon of simultaneous invention in electric power and lighting, Beniger's concept of the macro-scale control problems of industrial capitalism helps account for the massive investments in information infrastructure and in information technology research and development throughout the nineteenth and twentieth centuries.

Issues of Scale in the History of Information Technology

At this point I want to illustrate the implications of attention to scale in some of my own work on the history of computers. Electronic digital computers were developed for entirely modern purposes: code-breaking and ballistics calculations for military forces, calculation and data processing for giant corporations and governments, and numerical analysis for "big science." One of the most important episodes in early computer history was the construction of the largest and most grandiose single-purpose, centralized control system ever designed: the nuclear command-control system of the Cold War era. Few infrastructures could serve better as icons of modernity.

Ironically, within a few decades these same machines had evolved into desktop devices and embedded computers that distributed and dispersed control to a completely unprecedented degree. The present era, well characterized by Castells' phrase "the network society," looks very little like the subjection to large, panoptic systems characteristic of some concepts of modernity. It is thoroughly postmodern, yet it is also, as I mentioned earlier, in many ways antimodern. Indeed, the tensions between centralized, hierarchical forms of power on the one hand, and decentralized, distributed, networked forms of power on the other, are fundamental characteristics of the present moment. A great deal of evidence documents the relatively recent rise of networks as a major mode of sociotechnical organization, strongly facilitated (though not determined) by the availability of new information technologies (Arquilla and Ronfeldt 1997; Castells 1996; Held et al. 1999).

SAGE: The First Computerized Control System

The first important use of digital computers for control—as distinct from calculation, the chief purpose for which they were invented—arrived as a direct result of the Cold War. When the Soviet Union exploded its first nuclear weapon in 1949, well ahead of the schedule predicted by U.S. intelligence analysts, a nervous Air Force suddenly began to seek solutions to a problem it had until then been able to ignore: air defense of the continental United States.

Several different solutions were pursued simultaneously. All of them faced an extremely difficult communication and control problem: how to recognize and then to track an incoming Soviet bomber attack and mount a coordinated response that might involve hundreds or even thousands of aircraft. "Response," in that era, primarily meant interception by manned fighter aircraft. Limitations of radar systems, and the speed of then-nascent jet bombers, meant that the response would have to be mounted with only a few hours' warning at most. One warning system, the Ground Observer Corps, was labor-intensive; some 305,000 volunteers staffed observation towers along the entire Canadian border, reporting what they saw by radio and telephone. A second, the Air Defense Integrated System, proposed to automate some of the calculation and communication functions of the existing air defense structure using analog aids.

The third solution, proposed by engineers at the Massachusetts Institute of Technology (MIT), was radical. It involved using electronic digital computers to process radar signals, track incoming aircraft, calculate interception vectors for defensive fighters, and coordinate the entire response across the continent. The system concept included the abilities for the computer to send guidance instructions directly to the interceptors' autopilots, and even to control directly the release of air-to-air missiles. (The latter capability was never implemented.) All of these were real-time control functions; the computer, in other words, had to work at least as fast as the weapon systems (jet aircraft and others) it would guide. When the proposal was made in 1950, no digital computer could perform the required calculations at the necessary speed. Worse, electronic digital computers were extremely expensive, poorly understood, and highly unreliable. Containing thousands of burnout-prone vacuum tubes, their failure rates were enormous. In my book *The Closed World* (Edwards 1996), I argued that these issues made the choice of a computerized command-control system highly problematic, to say the least. Why did SAGE eventually win out?

With a colossal infusion of government cash, the technical problems were more or less resolved. The social problems—including resistance from some elements of the Air Force to a system that wrested control from individual pilots and placed computers in charge of command functions—were more difficult, but eventually they too were overcome. In 1958–61, after 10 years of research and development, the Air Force deployed the SAGE system across the United States. It was by far the single most expensive computer project to date. IBM, which built the system's 56 duplexed vacuum-tube computers, grossed $500 million from SAGE, its largest single contract of the 1950s. This was arguably among the chief reasons IBM came to dominate the world computer market by the early 1960s, since although it was not highly profitable, the project gave IBM access to a great deal of advanced research at MIT and elsewhere, much of which it introduced into its commercial products even before the SAGE computers were built.

SAGE consisted of 23 regional sectors. The computers at each sector's Direction Center communicated with neighboring sectors in order to be able to follow aircraft as they moved from one to another. Modems allowed radar data to be sent to the Direction Centers from remote locations and computer data to be shared. In a rudimentary sense, then, SAGE represented not only the first major computerized control system, but also the first computer network. Yet it was designed to permit hierarchically organized, central control of the nuclear defense system.

In a pattern entirely characteristic of infrastructure development (Bowker and Star 1999), SAGE piggybacked on other, existing infrastructures, relying on leased commercial telephone lines for intersector communications. Upon implementation, SAGE immediately spawned a host of follow-on projects with similar features. In the early 1960s, computers had already achieved a nearly irresistible appeal, far beyond what their actual capabilities then warranted. For example, intercontinental ballistic missiles (ICBMs) made the SAGE system obsolete almost before it was completed; the easily jammed system would probably never have worked anyway, and the co-location of SAGE Direction Centers with Strategic Air Command bases made them bonus targets.

Despite these glaringly obvious problems, literally dozens of computerized command-control systems, including the Ballistic Missile Early Warning System, the Strategic Air Command Control System, and the North Atlantic Treaty Organization's Air Defense Ground Environment (NADGE), were constructed in the following decade. Among the most ambitious of these was the World Wide Military Command Control System (WWMCCS), developed to automate planning for large-scale military operations across the globe.[10]

In short, computer-based command-control systems rapidly became a kind of Holy Grail for the American military. In 1969, General William Westmoreland, former commander-in-chief of U.S. forces in Vietnam, labeled this the "automated battlefield." The automated systems deployed during the Persian Gulf War and the recent Afghanistan conflict, though not nearly so perfect or so accurate as claimed, mark the near-realization of Westmoreland's vision.

Cold War–era nuclear command-control systems, all of them constructed on the model of SAGE, reflected the attempt to deal simultaneously with the imperatives of strategy, policy, technology, and culture. As the warning window shrank from hours to minutes with the deployment of ICBMs, constraints on command structures became extremely severe. The traditional hierarchical chain of command yielded to a "flattened," highly automated (but still hierarchical) version that reduced choices to a set of preprogrammed war plans for various "contingencies."

Military planners, attempting to reduce time delays inherent in the human command system, increasingly integrated computerized warning systems with weapons-release systems. Although the ultimate decision to launch nuclear weapons always remained in human hands, fears of nuclear war initiated by machine were far from groundless (Borning 1987). Soviet and American warning systems reacted to each other in an extremely sensitive way, producing a ratchet effect in which even sober analysts saw the possibility of "nuclear Sarajevos" (Bracken 1983).

Traversing Scales: "Mutual Orientation"

In *The Closed World*, I attempted an explanation of these developments that moved frequently between the macro- and meso-level constraints and enabling forces of strategy, policy, history, and culture on the one hand, and the micro- and meso-level worlds of individual inventors, work groups, and institutions on the other.

A process I call "mutual orientation" described the relationship between small groups of civilian engineers and scientists and their military sponsors, large institutions whose goals derived from the kinds of macro- and meso-scale imperatives discussed earlier.[11] In the early Cold War, most funding for research and development came directly or indirectly from military agencies. Very often these agencies did not know exactly what they were looking for. They could define general goals, but not a new means of reaching them. Generally speaking, military institutions of that era were inherently conservative, suspicious of innovation, and worried about "egghead" scientists taking over their traditional responsibilities. At the same time, WWII was widely perceived as "the scientists' war" (Baxter 1948). In the wake of radar, the atomic bomb, missiles, jet aircraft, and computers—all WWII products—American society credited scientists and engineers with almost superhuman powers. So, after the 1949 Soviet atomic test, the Air Force turned to them for help.

Here, as in very many other situations during the Cold War, the Air Force offered a general problem—continental air defense—and a set of existing weapons, such as airplanes. At the time, it was still integrating radar-based ground control into the cowboy pilot culture it had inherited from the days of dogfighting during World War I. It had no real concept of how to conduct air defense on such a scale, *nor did many believe such a goal was even feasible* (see Edwards 1996, chap. 3). In fact, the primary strategic policy of the period was "prompt use," or preemptive strike—one that left no role for a defensive force, since Soviet bombers would in principle be destroyed before they left their runways (Herken 1983).

The MIT engineers who designed the SAGE system, on the other hand, saw air defense as just one system control problem among others, solvable with the right equipment. Most of them had wartime experience with military problems (and sometimes with combat), but they were not military officers and they took a fresh view of the situation. The pieces of the puzzle as they imagined it were all in place—with the sole exception of the unfinished Whirlwind computer, which they were already building for other reasons and whose completion was their own primary, overriding goal. Making the computer fast and reliable enough to solve the Air Force's problem would also solve their own. The large implications of their concept were not lost on them.

In 1948, Jay Forrester and Robert Everett, later to become the chief engineers behind SAGE, had produced a comprehensive, compelling vision of computers applied to virtually every arena of military activity, from weapons research and logistics to fire control, air traffic control, antiballistic missile defense, shipboard combat information centers, and broad-based central command-control systems. They had written a plan for a crash 15-year, $2 billion program leading to computerized, real-time command-control systems throughout the armed forces, projecting development timetables and probable costs for each application (Redmond and Smith 1980).

The question here is why civilian engineers would spend their time working out a general systems concept for the military, which it had never requested and to which it was hardly (at that time) even amenable? The answer to this question requires understanding multiple factors and levels (for a full discussion see Edwards 1996, chapter 3). Among these factors and levels are Forrester and Everett's own backgrounds and interests; their personal relationships with foresighted specialists at the Navy Special Devices Center, which funded Whirlwind during 1944–49; other Navy elements which viewed Whirlwind as a white elephant and slashed its budget in 1949; and MIT's institutional response to this funding crisis. Seen in its full context, Forrester and Everett's plan for military computing represented not simply an engineering proposal, but more importantly a fundraising maneuver for a threatened project. When massive Air Force funding suddenly became available after the Soviet atomic test of 1949 and the outbreak of the Korean War in 1950, Forrester and Everett suddenly found themselves uniquely situated to bring digital computers to bear on a new kind of problem.

This multiscalar, many-dimensional history shows why a cowboy culture of pilots came to adopt a computer-based ground control infrastructure which it saw (initially) as a useless nuisance and anathema to the military ethos of battlefield responsibility. The civilian engineers oriented the Air Force toward a systems concept involving computerized control, while the Air Force oriented the engineers toward problems of very large-scale, real-time, high-reliability command. The SAGE engineers were system builders in the Hughesian sense: they perceived the control problem as the reverse salient, and devised a general-purpose solution that could be applied ad infinitum to other control problems. That particular reverse salient emerged simultaneously from technical, political, and cultural sources. Ultimately, U.S. geostrategic policies dictated the speed, reliability, and scale of SAGE, while a few engineers fascinated by then-nascent digital computers convinced the Air Force that the latter could be forged into a possible solution. The consequences of this interplay were profound indeed: a global command-control infrastructure based centrally on digital computers.

The concept of mutual orientation, I argue, characterizes quite broadly the general relationship between Cold War scientists and engineers and their military sponsors. In that era of swollen military budgets, sponsors did not need to direct research and development in detail. It was enough to *orient* scientists and engineers toward a general problem area. If even a fraction of the results proved useful for military purposes, that was enough, since cost was not the dominant concern. Even the most indirect value, such as pushing forward the high-tech economy (a.k.a. the "defense industrial base"), could be counted among the useful results of military R&D spending, within the totalizing vision of Cold War military planners.

Yet this was no conspiracy. Military sponsors relied in turn on scientists and engineers to generate applications concepts for new technologies. Grant writing—frequently viewed by scientists and engineers as a kind of make-believe, in which they pretended to care about military problems, while their sponsors pretended to believe in the military value of their work—looked quite different to military sponsors, who often took it quite seriously. This led to the weird (and often willful) near-sightedness of the legions of American scientists and engineers who consumed a steady diet of military money, yet claimed their research had nothing to do with practical military goals. They could be right, on the micro level, while being totally wrong about the meso-scale process in which they were caught up.

ARPANET History as Mutual Orientation

Another example of this process at work can be seen in the history of the ARPANET, which has developed a strange dual origin story. The version I described earlier holds that ARPA simply wanted to make links between its research centers more efficient and test some technically interesting concepts. A compelling part of this legend concerns the remarkable role of an anarchically organized group, consisting largely of graduate students, that developed the protocols for ARPANET message transmission. The nonhierarchical, contributory "request for comments" (RFC) process by which these protocols developed looks nothing like the hierarchical, specification-driven procedure held to characterize military operations. Indeed, the supposedly meritocratic, otherwise egalitarian culture of the ARPANET protocol builders has become part of the defining libertarian mythology of Internet culture.[12] Computer scientists themselves frequently recount this version of ARPANET history (Hafner and Lyon 1996; Norberg and O'Neill 1996). Note that this is a *micro-scale* story, both in time and in social organization: ARPA's tiny staff promoted the ARPANET, of course, but they did so as fellow travelers (most being computer scientists themselves, rather than military bureaucrats). For their part, the scientists involved pursued packet switching strictly for their own ends, and created their own, unofficial processes, such as the RFCs, to do so. There is an unmistakably gleeful tone in some of these recollections, a feeling that ARPA actually stood between computer scientists and the military, allowing the former to do exactly what they wanted while casting a smokescreen of military utility before higher levels of the Pentagon.

An entirely different ARPANET origin story takes the meso-scale approach. On this view, U.S. military institutions, seeking a survivable command-control system for nuclear war, were the driving force (see, for one of many examples, the widely distributed account by Sterling [1993]). This version begins in 1964, with a suite of RAND Corporation studies of military communications problems (Baran et al. 1964). One RAND proposal involved a "packet-switched" network. Digital messages would be carved up into small pieces, individually addressed, and sent through a network of highly interconnnected nodes (routers). Based on network load, every node would determine routing independently for each packet; in an extreme case, each packet might take a different route through the network, passing through many nodes on the way. Upon arrival, the message would be reassembled.

Packet switching meant that during a war, destruction of a few (or even many) individual network nodes would not prevent the message from reaching its final destination. This contrasted with the existing circuit-switched telephone network, in which two correspondents occupied a single circuit whose communication would be interrupted immediately upon destruction of any node in the circuit link. Packet switching was an express response to nuclear strategy, with its very high levels of expected destruction. In this second ARPANET origin story, the RAND studies fed directly into the ARPANET project. ARPA sought to build a packet-switched network for digital military communications. Whatever the research scientists believed, it was all along a deliberate strategy to build military applications.

Finally, a third, macro-scale story might also be told. This story would place the ARPANET against a larger background of the many other computer networking experiments already underway, some (such as Donald Davies' 1967 network at the UK National Physical Laboratory) having quite different social goals. Or it might situate the Internet against the long-term history of information and communication infrastructures, tracing it back at least to the telegraph, which used a "store-and-forward" technique remarkably similar to packet switching. Long-term studies of military command, control, and communication can now be re-read, seeking similarities among problems and solutions from historical periods predating the Internet (Bracken 1983; van Creveld 1985). Predictably, as scholars begin to explore Internet history, these macro-scale stories are rapidly emerging (Abbate 1999; Castells 1996; Rowland 1997; Standage 1998).

Multiscalar Analysis of ARPANET History

It is tempting to try to choose between micro-, meso- and macro-scale analysis to ask the question: Which version of this story is correct? A social constructivist view might opt for the micro level, holding that the actor perspective debunks the macro perspective. A modernity-studies approach might do the reverse, taking the meso-scale story as "true" and the micro as irrelevant or illusory. On this view, ARPANET history would be a typically modern episode in which huge forces and systems dominated individuals and prevented bottom-up social self-organization. Computer scientists and popular journalism frequently take the macro-level, functional view of the ARPANET, seeing it as one step in the continuous evolution of better, faster information infrastructures.

The concept of mutual orientation allows us to move among these scales and consider instead that *all three* stories are true. At the micro scale, scientists rarely if ever thought about the military communications problem; they had their own, private motivations for the work they did. Yet at meso scales of time and social organization, a packet-switched military communications network *was* a deliberate goal of military agencies (Abbate 1999). At a recent conference, a former high ARPA official told me: "We knew exactly what we were doing. We were building a survivable command system for nuclear war."[13] And indeed, within a few years (and with heavy ARPA backing) packet-switched networks had made their way into everyday military use (Norberg and O'Neill 1996). At this scale, the ARPANET's military backing explains not so much its particular structure as why it grew faster than other proto-types. Finally, the macro-scale view reveals deep, repeated patterns in infrastructure development. Military needs for speed, survivability, and remote coordination can be seen as ongoing functional demands that have shaped the form of communication infrastructures under many technological regimes; meanwhile, the constraints and enablements of varied communication networks have clearly shaped military capabilities (van Creveld 1985).

The subsequent history of the Internet also bears out all three stories.

On the micro level, as I pointed out earlier, by the early 1980s users had turned the Internet into a general-purpose communication tool. Hackers, largely working without pay and without a practical purpose other than invention for its own sake, played major roles in the Internet's development. The legend of Internet culture as a libertarian meritocracy—"on the Internet, no one knows you're a dog"[14]—is partly legend, but also partly true. The astonishing growth of the World Wide Web after 1993 was also strongly driven by the private purposes of individuals and small groups. The technical tools for website construction and web browsing (HTTP, Mosaic, Netscape, etc.) were by design free and open; the development model for HTTP was the Network Working Group that designed and managed Internet protocols.

On the meso scale, digital packet-switched command-control systems rapidly became the military norm, partly as a result of ARPA proselytizing (Norberg and O'Neill 1996; Reed et al. 1990; Van Atta et al. 1991). Pursuit of Westmoreland's (totally modern and modernist) centralized, electronic "automated battlefield" continues into the present. At a conference of the President's Commission on Critical Infrastructure Protection at Stanford University in 1997, an Air Force general claimed that "we are two years away from 24-hour, real-time surveillance and weapons delivery of any place on the planet." On a different meso-scale plane, corporate adoption of the Internet and the advent of e-commerce—especially pornography—were the decisive factors in turning the Web from a curiosity into a genuine global infrastructure.[15]

On the macro level, networking can be seen as a control problem along the lines posed by Beniger. The Internet explosion of the late 1980s would not have happened without a development entirely unrelated to the ARPANET, namely, the spread of personal computers (PCs) through the business world. As Gene Rochlin and James Cortada have argued, desktop PCs were initially adopted piecemeal by individuals and departments rather than by central corporate decisions. The effect of this pattern was to decentralize data (and therefore power) within corporations. Networking these many machines represented an attempt to reestablish central control, or at least coordination (Cortada 1996; Rochlin 1997). Until the later 1980s, most corporate networks were built without a thought of Internet connectivity. Yet they could easily be connected (because they generally used the same protocols), so that once the Internet began to become popular, many thousands of computers could be rapidly connected to it. This version of the story sees connectivity and control as functional directions of the economic system as a whole.

But the macro scale also allows us to observe a fundamental transition, one frequently connected with the end of modernity and the arrival of postmodernity. The distributed architecture of the ARPANET, Internet, and World Wide Web, and the open design processes that became their hallmark, made possible distributed networks of power and control. This effect is nearly opposite to the central-control purposes for which the ARPANET was built. Elsewhere I have argued that the Internet and other computer technologies have made possible "virtual infrastructures" which can be created and dismantled at will by constructing or destroying channels for information and control (Edwards 1998a). These virtual infrastructures are the foundation of Castells's "network society" (1996): a postmodern world not of systems but of constantly shifting constellations of heterogeneous actors of widely varying scale and form.

Conclusion

In this chapter I have argued that studying infrastructures on different scales of force, time, and social organization produces different pictures of how they develop, as well as of their constraining and enabling effects on social and individual life. Different scalar views also lead to different pictures of the solidity of the "modernist settlement" that separates nature, society, and technology.

Modernity studies typically approach technology as fundamental to a generalized modern (or postmodern) "condition," i.e., on the meso scale (Borgmann 1984, 1992; Harvey 1989). Meso-scale analysis typically takes historical time scales (decades to centuries) as the relevant frame. It describes large institutions—a typically modern form—as the dominant actors in infrastructure development. As large, force-amplifying systems that connect people and institutions across large scales of space and time, infrastructures seem like paragons of modernity understood as a condition of subjection to systems, bureaucracies, hardware, and panoptic power. The empirically observed meso-scale phenomenon of "technological momentum" explains the sense that infrastructures are beyond the control of individuals, small groups, or even perhaps of any form of social action, and that they exert power of their own. Infrastructures constitute artificial environments, walling off modern lives from nature and constructing the latter as commodity, resource, and object of romantic utopianism, reinforcing the modernist settlement.

Yet both micro- and macro-scale analyses challenge these constructions of technology and modernity. Macro-scale perspectives on force see infrastructures as imbricated within, rather than separate from, nature. The view from this scale emphasizes the role of infrastructure in creating *systemic vulnerabilities to*, rather than separation from, nature. It also underscores the metabolic connections between technology and nature, through fuel and waste. Here problems such as anthropogenic global climate change come into focus as the outcome of decade-to-century scale carbon metabolism. Macro-scale perspectives on time and social organization show infrastructures as solutions to systemic problems of *flow* in industrial capitalism: how to produce, transport, and sell increasing volumes of goods; and how to control the overall production-distribution-sale system (what Hughes might call the maximization of "load factor"). At this scale, their structure and form shift constantly. Particular technologies and systems are less important than the functions they fulfill. Thus infrastructures become, not a rigid background of overpowering technologies, but a constantly changing social response to problems of material production, communication, information, and control.

Micro-scale, social-constructivist analyses, especially those that study user activity, demonstrate that individuals and small, spontaneously organized social groups shape and alter infrastructures. In redeploying emerging infrastructures to their own ends, users participate in creating versions of modernity. Here too, the form and function of infrastructures shift and change over time, albeit for very different reasons than at the macro scale.

Thus, if to be modern is to live within multiple, linked infrastructures, then it is also to inhabit and traverse multiple scales of force, time, and social organization. My concept of "mutual orientation" describes one process by which micro-scale actors interact with meso-scale institutions; doubtless many other such processes await discovery. As for interaction between meso and macro scales, I have advocated describing infrastructures in terms of function rather than technology.

This multiscalar, empirical approach suggests problems with most conceptions of "modernity" itself, stemming from modernity theory's typically meso-scale perspective. Is there really a single condition describable as "modern"? Or is this a contemporary form of idealism, an abstraction to which reality corresponds only when viewed on a single scale? Micro-level, user-oriented approaches suggest that subjection and domination only partially describe actors' complex (and active) relationship to technology and institutions. Meanwhile, macro-scale approaches suggest a general trend toward infrastructural integration, facilitated by new information technology. But this integration seems to be leading, not only toward a shoring up of modernist state and corporate power and panopticism, but also toward a decentralized, rapidly reconfigurable "network society" whose postmodern dimensions are only beginning to be visible. Perhaps, then, "modernity" is partly an artifact of meso-scale analysis, to which the multiscalar approach recommended here might be an antidote.

I will close, *sotto voce*, with two important asides. First, the social constructivist approach currently popular in science and technology studies cannot generally, in practice, be distinguished from a micro-scale view (Misa 1988, 1994). Social constructivist approaches almost always explore the early phases of technological change, when technologies are new, salient, and controversial. This is also the point at which individual and small-group activity is most important. For example, user intervention in network design becomes decreasingly important and effective as standards are established and infrastructures become national or global in scope. The typical social constructivist argument is that if a technology was once controversial, it could become so again, and/or that ongoing social investment is required to maintain any given technical system. Constructivists tend to be skeptical of macro-scale explanations in any form, although they sometimes give attention to meso-scale actors.

My point here is that constructivist arguments not only depend upon, but actually function by, *reduction to micro scales of time and social organization.* Social constructivism is a contemporary form of reductionism analogous to the physicist's claim that all higher-order phenomena must ultimately be explained at the micro level of atoms and molecules. It is not that constructivist explanations are false; they have added enormously to our understanding of science and technology, and they offer a useful counterpoint to modernity theory's meso-scale view. But taken alone, without attention to meso- and macro-scale analysis, constructivism creates a myopic view of relations among technology, society, and nature.

Second, my multiscalar approach suggests a complementary reflexive conclusion. The present popularity of constructivism and other micro- and meso-scale approaches among academics may stem (in part) from meso- and macro-scale forces we too often ignore. As the academy's ranks swelled after WWII, institutions and disciplines responded by increasing scholarly specialization, thus allowing the creation of new niches (e.g. jobs and academic journals). This specialization (a modern condition?) drives scholars to focus on ever-smaller chunks of time and space. The discipline of history, for example, demands topics (and archival sources) that a historian can hope to master within a few years. Working typically alone or in small groups, historians are ill equipped to explore broad patterns and multiple scales. Similar points could be made about sociology, anthropology, and other empirical approaches to modernity. Today's scholars tend to sneer at genuinely macro-scale empirical studies, likely as they are to contain mistakes at the level of detail that occupies the forefront of specialists' attention.

Multiscalar analysis requires an enormous depth of knowledge—more than can be expected of most individuals. Social and historical scholarship has few precedents for genuine team-based approaches, which require a complex process of coordination, agreement on methods, and division of intellectual labor. It may be too much to hope that our disciplines will evolve in this direction, particularly given the present reward structures of most academic institutions. But if I am right that multiscalar analysis holds the key to an understanding of technology and modernity, we must at least make the attempt.

Notes

1. Most users of "computers" confront them, not in their essence as general-purpose programmable machines, but in their applications as special-purpose, preprogrammed systems: grocery store cash registers, rental car return systems, online library catalogs, Web browsers (Landauer 1995). Even more invisible to ordinary users are the ubiquitous "embedded" microprocessors contained in everything from automobiles to refrigerators.

2. Here I also want to acknowledge my friend and colleague Stephen Schneider, whose insistence on the importance of scale in climate science first led me to think about these issues.

3. Speed, which may be understood as the application of force amplification to the problem of human time, is another aspect of modernity produced through infrastructures. I lack the space to treat this here, but see, for example, Virilio (1986) and Rabinbach (1990).

4. The epochal character of these changes led Marvin (1988) to the correct insight that the perceived pace of technological change in the late nineteenth century was in fact faster even than today's (see also Kern 1983).

5. Small size does not always correlate with short duration. Families, for example, are a basic social unit that can endure coherently in time over extremely long periods. Nor does large size guarantee long survival.

6. For reviews of these literatures, see Friedlander (1995a,b; 1996).

7. In modern India and Bangladesh, microcredit programs are deliberately promoting a similar, community-centered telecommunications strategy. Village women receive cellular telephones from the Grameen Bank and other sponsors. They then sell call time to local customers. They earn money, but in the process they also become central to village life in a new and significant way.

8. Business users at first resisted general use of the telephone because it left no written record. Fax machines, piggybacking on the telephone system, serve this record-making function today.

9. Micro-level studies would certainly reveal systematic though subtle changes in the content and form of messages sent through each infrastructure for example, McLuhan's "hot" and "cold" media, or recent studies of differences between email and other communication forms in business organizations (Sproull and Kiesler 1991). Part of my overall argument is that these differences, too, could be seen as a matter of scale.

10. First operational in 1972, WWMCCS was replaced in 1996 by an updated version, the Global Command Control System.

11. This concept resembles, of course, other sociological ideas for relating actors and contexts of widely varying sizes and capacities, such as Giddens' dialectic of agency and structure (Giddens 1979, 1981) and actor-network theory (Bijker and Law 1992; Callon and Latour 1981; Callon et al. 1986; Latour 1987). I like to think that "mutual orientation" is a more directly descriptive and hence more useful term.

12. The term "mythology" here is intended in its full culture-defining sense, not as a contrast to a "true" history.

13. Because this comment came during a casual conversation, I omit this official's name. Suffice it to say that no one could have been in a better position to make this statement.

14. This was the punch line of a popular *New Yorker* cartoon, which shows two dogs working at a home computer.

15. As of 1998, 84 percent of registered Internet domain names were in the .com category, according to *The Internet Index*, vol. 24 (http://new-website.openmarket. com/intindex/99-05.htm). This figure probably presents a radically inflated view of the actual number of commercial websites, since many .com domain names are registered by speculators hoping to sell them later (or corporations trying to occupy a "name space"), and are not yet (and may never be) actually in use. Still, commercial and economic activity clearly became the dominant use of the Web in the late 1990s.

8

Creativity of Technology: An Origin of Modernity?

Junichi Murata

Technology studies are currently dominated by social constructivist approaches of many kinds: sociotechnical systems, social shaping, sociotechnical alignments, or actor-network approaches (see Grint and Woolgar 1997, chap. 1). Despite their differences, these approaches share a common stance against essentialist tendencies in one way or other. This characteristic can be found very clearly in the so-called social construction of technology (SCOT) approach (see Pinch and Bijker 1987 and Bijker 1995a), as well as in the actor-network approach of Bruno Latour (1987, 1999b) and Michel Callon (1995). Advocates of these approaches also argue against any determinism, whether it is a technological or a social determinism. That is, they do not presuppose a naïve distinction between the "technical" and the "social." They maintain that technological development is not determined by technical or social factors. These approaches emphasize the unique, contingent situation in which a sociotechnical network is developed and in which technological artifacts are correspondingly interpreted. Technological artifacts and their ways of working are considered to have no inherent and essential attributes and are subject to "interpretative flexibility."

While this nonessentialism makes discussions in technology studies intriguing, it also makes them at times very complicated and difficult, especially when the relationship between modernity and technology is under analysis. It is difficult to retain a nonessentialist view of technology when we consider technology to be one of the essential factors of modernity; it seems that we cannot but assume that there is an essential character of modern technology that marks it as different from traditional technologies. In fact, we have many conceptual schemes that orient our thinking in an essentialist direction; for example, Heidegger's concept of "*Gestell*" or Horkheimer's concept of "the domination of instrumental rationality" (see Feenberg 1991).

The use of these concepts to formulate questions concerning modernity and technology tends to presuppose that modern technology is essentially different from traditional technology. However, when we analyze concrete technological phenomena and search for criteria that distinguish modern technologies from traditional ones, these concepts are too abstract to be helpful. On the other hand, the newer approaches in technology studies have so far ignored the question of modernity and technology. While proponents of a social constructivist approach analyze how technological artifacts and their ways of working are constituted through sociotechnical networks, they seldom make any attempt to differentiate modern technologies from premodern ones. Perhaps for them this problem seems burdened by too many metaphysical or ideological factors that presuppose the essentialist way of thinking. We thus find ourselves in a difficult position when we try to deal with the relationship between modernity and technology.

Is there a way to deal with this relationship without taking an essentialist stance? How can we distinguish modern technologies from traditional ones while taking interpretative flexibility seriously? These are the questions I wish to address in this chapter.

The following section addresses the creative character of technology, which is rarely discussed in traditional philosophy of technology. In this section I draw upon concepts developed and elaborated by Kitaro Nishida, a preeminent modern Japanese philosopher. His philosophy can be interpreted as an attempt to develop a nonessentialistic way of thinking. According to Nishida, the creativity of technological phenomena can be described as "reverse determination," (Nishida 1949b) which is realized spontaneously in each historical situation and sometimes against the original intent of the designers and producers.

In the third section, I discuss case studies of technology transfer in late nineteenth-century Japan to illustrate the creative character of technology and to exemplify the idea of reverse determination. In the concluding section I suggest, based on several accounts of modernization in

Japan, a characteristic that differentiates modern technologies from traditional ones. If we focus on the creative function of technology, we could describe the distinguishing feature of modern technology as the institutionalization of creativity within a certain sociotechnical network, in contrast to a traditional technology, in which creativity remains a random phenomenon.

"Otherness" and Creativity of Technology

The Ambiguous Character of Technological Artifacts

One of the important and most general reasons we create technologies is to free ourselves from various types of work. However, if we examine this familiar aspect of technology more closely, its ambiguous character becomes apparent.

According to cognitive theories of artifacts, artifacts are considered to be not only the result of intelligent human work but also the cause of intelligent behavior by human beings. In order to solve a problem, such as keeping out of the rain, we make an artifact, such as a roof. Once we have made the roof, we can entrust the work of problem solving (keeping the rain off our heads) to the roof without worrying again about how to solve that problem. Gregory calls this role of an artifact "potential intelligence" (Gregory 1981: 311ff.).

From this cognitive view we can point out at least two features of artifacts and technology: (1) We use artifacts as instruments to solve certain problems. In this sense an artifact has a meaning only because human beings use it for a certain purpose. (2) But sometimes we are encouraged or compelled to use a specific means for a certain purpose, if we want to be intelligent and rational. Artifacts make our intelligent and rational behavior possible. In this way we can find in the most general characteristics of an instrument an ambiguous feature, which identifies a means as something more than a simple means.

I would like to call this surplus component—that which is "more than" a simple means—the "otherness" of technology, because it shows a component that cannot be reduced to a pure instrumental means and that sometimes motivates various interpretive activities corresponding to each situation. How can this ambiguous character be made clearer?

I think this problem is at the crux of the philosophy of technology. The kind of philosophy of technology we have depends on how we characterize this "otherness" of technology, or on which facet of the "otherness" of technology we focus.

Gregory focuses on the positive and active roles of technological artifacts that inspire intelligent thought and rational action by human beings. Gregory puts this role of instrument into a historical order by saying "we are standing on our ancestor's shoulders" (Gregory 1981: p. 312). When we emphasize the contemporaneous function of the ancestor's accomplishment, utilized during the process of problem solving, we could also say that artifacts play a role of "co-actor" in our intelligent and rational behavior. This co-actor role of artifacts has been focused on and impressively described in actor-network theory (Latour 1992, 1999b; Pickering 1995). According to their symmetry thesis, that is between humans and nonhumans, artifacts are regarded as hybrid actors or a material agency and play a fundamental role in constituting society. When we think about an artifact in our society, we can never neglect its actor element. In this sense the instrumental and co-actor roles of artifacts are inseparable and they must be considered to be two faces of one coin.

Surely it is important to characterize technological artifacts as co-actors, and surely it is important to see that the intelligence and rationality of human beings depends upon what kind of co-actors we have. It is especially important when we consider how to avoid designing inhuman environments and how to design "things that make us smart" (Norman 1993). On the other hand, it is also important to be aware that this active role of artifacts is only one element of the "otherness" of technology. In this perspective, artifacts are regarded as actors that function only according to the intention of the original designer, and there seems to remain no room for interpretative flexibility, which can be exercised in the interactive process between users and artifacts. In this sense, when we overemphasize this aspect of co-actor, there is a danger that we will adopt a perspective that is too rational and sometimes too deterministic concerning the relationship between human beings and technology.

For example, in principle it is possible not to use a roof in everyday life. But once a roof is made and widely used, it will be regarded as unintelligent, irrational, or even unhuman not to use it. Especially when artifacts are designed to be convenient and easy to use, this way of seeing them becomes unavoidable. However, exactly this character of artifacts (i.e., that artifacts determine the rational path of human action) constitutes the central core of theories embracing technological determinism. In this way, we can find a common ground between an instrumentalist view or a co-actor view and a deterministic view in which interpretative flexibility is neither sufficiently focused on nor highly prized. In either view, once the production process is finished, the artifact becomes a "black box" no longer open to various interpretations.

Creativity of Technology

We are frequently encouraged or even compelled to use a particular artifact in a particular way in order to solve a problem when we want to be rational beings. However, sometimes the situation is far from being well defined and is ambiguous enough that there is an opportunity to develop a new relationship between human beings and artifacts. For example, a hammer can be used not only to build a house but also as a murder weapon, a paper weight, or even an objet d'art (Ihde 1999: p. 46). Although this case seems to be a little extreme, every artifact has this kind of multidimensionality in some way or other, and the history of technology is full of cases of this kind.

In fact, in the history of technology it sometimes happens that invented artifacts bring us a new end-means relation in which problems and artifacts are reinterpreted and redefined for purposes far removed from the intent of the original designer. The Internet is a good example. Although originally designed for military use, it has now become a new form of communication in our everyday lives (see Edwards, chapter 7, this volume). Automobiles are another example. Before automobiles were invented, produced, and widely used, there was no urgent need to travel down a road faster than the speed of a horse or a horse-drawn carriage. In the beginning of the twentieth century, cars were not welcomed in the rural areas of America. They were called "devil wagons" and met a hostile reception from farmers (Kline and Pinch 1996). However, after automobiles became popular, traveling at the pace of a horse-drawn carriage became a "problem." In this sense new artifacts can be seen not only as problem solvers but also as "problem makers."

In addition to these cases, we can also find historical cases in which technological products are interpreted "negatively," contrary to the original intents of designers. Edward Tenner discusses various cases of this kind. Contrary to the prediction that making paper copies will become unnecessary because of electronic networking, offices are still full of paper. In another case, introducing cheaper security systems in a certain area caused malfunctions and user errors, which decreased the level of security. "Things seemed to be fighting back" (Tenner 1996: ix).

These cases impressively demonstrate the "otherness" of technology, which cannot be reduced to either a simple instrumental role or a co-actor role. This feature could be called the creativity of technology, because a new meaning for artifacts is realized, whether the new meaning is interpreted positively or negatively. What is characteristic in these cases is that the creativity is realized not in the design and production process, but rather in the interactive process between users and artifacts.

When it comes to the creative character of technology, we are often inclined to think mainly about the process of invention, design, innovation, and production, and not about users' reactions. Schumpeter (1961, chapter 2; 1950, chapter 7) emphasized the role of entrepreneurs in transforming technological changes into dramatic "innovations," resulting in economic development. Even social constructivists have tended to focus on the design and innovation process, in contrast to the process of diffusion to users. It is only recently that a designer-user distinction has been criticized along nonessentialist lines and the constructive role of users in finding creative new uses for artifacts designed for other purposes has been brought into focus (Fischer 1992; Kline and Pinch 1996; Kline 2000).

In order to clarify this creative role of the interaction between users and artifacts and also between producers and users, I would like to discuss Kitaro Nishida's philosophy, since his writings foreshadow the creative character of this interactive process.

Nishida emphasizes the creative character of our historical world and our experience of it. He describes the creative process of our historical

world with the phrase "from that which is made to that which makes"
(*tsukuraretamono kara tsukurumono e*):

Our concrete real world is a world which is a self-contradictory identity of one
and many and moves from that which is made to that which makes. That
means, our concrete real world is a historical world.

"From that which is made to that which makes" means being productive. The
historical world is the world of biological lives. But in the world of biological
lives there is no process of production. There is no process "from that which is
made to that which makes." In that world that which is made cannot be isolated
from the subject. That which is made does not become an objective reality. The
process is not that of active intuition.

There is no reverse determination there. It is not yet a world of true concrete
contradictory self-identity. (Nishida 1949b: p. 110)

According to Nishida, our real world has a feature that must be described with contradictory concepts, such as subject and object, one
and many, or motion and rest. Because our world always has contradictory characteristics, it cannot be stable; it moves incessantly and is always in a transformational process. This transformational process
cannot be characterized as mechanical or teleological because it is not
determined causally or planned or produced purposely, but arises spontaneously through the interaction of subject and object, of one and
many. The process is creative because a new situation is always incom-
mensurable with the old one from which it was formed. Because of
this transformational character, Nishida calls our world "historical"
and also "technological." Our world is technological because it is a
world of poiesis, a self-formative act that moves from the created to the
creating.

This transformation is an interaction in which subject and object are
inseparably connected but at the same time strictly differentiated. For
animals, the interaction is teleologically determined and not as contradictory as for human beings. In the case of human beings, the interaction is creative because the process has contradictory elements. The self
and the environment are so contradictory that the self is newly determined and produced complementarily by the object that the self makes,
and through it the self is brought to a new dimension. Cognition in this
process is called "active intuition" because the subject is not a passive
observer or a detached theoretician, but commits himself or herself to

and is co-constructed with an object. This cognition can be found in our daily experiences, or in the cognitive skills of artisans, artists, or experimental scientists.

Although Nishida himself did not develop the philosophy of technology in the strict sense of the word, I think we can develop his theses and apply them to concrete technological phenomena. Provisionally we can point out the following three features.

1. Nishida emphasized that the technological process does not end when the technological artifacts are produced and handed to users. When the products have left the hands of producers and become independent from them, they have a chance to acquire a new meaning and a new developmental direction through their interaction with users. In this sense, Nishida's view of technology is one in which interpretative flexibility can be found not only in the processes of design and production but also in diffusion and use. Nishida describes this creative process as "reverse determination." Perhaps this concept suggests that users instead of producers determine the creative process. But what Nishida emphasizes is that neither producers nor users alone have a decisive role in determining technological developments. Indeed, a creative process is possible only through an interaction between producers and users, both of whom stand in a contradictory relation. In this sense Nishida's philosophy of technology can be interpreted as a radical form of nonessentialism.

2. Concerning technology, Nishida does not emphasize its familiar instrumental role, by which our life is made convenient and stable; instead, he underscores the role of technology in radically transforming our historical world. According to him, because of this characteristic of technology, our life is always in the process of self-negating or self-creating and is therefore unstable. "Even in the simple process of building a house, things are not given only as material but as something which has a fateful significance for our action. In every action we stand on the brink of crisis in some way or other. Our world of everyday life is a world of true crisis" (Nishida 1949a: p. 70).

3. Nishida finds this self-negating creative structure in various levels of the historical world. Especially in his later years, he tried to define

the dynamic and critical structure of the world in the twentieth century. In a problematical essay written during World War II, he used the concept of "contradictory identity" to characterize the modern and global structure of the twentieth-century world in contrast to the eighteenth-century world. According to him, while nations and people in the eighteenth century were relatively independent and the concept of the world remained abstract, in the twentieth century the connections and the antagonisms among nations and people are so strengthened in a unified world that every nation is forced to transcend itself to fulfill its "world historical mission." "Today, as a result of scientific, technological and economic development, all nations and peoples have entered into one compact global space. Solving this problem lies in no way other than for each nation to awaken to its world-historical mission and for each to transcend itself while remaining thoroughly true to itself, and construct one 'multi-world' (*sekaiteki sekai*)" (Nishida 1950: p. 428). Although his description of the modern world remains abstract and problematical because of its political implications, it is certain that Nishida tried to characterize the modernity of the historical world with his idiosyncratic conceptual scheme (Feenberg 2000b).

Our next task is to explore the usefulness and the scope of Nishida's theses in the context of discussions concerning technology and modernity. How can we develop his abstract insights to solve the problems formulated earlier: dealing with the relationship between technology and modernity without taking an essentialist stance, and distinguishing modern technologies from traditional ones while taking interpretative flexibility seriously? In order to address this task, I would like to take up historical cases in which the relationship between technology and modernity became a central problem. The following cases relate primarily to the modernization of Japan, but I would like to compare this process with other technology transfer processes in different historical contexts. Through such a comparison it will become clear that western technology is "interpreted" and "translated" in different ways that correspond to different historical contexts. These are exactly the ways in which various types of interpretative flexibility and in this sense various "hermeneutical" experiences in the interaction of users and artifacts are realized.

Hermeneutics of Technology: Modernization of Japan

Radical Transformation

Impact of Civilization   In 1853 and 1854, the American Commodore
Matthew C. Perry visited Japan on warships powered by steam engines;
in their wake, as it were, a once-isolated Japan was opened to commerce
with the western world. The Japanese called these warships *"kurobune"*
(black ships), because they were painted black and raised a dense cloud
of black smoke. These powerful technological machines greatly impressed the Japanese people, who began to recognize, although reluc-
tantly, the necessity of cultural and technological exchange. Among the
presents from the U.S. president to the shogun, the magnetic telegraph
and a one-quarter-scale model of a locomotive engine especially stimulated curiosity in Japan. However, it was the ships' 10-inch cannons that
became the center of attention among Japanese officials, who fully understood the urgent need for introducing modern weapons in Japan to
prevent a third or fourth visit from Perry or other unwelcome visitors.
In fact, every effort to introduce and develop modern weapons was
made in the last days of the Edo period by the Tokugawa shogunate and
various feudal domains as well as after the Meiji restoration (1868) by
the new central government.

One of the main characteristics of the modernization of Japan in the
late nineteenth century was that the Japanese quickly understood that in
order to adopt modern western weapons, it was necessary to introduce
various industries connected with military technology. In order to build
and sustain those industries it would also be necessary to adopt the
western civilization that formed the background for those industries.

Even before the Meiji restoration, many samurai visited western
countries (illegally at first and then legally) and were greatly impressed
by the western world. Immediately after opening Japan to exchange
with foreign countries, the shogunate began to send various people to
America and Europe to study abroad and to negotiate treaties of commerce. In 1871, after the Restoration, a large mission was sent to America and Europe. The members of this mission consisted of 47 primary
officials of the new Meiji government. They spent more than 22 months
examining political, social, economic, and technological circumstances

in developed countries. The result of these observations was summed up in the famous political slogans of the new government: "Promoting enterprise and developing products"(*Shokusan Kougyou*) and "Enrich the country, strengthen the army"(*Fukoku Kyouhei*). In 1874, Ohkubo, the minister of home affairs, summarized the outlook of the new government: "The strength of the country depends on the wealth of people and the wealth of people depends on the amount of products. And while the amount of products depends upon whether people develop industry or not, its origin lies on whether the government leads and encourages the development" (Ohkubo 1988[1874]: p. 16).

These statements have been interpreted to mean that the highest purpose was in the (military) strength of the country, and in order to realize this purpose, the development of industrial technology was indispensable. Certainly this meaning was included in the sentence. But if understood in this way, the development of industry and technology can be regarded as only one means among others, and it is not clear why the Japanese wished to introduce the entire western civilization together with many kinds of technology so hastily. This consideration brings us to a slightly different interpretation.

I think the emphasis lay not on the military strength of the country but rather on the development of industry, so that it was understood in the following way: for the time being, the development of industry and technology was most important, because only through them could the wealth and strength of the country be realized. If we interpret the statement in this way, it clearly expresses an ideology of technological determinism, in that the development of industrial technology makes possible the wealth and strength of a country. This was exactly the response of the Japanese people to the challenges of western modern civilization. They fully understood that the engine of western modernity was industrial technology; from their viewpoint, technology and modernity were inseparable. How could the Japanese people have acquired such a point of view? In order to understand, we should look at how they arrived at this insight.

Technology as Instrument and Demonstration of Civilization   Stimulated by a telegraph demonstrated by Perry's crew, the Japanese began to introduce telegraph machines from various European countries, to learn this technology for themselves, and to make their own machines. As early as 1870, a public telegraph service began between Tokyo and Yokohama. Railroad service with locomotive engines began between these two cities two years later.

The rapid speed with which telegraph and railroad services were introduced was not in response to an urgent demand for them. Indeed, there was opposition to their hasty introduction because social and economic conditions in Japan were insufficient to support them, and in fact their economic results were disappointing. The many transplanted technologies, such as railways, telegraphs, shipbuilding, and iron manufacturing constituted a program of "industrialization from above" introduced by initiatives from the ministry of engineering.

Even if the process was an "industrialization from above," it did not meet a strong rejection from the grassroots or common people. Most of the people accepted and even welcomed with enthusiasm the modernization brought by these various technologies. In this context, I would like to focus on the demonstrative character of technological artifacts.

Although modern transplanted technologies such as steam locomotives and railway systems did not always function successfully in the sense of instrumental rationality, they had a great expressive meaning as a demonstration of western civilization in the early Meiji era. Tetsurou Nakaoka, a historian of technology, describes this characteristic of technology in the following way:

Enterprises of industrialization in the early Meiji era proved to be not directly useful for the industrialization per se. In a sense they could be considered to be a waste. But what I want to say is that they have played a significant role for the industrialization in reproducing the "impact of civilization" in the mind of people, although this role was indirect. Only when we take this role into consideration, [can we] understand why grassroots people have shown such an extraordinary active response to the industrialization. Through the understanding of this role, we can also come to understand what an important role exhibitions have played in the Meiji era. (Nakaoka 1999: p. 165)

In fact, during the Meiji era domestic industrial expositions were held regularly, and when the fifth exposition was held in Osaka in 1903, more than four million people visited it. This fact alone shows how much interest people had in modern technologies. Modern technical artifacts introduced into sociotechnical networks that were already present played not only an instrumental role but also an expressive role. A train pulled by steam locomotives could be viewed as a running advertisement for modern western civilization; people could see the modern western world "through" a train.

The situation was not radically different later in the twentieth century. After the bitter defeat of Japan in World War II, cars imported from America symbolized western civilization in Japan. Cars were seen as an artifact embodying the American dream, and their acquisition and use symbolized the acquisition of a most advanced civilization. Modern technology was never considered to be a neutral instrument in modern Japan, from grassroots people to government officials. Rather, it has been considered something that is always value laden and cannot be detached from its original sociotechnical network.

Contrast between Japan and China However, there is a famous proverb, "Japanese spirit and Western technology" (*Wakon yousai*), that seemingly contradicts this view. According to this proverb, western technology can be detached from its original context and introduced without changing Japanese culture. Sometimes the proverb is interpreted to show the real characteristics of the modernization process of Japan, and sometimes to explain its "success." While some intellectuals in the Edo and the early Meiji era emphasized the necessity of this thesis in order to introduce western science and technology without conflict, others criticized the one-sidedness and distortion of the "success" of the modernization process in Japan. On the assumption that the Japanese successfully introduced and developed science and technology detached from their original contexts, Steve Fuller has recently maintained that their success demonstrates that the "uniqueness" of western science is only a matter of contingency. According to Fuller, "the Japanese were bemused that modern Europeans could believe in such a superstitious sense of [Eurocentric] historical destiny" (Fuller 1997: p. 127).

Considering science at the time, that assertion has limited validity. During the nineteenth century, science experienced a "second revolution," became more institutionalized, and the connection between science and technology strengthened. But this does not mean that science became separable from its context, but rather that science was embedded more fully in its sociotechnical network. In this sense it became even more difficult to separate science from its context.

The proverb "Japanese spirit and Western technology" actually originated in China, where the Chinese followed more faithfully the thesis of adopting western technology but not western culture. However, the result was disastrous for them, at least at the end of the nineteenth century. Barton Hacker describes the contrast between the Chinese and the Japanese responses to Western technology:

The crucial issue, and the point from which Chinese and Japanese response sharply diverged in the 1860s and later, was how much of Western culture was attached to the hardware. China and Japan found different answers. . . .

In a deeper sense, China's defeat [in the Sino-Japanese War of 1894–95] was rooted in a fundamental miscalculation. Self-strengthening assumed that China could defend its traditional society against the West with Western weapons, that the West's military technology could be detached from Western culture as a whole. . . .

The Meiji Restoration of 1868 was so named from the presumed return to the emperor of his former power, usurped in recent centuries by the shogun. The rhetoric of imperial rule and a return to time-honored forms disguised far-reaching changes. Younger samurai had played key roles in toppling the Tokugawa regime. Deeply impressed by the West's military technology, they assumed their new government posts determined to sustain Japan's independence with Western weapons. But they accepted, as their Chinese counterparts did not, the price of that technology, which involved not only a complete revamping of the military system but also large-scale industrialization and all it implied. (Hacker 1997: pp. 283–286)

An important point is that the whole scale of modernization was not regarded as a necessary price by most Japanese people but welcomed by them. Perhaps the proverb "Japanese spirit, Western technology" also played a certain role in Japan. But if we think it did, its function must be considered to belong to an ideological dimension. If we pretend to believe it, it is possible to develop a radical cultural change under the guise of this ideology, avoiding, or at least decreasing the conflict between traditional culture and modern technology. While in the ideological dimension the thesis that technology is a neutral instrument played a certain role, in the material dimension everything was changed, continuously responding to and accepting the modern technology.

Radical Translation

The story about the hermeneutical process of modernization in Japan is not yet complete. In order to highlight the characteristics of this process, I would like to go back to another type of encounter that took place a few centuries before the above-mentioned story.

Medieval and Early Modern Age of Europe   "The clock, not the steam-engine, is the key machine of the modern industrial age" (Mumford 1934: p. 14). This famous statement by Lewis Mumford identifies the clock as the icon of modern machinery. Mumford did not tell a deterministic story concerning the relationship between technology and modernity. Rather, he emphasized social factors, such as the discipline and regularity of the monastic life, which constituted the background of the invention and diffusion of mechanical clocks.

Recently historians have suggested that we should not overemphasize the mechanistic image of the monastery and that Mumford's thesis has only limited validity. Certainly it is misleading to talk about the machinelike rhythm of monastic life because "life according to the Rule was bound in a very high degree to natural time givers, daylight and the seasons, and was by no means marked by ascetic resistance to the natural environment" (Dohrn-van Rossum 1996: p. 38). Although the Christian church played an important role in the growth of interest in time measurement and timekeeping, and also in the development of mechanical clocks, it was only one factor among others. The new source of demand for mechanical clocks came from "the numerous courts—royal, princely, ducal, and episcopal" and "the rapidly growing urban centers with their active, ambitious bourgeois patriciates" (Landes 1983: p. 70).

I wish here to emphasize the role of clocks that is essentially connected with technical functions but includes more than these. For a long time, clocks have been used as a metaphor for the mechanical world-view. We find this even in the early history of clocks. "It is in the works of the great ecclesiastic and mathematician Nicholas Oresmus, who died in 1382 as Bishop of Lisieux, that we first find the metaphor of the universe as a vast mechanical clock created and set running by God so that 'all the wheels move as harmoniously as possible'. It was a notion with a future: eventually the metaphor became a metaphysics" (White 1962: p. 125).

During the change from the medieval to the modern age, clocks influenced (and were influenced by) dynamic changes in sociotechnical networks. But more than that, clocks also played a decisive role in the radical change of the worldview. We could describe this as a creative role of technology, although it does not have a direct relation to certain technological innovations. What is different in the case of clocks is that clocks had no strong social or technological networks by which their creative function could be transferred and realized, so that they were "interpreted" in radically different ways. We can clarify this point by contrasting the introduction of western mechanical clocks into China and Japan, which began in the late sixteenth and early seventeenth centuries.

China In the seventeenth century, many Christian missionaries from Spain or France visited China to propagate their faith, and in the eighteenth century, many Europeans rushed to establish commercial ties with China, to meet the soaring demand in western markets for China's silk, porcelain, and especially tea. This cultural and commercial exchange between China and Europe remained unbalanced or even one-sided for a long time because the Chinese found nothing interesting in what Europeans brought, while Europeans wanted to import various things from China.

One of few things in which people in China expressed an interest was the mechanical clock. Chinese emperors showed great interest in mechanical things and collected many kinds of western clocks. Father Valentin wrote in the 1730s, "The Imperial palace is stuffed with clocks, . . . watches, carillons, repeaters, organs, spheres, and astronomical clocks of all kinds and descriptions—there are more than four thousand pieces from the best masters of Paris and London, very many of which I have had through my hands for repair or cleaning" (Landes 1983: p. 42; Tsunoyama 1984: p. 42). Clocks were displayed together with pictures, porcelains, pottery, and many kinds of playthings in palaces and enjoyed by people in court. There was even a factory in which clocks were made and repaired by artisans, instructed by Christian fathers who were specialists in the technology. In spite of this interest in mechanical clocks, the Chinese did not use them as an instrument for time measurement and timekeeping.

Why didn't the Chinese use mechanical clocks in their everyday life? Why did they not develop the technology of mechanical clocks when in the tenth century they had invented a splendid mechanical device that expressed astronomical movement and was used for measuring time? The simplest answer would be "because they were useless in a society in which timekeeping had no decisive role." A more insightful answer would be the following: While the Jesuits wished to persuade the Chinese people that a civilization that could produce a manifestly superior science and technology must be superior in other respects, especially in the spiritual realm, the Chinese had seen a dangerous element embodied in the European mechanical clock, which made an assault on China's self-esteem and could not be reduced to a neutral instrument. The Chinese people were deeply disappointed by the western worldview, in which China was located not in the center but only in a small and peripheral part of the world. In this sense, we could interpret the response of the Chinese to western clocks as a deliberate rejection (Landes 1983: pp. 44–47). In any case, until the middle of the nineteenth century, clocks were considered mainly to be interior decoration or playthings for emperors and high officials.

If we say that aesthetic meaning is not the main function of a clock and the use of a clock as an objet d'art is irrational, we presuppose what the main purpose is and what the side effects are. However, this distinction between purpose and side effect is always constructed in a cultural context, and side effects are well known to sometimes play a creative role in the development of technology. When we remember the windmill in the medieval age, the interpretation of clocks as aesthetic rather than functional objects in seventeenth- and eighteenth-century China could be recognized as a typical case of a hermeneutical experience concerning technological artifacts. According to Lynn White, "In Tibet windmills are used only thus, in the technology of prayer; in China they are applied solely to pumping or to hauling canal boats over lock-sides, but not for grinding grain; in Afghanistan they are engaged chiefly in milling flour" (White 1962: p. 86).

Japan   During the same period, very few western clocks were imported into Japan. Japan used a variable-hour time system so Japanese artisans adapted newly introduced western clock mechanisms to move according to the Japanese time system. In adapting the original mechanisms, the artisans invented complex mechanisms of their own to correspond to the complexities of the Japanese time system, in which daytime hours were longer than nighttime hours in summer and shorter in winter. "Some clocks had several interchangeable face plates with different spaces between the markings for the hours. On others there were sliding weights which had to be adjusted manually at sunrise and sunset to slow down or speed up the working of the mechanism. Others again had a double verge-and-foliot system which marked and measured the elusive flow of time" (Morris-Suzuki 1994: p. 52). In effect, Japanese artisans developed many original types of clocks.

The development of these "traditional Japanese clocks" (*wadokei*) can be seen as unique and original in the history of clocks, but as soon as the western time system was introduced after the Meiji Restoration, these clocks became useless, abandoned, and forgotten. Sometimes the Japanese pattern of clock development, adapting western technology to a Japanese time system, is considered to be a degeneration of technology.

However, there is no need to regard this adaptive process as degenerative and the western way as progressive. Rather, one could view traditional Japanese clocks as successful accomplishments of instrumental rationality, which supports the thesis of social constructivism of technology. Japanese artisans opened the black box of a western clock mechanism and redesigned it to correspond to the needs of Japanese social groups. In this way they showed the interpretative flexibility of technology across different cultures.

Thus we can find three types of interpretations of clocks. In the first case, clocks were interpreted as something *more* than technical; in the second as something *other* than technical; and in the third as something *simply* technical. In this sense, the Japanese reaction could be considered to be the most rational and enlightened on technological grounds in the narrow sense of the word.

In contrast to Japan, in Europe clocks were seen as embodying a metaphysical meaning, and people did not perceive clocks alone but perceived the world "through" clocks (Ihde 1990: p. 61). Here we can find a similar relationship between the artifacts and the meaning embodied in them, as in the case of the introduction of modern technology into Japan in the late nineteenth century. In both cases, modern technology was not regarded as merely a neutral instrument, but as something more. It is not the case that because modern machines are considered to be useful in a pregiven society, they are introduced into it. Rather it is because they attract people as something more than a simple instrument that they are introduced and accepted as a useful instrument. The meaning embodied in artifacts varies and depends on historical situations. In any case, modern characteristics of artifacts cannot be reduced to their instrumental or co-actor role, and they cannot be fully understood without taking their surplus component into consideration, which is what motivates people to accept them, whether it belongs to a metaphysical or an ideological dimension.

Why did the Japanese show such an enthusiasm for western technologies in the late nineteenth century, while they were so "rational" about western clocks earlier? In other words, why did the surplus component embodied in modern machines in the late nineteenth century not remain in the ideological dimension, but in fact have a material influence on Japanese society? Why weren't modern machines detached from their (western) sociotechnical network, as in the case of the clocks in the seventeenth or eighteenth century? Certainly there were many reasons that must be clarified through empirical studies. But in order to find an answer to this question, I would like to go back again to the modernization of Japan.

Mediated Transformation: Continuity and Discontinuity

The Japanese cases demonstrate contrasting types of technology transfer between cultures. In the case of clocks in the seventeenth and eighteenth century, the new artifacts (western clocks) underwent a radical translation, as Japanese artisans developed an efficient instrumental rationality to fit them into a traditional network. In contrast, the encounter between modern technology and the Japanese in the late nineteenth century produced a radical transformation of the sociotechnical network, and as we have seen, the conception of technological determinism accompanying this transformation was the result of the interpretative activities of Japanese people. Despite the striking contrast, there is an interesting relation in the two types of technology transfer. In order to clarify it, I would like to follow the story of the artisans who developed Japanese traditional clocks.

The artisans who developed and produced a Japanese style of clock were closely connected to another innovation in the Tokugawa Edo period: the automaton (*karakuri*). The introduction of clockwork provided the opportunity to make a more realistic representation of human behavior possible. One of the most famous artisans in this technological tradition was Hisashige Tanaka (1799–1881), who built a very impressive astronomical instrument in the Edo period. Immediately after the arrival of Commodore Perry's fleet of ships, Saga Domain invited Tanaka to advise on technological modernization of steam engines of ships and guns, among other things. In 1875 Tanaka established a private machine-making firm, which later became part of the twentieth-century manufacturing giant Toshiba (Morris-Suzuki 1994: p. 53).

The connection between Japanese clocks, automatons, and advanced technology was not direct, but was rather complicated. The gap between traditional Japanese technology and more advanced western technology was huge at the time. In the iron or railroad industries, for example, almost every machine part was imported during the early adoption phase in Japan, and many foreign engineers and artisans (*Oyatoi gaikokujin*, literally "hired foreigners") were invited to build factories and teach and advise Japanese engineers and artisans. However, few of the transferred technologies took root easily in the new context; only after Japanese engineers and artisans worked hard to translate those technologies into local terms did they function successfully in the Japanese context. In this sense we must confirm that the rapid and radical transformation of the Japanese technological network depended on the support work of Japanese technicians. This was a decisive point in the modernization of Japan, as it is in many other cases: skilled artisans and domestic engineers played a critical role. I would like to especially emphasize the role of *traditional* artisans in adapting the new technology to the environment and preparing a suitable environment for the new technology. While machines and factory systems introduced during the industrial revolution are often thought of as deskilling laborers and leading to the disappearance of traditional artisans, many economic historians emphasize the important role of artisans in the innovation and development of industrial technology.

As for the role of artisans and skilled workers in the process of industrial revolution and the modernization of industry, there continues a debate (Sabel and Zeitlin 1985, 1997; Odaka 1993). The concept of *artisans* itself is sometimes ambiguous, and the situation around artisans and skilled workers is different in different countries and dependent on historical conditions. But at least in the case of Japan, almost all historians seem to agree that traditional artisans played an important role in the early phase of the industrial revolution in Japan.

According to Rosenberg (1970), for example, a capital goods industry plays an important role in the development and transfer of a technology, by creating an appropriate environment for repair and maintenance and successful performance of the machines. Rosenberg also emphasizes the aspects of technology that are incorporated by skilled personnel and are not explicitly codified. The transfer of these people played a decisive role in the process of technology transfer in many kinds of machine-making industry in the nineteenth century.

But in making new products and processes practicable, there is a long adjustment process during which the invention is improved, bugs ironed out, the technique modified to suit the specific needs of users, and the "tooling up" and numerous adaptations made so that the new product (process) can not only be produced but can be produced at low cost. The idea that an invention reaches a stage of commercial profitability first and is then "introduced" is, as a matter of fact, simple-minded. It is during a (frequently protracted) shakedown period in early introduction that it becomes obviously worthwhile to bother making the improvement. Improvements in the production of a new product occur during the commercial introduction.

Alternatively put, there has been a tendency to think of a long pre-commercial period when an invention is treated as somehow shaped and modified by exogenous factors until it is ready for commercial introduction. This is not only unrealistic; it is a view which has also been responsible for the neglect of the critical role of capital goods firms in the innovation process. (Rosenberg 1970: p. 569)

In the capital goods industry, various machines and parts for machines are invented, designed, and produced in order to solve problems that occur in the interactive process between producers and consumers. In this sense, the capital goods industry plays a necessary role in preparing an environment in which a new technology can be realized or transferred. We could also say that capital goods industries are an institutional foundation that constantly makes possible technological innovation and transfer by mediating between the producer of machines and their users.

We can find many parallels between the capital goods industries cited in Rosenberg's cases and the roles Japanese artisans played in the modernization process in Japan. Surprisingly, after the Restoration, the central Japanese government took into consideration this role for skilled personnel. When the new government sent a mission of artisans and high officials to the international exhibition held in Vienna in 1873, several of them remained for two years after the exhibition to continue learning various technologies. Most of the technologies that they brought back were not directly connected to advanced technologies, but to *traditional* ones. These were more readily accepted and this allowed them to introduce new inventions and innovations very rapidly. In addition to the international exhibition, regularly held domestic exhibitions provided occasions for the rapid and wide exchange of information about new inventions and technical know-how (Nakaoka 1999: 169ff.). In this case, exhibitions played a role in an instrumental dimension, rather than in an ideological or demonstrative dimension. Within the radical technological change in the realm of advanced technology, there was a relatively continuous and gradual transformation in the field of traditional technology.

Thus we find a material background for the rapid introduction of many types of Western technology in the late nineteenth century and a technological foundation for the enthusiastic response of Japanese people at that time. Without this, the ideology of modern civilization incorporated in various machines would have remained only an ideology. The ideology of technological determinism at the core of Ohkubo's proposal for the government to promote enterprise and industrialization would also have remained merely ideology. A number of historians of technology support this view of modernization in Japan.

Rosenberg indicates that the subcontract structure common in Japan contributed to the success of modern Japan. Traditional technology, constituting a lower level of this dual structure, played the role of a capital goods industry, making possible the interaction between producers (in this case, European advanced industry) and users (the Japanese industrial system). In addition, Rosenberg attributes many *un*successful technology transfer projects to the absence of such appropriate conditions (Rosenberg 1970: pp. 565, 570).

Jun Suzuki indicates the importance of gun smithery, which remained at a certain developmental level during the Edo period. Guns were introduced into Japan in the mid-sixteenth century, adopted very quickly (with innovations), and used widely for the next 100 years. After the establishment of the Tokugawa Shogunate, there were few occasions when guns were used in war, but guns have been produced on a limited scale ever since. According to Suzuki, traditional gunsmiths played an important role in the effort to modernize guns and cannons in the last days of the Tokugawa shogunate, and after the Meiji Restoration these gunsmiths moved into manufacturing industries just as the clock-artisan Tanaka did (Suzuki 1996: chap. 1).

Konosuke Odaka indicates the difficulty that Japan would have had in promoting an iron and machine industry if there had been no skilled mechanics at the beginning phase of industrialization. "If there had not been these artisans and their tradition, the process of iron manufacturing and machine making would have remained a 'black box' for Japanese people, which could not be understood for a longer time and the domestication of this process would have proceeded (even if it should succeed) much more slowly" (Odaka 1993: pp. 239–240).

Tessa Morris-Suzuki emphasizes more clearly the role of the traditional technology developed before the Meiji Restoration and describes the course of development of industry and technology with the concept of a social network:

The upheavals accompanying the transition from the Tokugawa political order to the centralized Meiji state resulted in reshaping of this network. The new system bore traces of its pre-Meiji heritage, but was at the same time distinctively different both in its structure and in its implicit objectives. In the first years of the Meiji era, the technological initiatives of local, grassroots groups were relatively far removed from the ambitious modernization schemes of the central state. While central government laid the foundations of a modern industrial infrastructure, with railways, telegraph and imported mining, factory, and military technologies, regional institutions encouraged incremental innovation and the incorporation of simple foreign techniques into existing production system. By the end of the century, however, center and periphery were beginning to be woven together into a multiple-layered hierarchy of connected institutions which proved an effective means of spreading technological information. (Morris-Suzuki 1994: pp. 103–104)

The characteristics of modern technology are sometimes considered to be universal and context independent, in contrast to traditional technology, which is considered to be embedded in a local cultural context. However, without an environment provided by traditional technologies, modern technologies cannot be transferred and introduced into other contexts. In this sense, we could say that it is the developmental processes, mediated translation, and transformation processes of *traditional* technology that make the *modernity* of technology possible. Without support from traditional technologies, the ideological character of modern technology could not be transformed into reality. Modernity without the help of tradition would remain only an ideology.

Conclusions

What can we conclude from these stories about the modernization process in Japan? One of the most conspicuous characteristics of this process in Japan is the dual structure of its sociotechnical network, with an advanced sector of modern technology and a parallel domestic sector of traditional technology. The advanced sector functions as if transferred technology guides and determines the direction of modernization. In reality, however, the advanced sector interacts with the domestic sector, where traditional technology plays a role of instrumental rationality, decreasing the gap between the two sectors sufficiently that advanced technology is adapted to local circumstances. Through this interaction, the scope of possibilities is restricted; the process is channeled in a certain direction; and rapid and continuous adaptation and development of technology becomes possible.

What made the modernization process in Japan possible were these seemingly contradictory yet inseparably connected technology sectors. If these factors had been too contradictory, there would have been no successful process, as was the case during an encounter between China and western civilization in the late nineteenth century. If they had not been contradictory enough, there would have been no radical transformation, as happened in the encounter between Japanese artisans and western clocks in the early seventeenth century. The encounter between the Chinese people and western clocks in the seventeenth and eighteenth century can be considered to belong to the latter kind of case because the two sectors remained indifferent and no contradiction developed between them. In this sense, the manner in which the creativity of technology is realized depends on each historical context; it is thoroughly contingent, and we cannot generalize the lessons of the Japanese experience. What we can say is that modernity does not exist in a universal sense, but in modernity there is always a dual structure of modern factors and traditional factors. In this sense there are always various modernities (in plural) together with various transformational processes of tradition.

On the other hand, we have developed a relatively general and formal structure of modern technology in which the capital goods industry plays a decisive role. In this structure, a capital goods industry is an environment in which the interaction between producer and user is constantly made possible and the "reverse determination" initiated by users can be realized. What about the role of artisans, which we have confirmed in the case of the early stage of the industrialization of Japan? Does the argument still hold concerning the later stage of industrialization? Even scholars who emphasize the role of skilled workers who made flexible industrialization possible, confirm that by the 1920s (by the 1960s in Japan) the dominance of mass production became irreversible and the role of artisans declined.

Surely artisanal skills have changed greatly and most traditional artisans disappeared by the middle of the twentieth century. Especially after computer-operated machine tools were introduced, many types of knowledge became obsolete and disappeared. However, we cannot neglect the fact that while old skills might disappear with the introduction of new machines, new skills become indispensable in adapting new machines to new circumstances. With advanced technologies, these translating and mediating roles are no longer filled by traditional artisans, but by engineers who have academic training. In spite of these circumstances, the knowledge necessary for accomplishing the work cannot be reduced to codified scientific knowledge, but still requires skill and intuition gained through experience, just as the knowledge of traditional artisans did (Ferguson 1992: chap. 2). Only through the application of this kind of knowledge by skilled people is flexible and rapid mediating work possible.

We find a similar structure in many twentieth-century technologies as well. Paul Rosen, for example, finds a flexible feature of technology in the development of mountain bikes in the United States and England since the 1970s (Rosen 1993). In contrast to the design of the standard bicycle, which has been mostly stable for the past hundred years, the design of mountain bikes has changed constantly since their invention.

Stabilization in mountain bikes has occurred, at a certain level. The features that distinguish mountain bikes from road bikes [. . .] continue to hold true. However, closer investigation of the technological details shows constant shifting in the design of frames and components, which means that since their inception, mountain bikes have been moving further and further away from being a stable artefact. They are in a constant and irresolvable state of interpretive flexibility (Rosen 1993: p. 505).

Rosen calls this type of industry post-Fordist and labels mountain bikes as "a technological artefact of postmodern society" (Rosen 1993: p. 494). This kind of continual innovation of mountain bikes has become possible because the base of production has been transferred and almost all components are produced in Taiwan. Taiwanese companies have the capacity to fulfill continually changing requirements from trading companies in England and the United States. These trading companies only assemble the imported components. In this sense, Taiwanese companies play the role of a capital goods industry.

In the history of many technologies there is a reciprocal interaction between producer and user on the one hand, and the capital goods technology that supports such an interaction on the other hand. Through the processes supported by this institutional structure, new values and new problems are constantly created. This type of creation is held in high esteem in "our" modern society, while it was not in premodern traditional society. If we can think in this way, and include the concept of "postmodern" in the concept of "modern" in the broad sense of the word, we can arrive at a distinction between "traditional" and "modern" in the realm of technology.

This distinction, then, lies in the way in which creativity is realized differently in modern and traditional technologies. The creative process can be found in any course of technological development since the beginning of the history of human technology. What is distinctive in the modern age is that this process is not a random phenomenon, but is institutionalized in a sociotechnical network that has a particular dynamic in which technologies are continually transformed. Since the latter half of the nineteenth century the international connections between different counties and different cultures have strengthened, and the global character of the world has begun to become conspicuous. While capital goods industries support this global tendency by accelerating the interactions between producers and users in various fields, they are also supported and oriented by this tendency (Feenberg 2000b). Different and heterogeneous parts of the sociotechnical network of the modern world are not indifferent to each other and are always involved in a contradictory, interactive process that occurs between them. In this way, the interaction between producers and users does not remain stable, but is always part of a transformational activity where, in the words of Nishida, "reverse determination" leads to conspicuously "creative" results.

III

Changing Modernist Regimes

9

The Contested Rise of a Modernist Technology Politics

Johan Schot

This chapter explores the idea that as part of a modernization process that gained speed in the nineteenth and twentieth century in the western world, a typical modernist practice of technology politics emerged.[1] The concepts of modernization and modernity need to be handled with care, of course, since their use may easily lead to an identification with modernizers, actors who have invented and used these labels to advance their cause. In addition, using these concepts for analysis might lead to finalism, as if past developments have led right up to the present. When these two pitfalls are avoided, the concepts of modernization and modernity are useful categories to discuss various structural changes in western societies since the eighteenth century. The concept of modernization refers to a new mode of social organization, a new social order, and a discontinuity in history (Wehler 1975; Giddens 1990). It is best understood as a process associated with a specific time period (eighteenth century to the twentieth century) and geographical location (the western world). The concept of modernity furthermore refers to a specific mode of thinking in which technology is identified as the main way of advancing the modernization process. Technology has been far more central to the making of modernity than is usually recognized (Brey, chapter 2, this volume; Hård and Jamison 1998; Latour 1993).

The modernist politics that slowly emerged consists of separating the promotion of technology from the regulation of technology. In this practice, technology development is perceived as a neutral, value-free process that needs to be protected and nurtured (because it creates progress, material wealth, health, etc.). Special "free places," often called laboratories, are created where engineers, inventors, and other technology developers can focus on solving technical problems. If these problems are solved, technologies begin their journey to the "real world." Fitting technologies into a market is the business of entrepreneurs (innovators).

Sometimes, as the modernist politics recognizes, these technologies will have undesirable impacts for society. To help societies deal with these impacts, government or other bodies put into place regulations to protect and if necessary compensate citizens. These undesirable impacts, in the modernist view, are unrelated to the choice of a technology. The modernist view does not recognize an important feature of technical change, the co-production of technology and its effects. The social effects of any technology depend crucially on the way impacts are actively sought or avoided by the actors involved in its development. In the modernist view, impacts are perceived as acceptance problems. Hence, technology promoters devote substantial resources to persuading the public to adopt a "better understanding" of the issues at stake. Technology promoters do test their innovations and if necessary modify them to fit with the regulatory system and the worldviews of the public. However, the modernist style of regulation does not require technology developers to consider impacts and "impact" constituencies systematically, let alone at an early stage, while technologies are undergoing development and taking on their durable forms. The emergence of this modernist technology politics went hand in hand with the development of a dichotomized discourse on technology. Reinforcing the modernist practice of promoting and regulating new technologies was the emergence of two dominant perspectives: an *instrumental* one in which technology is a neutral means toward an end, to be defined outside the technical area and, by contrast, a *strong critique* asking for (regulatory) limitations on technical action.[2]

This essay first explores the rise of this modernist technology politics, spotlighting key turning points from the early nineteenth to the midtwentieth centuries, and then suggests ways to go beyond such a dichotomous politics. My ultimate aim is to identify ways to open up space for the actual shaping of technology and for discourses on how to manage technology in society. In my discussion of the rise of the modernist technology politics, I particularly focus on episodes of resistance to technology. There are both substantial and methodological reasons for doing so. The emergence of the modernist regime of technology management was highly contested and it is important to make this contested process visible, particularly because the notion of modernization can easily lead the author (and reader) to the pitfall of finalism and the writing of whiggish history. Resistance is also interesting for methodological reasons because various kinds of positions can be more easily found in source material.

This essay is an attempt to construct a plausible account of a modernist regime of technology management. It is a broad-ranging and interpretative attempt to bring together diverse material to form a meaningful and coherent story. It can also be read as an attempt to bring together my background in social history, sociology of technology, and policy studies, together with my practical experience in several technology-policy networks.[3] It draws on systematic reflections resulting from circulating in various networks and disciplines. The argument is, therefore, speculative, but a starting point for further research and discussion on the relation between modernity and technology.

Politics and Innovation in Early Modern Europe

In the early modern period, a distinct technological domain did not exist. Technological development was embedded in religious, economic, and social practices, and it was assessed against social norms. The assessment processes, which were often informal, took place in guilds, for example. While guilds often slowed down specific innovations, they were not against all forms of technological development; they hindered only those technologies that were contrary to their ideas about the "good society," for example, machines that would threaten skill or employment. Technological development was heavily influenced by the regulatory (and evaluative) practices of guilds (Mokyr 1990: pp. 258–289). It was also shaped by a variety of protests, such as organized demonstrations, petitions, threats to inventors and entrepreneurs, and breaking machines (Rule 1986).

The destruction of machines is associated with the acts of the Luddites, the English workers who destroyed textile machines in the early nineteenth century.[4] For decades, the Luddites were held up as irresponsible if unwitting technophobes. Historians once viewed them as the victims of progress, who saw no other recourse than taking out their aggression on the machine. Often, it was added, every new technology is resisted because of vested interests, but that resistance eventually subsides. Hobsbawm (1952), Thompson (1963), Rule (1986), and Randall (1991) have corrected this mistaken image of the Luddites. According to their research, organized machine breaking had been a rather popular and successful form of protest since the seventeenth century. It was more effective than striking because employers could not employ scabs to keep the machines in operation. Hobsbawm called it "collective bargaining by riot." In saving the Luddites from modernistic criticisms, these revisionist historians have sometimes argued that the Luddites' protests were not directed against technical change or machines. I would like to argue, however, that their protest did entail a strong criticism of technology. Their critical stance was not based, however, on disdain for technology in general. On the contrary, it was directed at particular machines. The only machines the Luddites destroyed were the ones against which the workers had particular grievances. Other machines, even in the same factory, were left unscathed. A crucial point that is often lost in the popular image of the Luddites as an uninformed antimachine mob is that most Luddites were skilled machine operators in their own shops.

Moreover, I would like to emphasize that the Luddites' resistance ran much deeper than the rejection of particular machines. It concerned the rise of a new kind of society, embodied in a new set of specific machines, in which employers had the right to introduce machines that made workers redundant, produced unemployment, and lowered the quality of the products and the quality of society. Randall, who carefully analyzed the discourses used by various workers, argues, rightfully, that the workers were not just trying to restore an old situation (Randall 1991). Rather, they acted proactively to develop their own view of the future, a future that in their time was a genuine and feasible alternative. It was a struggle between rival models of how to organize society. The Luddites demanded that those who introduced new machines should anticipate their social effects. One of the Luddites' proposals was a machine tax intended to create fair competition between the power loom and the hand loom (see Berg 1980). In other cases, workers asked for a negotiated introduction of machinery. They proposed an experimental period to assess social costs and social benefits (Randall 1991: pp. 72–74). Some evidence exists that attempts were made to construct "intermediate" machines, which would need *more* hands and skills; in addition, certain machines were available for small-scale domestic manufacture. Two such cases from the cotton textile sector, which would benefit from economies of scale, are James Hargreaves's "jenny" and Richard Arkwright's water frame, which was deployed on a large factory scale because of patent-law considerations even though it had been developed initially for small-scale domestic use.[5]

To the employers and entrepreneurs, as well as the politically dominant classes in Britain, the Luddites were criminals. Labeling machine breaking a criminal act was, however, part of the struggle of developing a specific kind of industrial society. Initially the Luddites had English law on their side, for machine breaking as a form of protest was legitimated by the common law. Only in 1769 did the Parliament pass a new law against machine breaking. Luddites were not alone in their dissent. They were supported by craftsmen, small-time entrepreneurs, and conservative politicians (Randall 1991), the last of whom were strongly influenced by early Romantic authors such as Carlyle and Southey (Berg 1980: chap. 11). Finally, Luddite resistance must be seen against the background of the national debate on the "machinery question." This debate centered on the sources of technical progress and the impact of new technologies on the economy and society. It spurred the development of a new discipline, political economy (Berg 1980).

The Luddites lost their battle in the end, partly as a result of strong state intervention. During a wave of protests in 1811–13, some 12,000 soldiers—a force much larger than Wellington's army then fighting Napoleon at Waterloo—were sent against the workers to "restore order" in the textile regions of England. While the Luddite movement was destroyed, it can be argued that it slowed the introduction of a number of machines, particularly in the woolen industry, and the threshing machine in agriculture (through the so-called Swing riots). In this way the workers bought time to adjust to the changes.[6] However, the main outcome was the emergence of a new ideology and practice

that granted inventors and entrepreneurs near-complete freedom to introduce new machines into society without having to think about their
effects.

The replacement of the early-modern order by a new industrial order,
including a new relationship between politics and technology, was an integral part of industrialization in many western European countries.
Ken Alder has argued that in France during the French Revolution engineers pioneered and founded new institutional structures to control and
discipline the productive order (Alder 1997, see especially the introduction and chap. 8; for the French case see also Rosenband 2000[7]). The
French Revolution was not initiated in the name of the factory, but it
was supported by engineers seeking to create institutional forms to regulate production, especially to enforce forms of industrial and factory
production. As in the case of England, these attempts met fierce resistance from labor and petty commodity producers. For example, in
Saint-Etienne in 1789 a crowd of armorers, with the municipality's con-
sent, destroyed a factory that aimed at mechanized barrel forging with
trip-hammers (Alder 1997: pp. 214–215). When the Revolution turned
violent, engineers, to keep their heads attached to their bodies, learned
to position themselves as neutral, not involved in politics.[8] (Historians
largely accepted this view in subsequent decades, obscuring the relationship between the industrial and political revolutions in France.) This
neutral position led to the development of a new strategy, one in which
engineers became licensed experts of the state responsible for controlling
the productive order.

In many European countries persistent resistance to new technologies
became obsolete, condemned, and perceived as reactionary. Romantic
thinkers, who had struggled to construct a political vision that allowed
innovation while protecting society against some of the impacts of new
machinery, made a utopian turn after the French Revolution and the
dreadful experience of the English industrialization (Sieferle 1984). The
machinery question was "solved" through the gradual acceptance of the
instrumental vision by all parties during the course of the nineteenth
century. Leading spokesmen of all major political parties and most interest groups agreed on a consensus in which technological innovation
was acclaimed as a progressive force. Even radical reformers (such as

Owen), and later Marxists and socialists, came to share this instrumental vision. These radicals argued that social problems must not be associated with the machine itself, but with the machine's use in a capitalist context. Owenites argued, for example, that machinery used in a cooperative social context would benefit labor since the productivity gains following mechanization could be redistributed to the working class (Berg 1980: p. 270).[9] By the end of the nineteenth century, modernist technology politics was firmly in place. The elite (the right wing as well as the left wing, employers as well as unions and intellectuals) almost automatically condemned resistance to new technology as reactionary.

Testing and Celebrating Modernization

One of the few violent outbursts of resistance to the machine in the early twentieth century took place in the Netherlands. In 1905 grain elevators (unloaders) were introduced at Rotterdam harbor. These elevators were large suction devices that conveyed grain from one ship to another almost without human intervention. Thousands of dockworkers, who had worked carrying sacks of grain, were to lose their jobs. When the first grain ships were unloaded, the automatic weighing did not work; its indications were too high. In the 6 weeks it took to repair this, the dockworkers organized themselves. When the unloaders were ready to start working again, they called a strike. This strike was a great success, blocking almost the entire grain transshipment. When the German grain importers heard about the strike, they negotiated a contract with the labor leaders and the factors, the importers' local representatives in the harbor, to accept only grain that had been weighed by hand. As a result, the unloader company could not find enough work for their two unloaders. For two years the elevators remained dormant.

In 1907 the unloader company began once again unloading grain ships with the elevators, provoking another strike, but this time the workers, with their strike funds depleted, could not win. The employers, including the factors, had united and had recruited strikebreaking scabs from all over the Netherlands and from Germany. Rotterdam's mayor proclaimed a state of siege; warships appeared in the harbor; and military troops were called out to preserve order. More elevators were introduced and many jobs were lost. By 1912 sixteen unloaders were on duty, handling 90 percent of the harbor's flow of grain.[10]

In this pitched conflict, conducted not only by striking workers but also in a wide public debate, it is curious that the obvious technology choices involved in elevator design did not come under discussion.[11] Union and socialist leaders embraced the new technology and argued it would bring progress to the harbor. Machine breaking, condemned as Luddism, was not on the agenda. The union and socialist leaders endlessly repeated the message, familiar from Marx and Owen, that any problems were not due to technology but to its uses under capitalism. In the new socialist society, the tremendous productive forces built up under capitalism would be employed for the benefit of all: "our watchword should not be 'away with machinery' but 'away with the capitalists and capital to the workers' " (see van Lente 1998a: pp. 93–94). One prominent socialist leader even argued that losing strikes against new machinery was in the best interest of the working classes.

Representatives of the broad-based anarchist movement, probably representing a larger part of the laborers, however, denounced the technological determinism implied in the views of the union and socialist leaders. Much like the Luddites a century before in England, the anarchist movement viewed the harbor as a community in which the employers had no right to impose, without negotiation, a machine that would deprive hundreds of workers of their daily bread. They also denied the economic necessity of the unloaders, without rejecting labor-saving machinery in general. Research in the minutes of the meetings of grain traders has proven that this view was, remarkably enough, shared initially by a number of grain traders and employers (see van Driel and Schot 2001). However, when the conflict hardened, the grain traders redefined the conflict into one about who controls the harbor and the introduction of new machinery, and closed ranks with those arguing for the economic necessity of elevators.

The consensus among the Dutch elite and part of the labor force on the instrumental role of technology in society was certainly challenged by labor in the elevator conflict during 1905–7. Yet the instrumental view emerged stronger than ever. In this period, the instrumental vision was also challenged in different ways in a number of European countries and the United States (Hård and Jamison 1998). After 1870 in a number of European countries it was impossible to ignore some of the problems associated with the introduction of new machines, such as bad hygiene in cities, child labor, and accidents involving machinery. "New liberals" started to write extensively about the social consequences of industrialization. They argued that these problems should not be attributed to industrialization itself, but to human ignorance and immorality, obsolete institutions, and outmoded laws. Social legislation could solve these problems. These issues were part of the "social question," which would dominate discussions.

In Germany the social question took the form of a machinery question, partly as a result of the dreadful experience of World War I. In Germany, one had to come to grips with wartime chaos and postwar depression. Technology became a much-debated issue (see Hård 1998; Dierkes et al. 1990; Herf 1984). To summarize, technology was seen as important for creating order and control, but only in a modified form. Technical change needed organization and control and regulation by the state, and the creation of domestic monopolies to guide its implementation. For example, Sombart argued that the government must appoint a body to decide what new inventions should be developed. He also argued that the police must prohibit the use of technologies with negative consequences for citizens and workers. He approved the decision of the Swiss canton of Graubünden to ban the use of automobiles and motorcycles (Hård 1998: p. 62). These modifications would make technology part of the German *Kultur*. Whereas U.S. technology was part of corrupt western *Zivilisation*, German appropriation would transform technology into an order-bringing and *Kultur*-enhancing mechanism. A number of influential authors (Schweitzer, Sombart, Rathenau, and Spengler) argued, in various ways, for a German *Sonderweg* (loosely, "alternative path") in technical change.

Generally, participants in the German debates considered modernization to be desirable, but thought that its consequences should be controlled and regulated, either by engineers or sociologists. Modernization could thus become controlled modernization. In the debates, it is clear that for Sombart and others, technology was not an autonomous realm of society; it could be shaped and fitted into the German context.

Sombart used the notion of "cultural carpet" to analyze the relationship between technology and other spheres of society, suggesting that Germany could develop its own style and combination of technology and society (Hård 1998: p. 58).

Even though some figures proposed a new kind of technology politics (Dessauer 1958) that would exert more control over technology, the outcome of the German debates reconfirmed the modernistic technology politics: state intervention might accommodate the embedding of technologies in society (for example, with safety regulations), but there could be no direct intervention in the innovation process itself. Even Sombart eventually accepted the instrumental view. He argued in 1934 that "technology is always culturally neutral and morally indifferent; it may serve either the good or the bad," a definition that, rather jarringly, does not fit his earlier use of the notion of cultural carpet (Hård 1998: p. 63).[12]

A wave of technological enthusiasm in the early twentieth century stiffened the modernistic consensus about the *a*political role of technology in society. People started to refer to "technology" in the singular—an independent and abstract phenomenon that transcended its many individual fields of application (Marx 1994; Oldenziel 1999). Technology became the very symbol of modern society. The belief in the technical fix, in shaping a new society by means of modern technology, assumed unprecedented proportions. Social and cultural advances through technology appeared limitless. This belief became visible in several technocratic movements in many western European countries and the United States. Their objective was to promote the prosperity of the people, through the use and implementation of technology. In art and architecture, the new belief in technology led to the emergence of new movements, such as Futurism and *De Stijl*. These movements celebrated the coming of the machine as a new joyful age. Theo van Doesburg, one of the leading figures in *De Stijl*, heralded the new age as follows:

You long for wildernesses and fairy tales? I will show you the order of engine rooms and the fairy tale of modern production methods. Each product is a real miracle. You long for heaven? I will show you the ascension of the aeroplane with its quiet pilot. You long for nature? Her dead body is at your feet. You have beaten her yourself. Your high mountains have changed into skyscrapers. Your windmill is no longer turning—a chimney has taken its place. Across the place once occupied by your stage-coach, now an automobile is zooming along. (Quoted in Anbeek 1994: p. 123)

Technological enthusiasm was pervasive in this period (Hughes 1989: chap. 7). This enthusiasm was not only widespread among the elite of engineers, scientists, architects, and artists, but also in the world of business, social organizations, and among citizens. The enthusiasm was embodied most clearly at the New York World's Fair of 1939, which presented "The World of Tomorrow" (see Nye 1990: pp. 368–379). In this world technology was presented as the key instrument of a better society. The fair was explicitly and consciously concerned with selling the vision of a technology-driven and technology-based future. Technology would fix many of the world's problems, including hunger, disease, scarcity, and war. That 45 million people attended this fair indicated how much the instrumental (and enthusiastic) view of technology had captured the feelings of a larger part of the American people.

The Coming of Reflexive Modernization

Although World War II showed again that death could be efficiently mass produced by technology, technological enthusiasm prevailed for at least two decades after 1945, in Europe as well as in the United States. These were the decades of Big Science, and after two decades of hardship during depression and war, consumer society finally became a reality for all, including labor and Europe. Science and technology were seen as the key to American prosperity, the rebuilding of Europe, and the future of the world. In the 1960s, however, people began to find, somewhat to their surprise, that new products can have serious problems, so-called unintended consequences. Various citizens' groups, nongovernment organizations (NGOs), and intellectuals, such as Commoner, Ellul, Mumford, Nader, Marcuse, and Roszak, started to challenge the promise that science and technology could solve any problem (see, for example, Nelkin 1979; Hughes 1989: chap. 9; Eyerman and Jamison 1991; Bauer 1995).

Overt resistance against new technologies, especially nuclear energy, flourished, effectively frustrating its further development in the 1970s. New social movements reversed modernism's trust in technology by issuing critical calls for values such as quality of life, wholeness, small-ness, care for nature, and concern for future generations. The attempts by governments and companies to improve "public understanding" were seen as defensive and self-serving. This distrust was not only fueled by accidents and other impacts of technology that became visible, but also by a critique of centralized large-scale technologies. The various controversies and disputes were not merely about the impacts them-selves, but also were about wider social and moral preferences and val-ues (see Irwin 1995; Irwin and Wynne 1996).

Not only was resistance against new technologies reinvented, the idea of developing alternative systems and technologies consonant with the new value system became popular. In 1973 E. F. Schumacher published *Small Is Beautiful: Economics As if People Matter*, in which he advo-cated a latter-day "intermediate technology." In 1977 appropriate tech-nology in the United States received official sanction in a new National Center for Appropriate Technology. The Army Corps of Engineers was ordered to identify dams that might be retrofitted to low-head hydroelectric production. Many programs for research and development on renewable energy were set up (Pursell 1995: chap. 13, 1993). Particularly in Denmark, small-scale wind energy was developed and used successfully (Jørgensen and Karnøe 1995). Many examples of so-called clean technologies emerged during these years (Green and Irwin 1996).

In addition to the development of more appropriate and cleaner tech-nologies, western societies since the 1970s have witnessed an explosion of new governmental regulations as well as a huge increase in knowl-edge about environmental problems and solutions. The consequences of new technologies have been increasingly assessed, monitored, and regu-lated. Also, these consequences (dangers, risks, impacts) began to domi-nate public and political debates. For this reason Ulrich Beck has argued that we have entered a new phase in the modernization process, a phase of reflexive modernization in which industrial society confronts its own problems (Beck 1992, 1994). Thus "reflexive" does not refer merely to reflection, but foremost to self-confrontation.

Still, although western societies seem to recognize their problems, a solution to them is not at hand. Alternative technologies, such as wind energy, organic farming, and electric vehicles occupy only small market niches, while many "clean" technologies, from the chimney filter to the catalytic converter, do not solve the problem, but only displace it and create new problems elsewhere. Regulation is often not very effective, while the promotion of new, risky technologies such as genetic engineering continues, and many problematic old technologies such as gasoline-fueled automobiles flourish. No clear picture has emerged on how to effectively handle even widely recognized problems. This leads for some to an uneasiness; for example, people still drive automobiles but feel a bit guilty about it. For others, the intractability of these problems leads to apathy and indifference.

The case of the expansion of the Amsterdam Schiphol Airport serves here to illustrate the strains of reflexive modernization and the persistence and limitations of modernist technology politics. In this case the dual-track approach of separating promotion and regulation was clearly articulated and codified in official policy, even as the defects of this policy became clear to many parties involved.

In 1969 the director of Schiphol Airport made a plea for a large expansion of the airport, particularly the construction of a fifth runway.[13] A long battle ensued between the national government, the provincial government, and various local municipalities, against a background of organized resistance by a variety of local communities and environmental groups. During this battle, the number of flights at Schiphol increased dramatically and the airport itself was expanded, but permission to build a new runaway was repeatedly delayed until February 1995. In these years, many studies—including a so-called integral environmental impact statement—were done to explore, determine, and calculate all the impacts. The national government's decision to allow the construction of a fifth runway was part of a broader policy for the airport. According to this policy, Schiphol would be allowed to grow, albeit within certain limits set by noise standards. The number of residences to be affected by serious noise pollution was set at a maximum of ten thousand. This policy was labeled a "dual decision," and defended as a policy that would achieve competing economic and environmental goals.

This "dual decision" was developed by a project group that included the airport managers, municipal administrators, and various national ministries, who arrived at that consensus before commencement of the formal decision-making procedures, including a public inquiry. Consequently, citizens' groups and various NGOs distrusted the ensuing process of public participation from the start. The "dual decision" has dominated the political debate since 1995. Discussions range from such issues as how to measure noise effects to which types of runway configuration would allow steady growth with the least noise. Resistance also continued, as citizens and NGOs tried to slow down the process of building a fifth runway. These efforts have met with some successes; namely, court appeals and other actions such as the refusal to sell land needed for the expansion of the airport (which was preemptively bought up by activists before the airport started to buy the needed land).

The drive to expand the airport cannot be understood merely in terms of a growing need for air travel. The expansion of Schiphol is a part of the story of the Netherlands as "the Gateway to Europe," distributing goods and people. This story is particularly forceful in the Dutch context because it reconnects the present to the Golden Age of the seventeenth century, when Holland and especially Amsterdam was the hub of international trade. In this storyline, resistance to growth and a growing transport sector is viewed as resistance to progress, a sound economy, and to a core cause of Dutch prosperity.

In the debates since the end of the 1960s, environmental groups and local communities have hammered home the adverse environmental effects of expansion and trivialized the appeal to national economic interest. These critics pointed at airplanes contributing to the greenhouse effect, overuse of space, noise production, congestion of automobiles around the airport, and safety problems. (In 1992 an airplane crashed into an Amsterdam neighborhood, killing 47 people.) They called for stricter norms and limits to growth. At the same time, they attempted to develop alternatives. For example, they proposed a much smaller airport that would not accommodate so many transit passengers flying to the rest of Europe (such transit passengers, it was argued, contribute little added value to the Netherlands). A fierce debate among economists has persisted over the calculations of the added value of the airport expansion. NGOs developed the idea of a "railport," whereby passengers bound for Frankfurt, Paris, and other European cities would be forced to continue travel by rail. They also hinted at options for integrating air transport into a broader, multimodal transport policy.

Other critics argued that the only way out is *not* to start with technology. Real solutions will only come from social and cultural change, to be enforced through regulation. The way forward, in this view, is through restricting mobility (through price mechanisms that make jet fuel much more expensive, creating a new tax on flying, or enforcing mobility quotas). The management of Schiphol Airport hardly responded to these ideas, other than by pointing at the growth in the number of flights and the competition among European airports, forcing Schiphol to grow as fast as possible. At the same time, airport planners did incorporate a train station in the construction plans for an expanded airport. Also, a number of successful measures were taken to reduce the airport's energy use.

In the prolonged Schiphol controversy, economic growth was discussed simultaneously with risk production and risk distribution. Risks were made visible, and attempts were made to measure and predict them; this is a key element of reflexive modernization. Also, the two tracks of promotion and regulation—identified in this essay as the modernist way of handling technology in society—were explicitly labeled in the "dual decision" governmental policy. However, the attempts to integrate the risks into a policy did not lead to a viable solution that was acceptable to the range of actors. This suggests limits to the modernist technology politics. How can we explain this lack of room for negotiating a solution?

The Schiphol case is an exemplar for many other "risk issues" (BSE [so-called mad cow disease], food toxins, nuclear threats, global warming). The failure to resolve these issues, it seems, deepens the distrust and alienation experienced by many citizens. Following my analysis of the rise of a modernistic technology politics, we can see two phenomena at play. First, no space or arena for collaboration, discussion, and mediation on how to deal with the impacts of technology was available.[14] Second, no discourse was readily available to the participants so they could understand the relationships between technology choice and technology impact. A feeling of shared responsibility between producers of new systems and those who use or are affected by them cannot emerge in such a situation. Typically, only those acting to promote it have any access to decision making about a future technology, system development, and the attendant impacts, leaving ample room for viewing promoters as the "bad guys" seeking only profit.

In the Schiphol case, public participation, which is often held out as a robust solution to such conflicts, did not result in any substantial access or choice of technology. The national government tried several times to create a "roundtable" to discuss the future of the airport with varied actors, but these attempts failed because of the airport's low institutional credibility and the lack of common ground for discussion. The airport management continued to perceive a binary choice—Schiphol could remain a regional airport or it could become a huge international one, which would require a fifth runway. Opponents of the airport expansion viewed airport growth and the construction of a fifth runway as the problem. For them, system growth needed to be curtailed through strict regulation that might change the travel patterns of passengers.

Solutions to the Schiphol impasse were thus sought in either a "technology" fix or a "regulation" fix. As I have argued, both approaches are deeply embedded in our culture and dominate the debate about many technological systems. The key issue my analysis raises is whether modern societies are indeed trapped within these two conflicting positions. Would it be possible to conceive of a modern culture able to discuss contending social and cultural issues in relation to technology? This leads to the related question of what conditions would encourage and allow actors to work on both the technical *and* the social simultaneously, in a related way.

Contours of a Constructive Technology Politics

The core of modernist technology politics, as I have argued, lies in the separation of technology from its social effects. The separation emerged in the early modern period and was a defining characteristic of modernity. I have interpreted resistance by the Luddites in the eighteenth and nineteenth centuries as resistance to that separation. They demanded that those who introduced new technology anticipate its social effects. To the Luddites and their sympathizers, technology did not inhabit a realm separate from its social, cultural, and political effects. This was also the case for a larger part of the Rotterdam dockworkers at the turn of the twentieth century. The socialist leaders and other members of the elite, however, viewed technological developments as unavoidable and could not perceive viable alternatives. Environmental groups and other protesters against the prospective expansion of Schiphol were more ambivalent. By attempting to formulate alternatives, they did not define the contemporary plans for Schiphol's expansion as unavoidable. But their efforts were hampered by the absence of a language and space to create alternative designs for Schiphol.

These social, cultural, and institutional liabilities make it clear why, under the modernist regime, the technical is kept separate from the political. No wonder that it is so difficult to develop a new relationship between technology and the political realm. In this last section of my chapter, I develop some ideas about how to overcome the bias of modernist technology politics that separates the technical and the social. In doing so, my tone will become less descriptive (aiming at diagnosis) and more prescriptive. Indeed, I aim to prepare intellectual ground for a new kind of modernist technology politics, one that could be called "constructive technology politics."[15]

To achieve such a constructive technology politics, it will be necessary to nurture a new set of institutions and discourses that aim at broadening the design of new technologies to include societal actors and factors. When such institutions proliferate, design processes will happen in new networks and circumstances. Ultimately such a development would allow for the constructive experimentation of technology and society.[16] It is not constructive in the sense of avoiding conflict. Power games will still be played; however, these will be partly displaced to other arenas, and here affected persons and institutions will be in a position to take responsibility for the construction of technology and its effects. By institutionalizing negotiation spaces (or nexus), both proponents and opponents will become responsible for giving meaning to technology and its effects.

The view that design processes must be broadened is not based on any presumption that social effects play no role in present design processes. On the contrary, they are present in the form of (sometimes implicit) assumptions about the world in which the product will function (Akrich 1992, 1995). The effect of broadening is that the designers' assumptions or "scripts" concerning their technologies[17] are articulated as early as possible to the users, governments, and other parties who will feel the effects of the technology, and have their own scripts. At present, there is no space for such an early exchange of contending scripts.

If the design process is broadened, it could acquire three beneficial features: anticipation, reflexivity, and symmetrical social learning.[18] In the first feature, actors would organize the anticipated impacts on a continuous basis. Through reflexivity[19] they would have the ability to consider technology design and social design as an integrated process and to act upon that premise. Finally, through symmetrical social learning, the actors would learn about all aspects of a new technology simultaneously. The vision is of new technologies evolving through a mutual learning process: technological options, user preferences, and necessary institutional changes are not given ex ante, but are created and modified along the way. Many historical and sociological studies have shown how user demands and regulatory requirements are articulated and expressed during the development process itself, in interaction with the technological options (Clark 1985; Green 1992). Producers gain new perspectives on their technologies from their customers and in response modify their designs.

In current design processes, mutual learning rarely takes place because of a prevailing tendency to optimize technology first, then check for user acceptance, and finally examine regulatory fit. Of course, no design process is strictly linear, and most design schemes include planned feedback. Feedback also arrives unexpectedly as problems discovered during application force redesign. However, such adjustments rarely change the pervasive assumption that design and development have to focus first on optimizing a technology before specifying markets and detailing social effects.

Incorporating reflexivity, symmetrical learning, and anticipation in design is not directed at substantive goals such as the reduction of environmental pollution or the creation of more privacy. It does not even lead to an argument about the desirability of such goals. The purpose of incorporating these features in design processes should be to shape technological development *processes* in such a way that social and technical aspects are symmetrically considered. However, it can be argued that when design processes assume these features, fewer undesired (and more desired) effects will result. By incorporating anticipation, reflexivity, and social learning, technology development becomes more transparent and more responsive to the wishes of various social actors. They will address the social effects that are relevant to them. Furthermore, in a society where these new development processes have become the norm, technology developers and those likely to be affected by the technology will be in a position to negotiate about the technology. An ability to formulate sociotechnical critique and contribute to design will become widespread. Resistance to specific social aspects will not be viewed as "technophobia," but as an opportunity to optimize the design to achieve a better fit in society.

The effect of breaking away from modernist management patterns will not be to bring technology "under control" so that it plays a less dominant role in society. Technology is not out of control. What will change is the form of control and how technology development is played out. The goal is to anticipate effects earlier and more frequently, to set up design processes to stimulate reflexivity and learning, and thus to create greater scope for experimentation. Winner (1977) asked for more space for experimentation. His concrete proposals are a bit disappointing, however; they consist of negative experimentation, that is, *not* using a number of technologies. My proposal is for constructive experimentation. Technologies need to be nurtured, but in a design process that allows various actors to become engaged (see Smits 1997).[20]

Eventually this change will make technologies more open and more flexible so users can easily control them. Technological development will also become more complex. The variety of technological designs probably will increase, for more groups will be addressed in their capacity as knowledge producer and technology developer (Verheul and Vergragt 1995). More coordination and new competencies will be required. In some cases technical change processes will slow down. New institutions will emerge to encourage negotiation among developers, users, and third parties. Should design processes acquire the features of learning, anticipation, and reflexivity, technologists will not suddenly see their work disappear or have it constantly evaluated by all sorts of commissions. Most of the incremental design changes will not require special negotiation at all, since social aspects will be included on a routine basis. They will be part of the technological "regimes" that orient design and use (Rip 1995; Rip and Kemp 1998).

My call for new design practices extends beyond changing and/or improving the design processes surrounding individual technologies. The point is ultimately to change the way design is done in our modern society. This change does not imply that the design activity itself needs to be put up for discussion. Modern society—a society where there is room to innovate and to create stable artifacts and networks—is accepted. Only the design process is the object of change. To make that change, inspiration can be sought in early Romantic thought, in which technology and society are not pulled apart and in which individual autonomy and the relevance of different rationalities held by different groups is accepted and used as a resource (Blechmann 1999; Schwarz and Thompson 1990). The design process must make way for confrontation, power struggles, ideological criticism, and the exchange of various rationalities. Only then will anticipation, reflexivity, and social learning be well served.

Notes

1. I would like to thank Mikael Hård for his comments on an earlier version of this paper and Tom Misa for his suggestions on revisions.

2. In recent philosophy of technology, it is common to make a distinction between instrumental and substantive positions; both result, however, in an analysis that emphasizes that technical change is ruled by itself, that is, by norms of efficiency and gradual and linear improvements to better systems (see Feenberg 1999a and Achterhuis et al. 1997.)

3. My technology-policy activities include working as an analyst for the Netherlands Organisation for Applied Scientific Research – Centre for Technology and Policy Studies (TNO-STB), many consulting jobs for government agencies and NGOs, as well as founding (along with Kurt Fischer in 1991) and participating in the Greening of Industry network <www.greeningofindustry.org>.

4. The workers later were given the name "Luddites" after their legendary leader Ned Ludd, who signed messages in the name of the workers. My interpretation of the Luddites was first formulated in my thesis (Schot 1991). Recently, a similar argument has been put forward by Nuvolari (1997). His paper has helped me sharpen my arguments and views.

5. See Nuvolari (1999), who cites R. L. Hills, "Hargreaves, Arkwright and Crompton: Why Three Innovators?" *Textile History* 10 (1979) 114–126; compare Berg (1985: pp. 236–239, 243).

6. Nuvolari (1999: p. 7) cites Randall (1991: pp. 82–83) and Rule (1986: p. 365) for the woolen case and the seminal work of Hobsbawm and Rude (1969) on the Swing riot, *Captain Swing*.

7. See, for example, Rosenband (2000: p. 50): "Put another way, the state technicians' and the entrepreneurs' search for unfettered space, in which they could manipulate technique freely, placed them on a collision course with the workers' custom and the skills that undergirded it."

8. Alder (1997: p. 302) even calls the decision to become neutral "the *ur*-event in the relations between science and politics in the modern era. As many historians have noted, science as a profession and politics as a public activity both came of age in France at the end of the eighteenth century. Yet after a brief period of intense involvement, scientists (with very few exceptions) have generally shied away from formal party politics."

9. Notwithstanding the emerging consensus on an instrumental vision of technical change, various countries, regions, and industrial sectors did not follow an identical industrialization path; industrialization was a varied and complicated experience. In Britain and elsewhere craft production was highly innovative and contributed to economic growth (Berg 1985). France followed a specific route toward industrialization that was based more on skilled flexible small-scale industries producing a varied assortment of goods for large but constantly shifting markets. This other route was thus a result of technology choices (see Sabel and Zeitlin 1985; Mokyr 1990: pp. 113–148, 256–261). The same thing happened in the Netherlands, where in a number of industries small-scale solutions were preferred above mass-production technology. The Netherlands followed its own distinct path too, mixing craft and mass production in a Dutch blend (Schot 1995, 1998). For the United States, typically cited as the Mecca of mass production, Phil Scranton (1997) has emphasized the importance of smaller and medium-scale enterprises (over mass-production formats).

10. For sources on the elevator controversy, see van Lente (1998a,b) and van Driel and de Goey (2000, pp. 38–42).

11. Examples of such choices were elevators that would permit trade in sacks of grain instead of large "bulk" grain loads; smaller elevators that allowed slower discharge; elevators that allowed manual instead of automatic weighing. In addition, combinations of an elevator regime and a manual transshipment regime were conceivable.

12. The notion of the need to regulate technology, albeit in a different way, is also visible in the work of a number of philosophers who developed a substantivist critique of technology: Adorno, Horkheimer, and Habermas. They do not question the instrumental definition of technology, but only want to limit its application.

13. For a detailed history of the Schiphol expansion, see Schot (1995) and Bouwens and Dierikx (1996).

14. Such mediation forums are available for integrating constructed user needs into design (see Schot and de la Bruhèze, forthcoming.)

15. In other publications, this kind of politics of technology is labeled "constructive technology assessment" (Rip et al. 1995; Schot and Rip 1998).

16. See Weber et al. (1999) and Hoogma et al. (2002) for ideas on how to design constructive experiments.

17. I use a broad definition of designers, namely, technicians, managers, and workers who are directly involved in the design process. In this respect Staudenmaier (1989) has termed it a "design constituency."

18. Note that these three features can also be transformed into management criteria; that is, particular agencies could use them proactively in assessing and upgrading design processes. For a elaboration of these criteria, see Schot (2001).

19. My notion of reflexivity is very different from the one used by Beck.

20. Such constructive experimentation is a form of neo-Luddism, but the nature of this action is quite different from that implied in its use by a number of other authors. Kirkpatrick Sale (1995) invites individuals to scrutinize any new technology for possible harm and if needed, reject it. No proposals are made to change the nature of the design process itself. It is my conviction that this is precisely what the Luddites were after, while individual technologies per se were not their main concern.

10

Technology, Medicine, and Modernity: The Problem of Alternatives

David Hess

The problem of technology and modernity is both analytical and normative: as researchers we seek not only to understand dramatic social and ecological changes but also to channel the changes into more sustainable, just, and democratic pathways. The medical field provides a particularly important site for the problem because biomedical conflicts tend to magnify some of the issues of technology and modernity, and also because health policy occupies a central place in the political and normative discourse of late modernity. As biomedicine has become increasingly driven by science, technology, and industry on a global scale, it has also encountered growing countermovements that pose fundamental questions about technology, modernity, and the nature of health.

Three frameworks, drawn in part from anthropology, for analyzing the problem of technology and modernity are utilized here: cultural ecology, cultural values, and political economy. The three frameworks stand somewhat outside the body of technology studies literature that focuses on networks, technological systems, and micro-meso levels of analysis. As in many of the essays in this volume, the frameworks here bring into play macrostructural categories of analysis, in this case the environment, cultural values, and global capital. Furthermore, as discussed in other essays in this volume (e.g., Brey, Slater), the macrostructural categories are assumed, *pace* technology studies, to be mutually shaped by technology.

In each of the three framings of the problem, the analysis assumes a broad definition of technology as material objects that are used to modify the social and/or material world.[1] Specifically, in the medical field, technology will be understood as more than diagnostic and treatment equipment; medical technologies are assumed here to include therapeutic and preventive interventions such as drugs and nutritional supplements. However, mind-body techniques will be classified as a somewhat separate category of "psychotechnologies." The empirical materials presented here regarding medicine and technology are based on fieldwork and semistructured interviews in various alternative scientific, medical, and religious communities in Brazil (Hess 1991) and North America (Hess 1993, 1999) during the 1980s and 1990s.[2]

The introductory essays in this volume point to the vagueness of the concept of "modernity," which has been located as far back as technological developments in the fourteenth century and as far forward as the Enlightenment, French Revolution, and industrial revolution. Although the choice of centuries varies, most discussions of modernity share a common ground in their analysis of major political, scientific, religious, intellectual, and other events that (1) were located primarily in western Europe, (2) took place during the last half-millennium, and (3) continue to ramify and be modified in a continuing process of modernization through the present era of informational capitalism, globalization, and "development." Technology studies sharpen the importance of the role of material culture in modernity and add an analysis of mutual shaping that questions architectonic formulations of modernity. Although the problem of modernity has obviously generated a rich literature and deep insights into the present era, this essay places the problem of modernity and its relationship to technology in a somewhat wider framework, both temporally and comparatively.

Cultural Ecology

In the long-term perspective of human history and cultural evolution, it is potentially valuable to compare the technological and social changes of the modernizing West with other major shifts in material and social culture, beginning with the first major technological revolution, the domestication of plants and animals. With technological innovation, the natural ecologies to which human societies must adapt became increasingly modified; indeed, much of the recent work in cultural ecology has documented the extent to which societal adaptation involves modifying an already modified natural environment.[3] Resource depletion and other adaptational problems probably played a significant role in the collapse of civilizations such as the Mayan, Roman, and those of the Indus Valley (Tainter 1988). As such, the ruins of at least some of the major human civilizations stand as an ominous warning to the emergent globalized world system (Price 1995). One may speak today of designing "green" technologies and sustainable industry, but from a long-term perspective, "sustainability" is an elusive concept. It can be argued that no type of human society is sustainable, with the possible exception of a hunting-and-gathering mode of production. However, few moderns would voluntarily return to a hunter-gatherer existence, and over the millennia even hunter-gatherers apparently produced substantial environmental changes, mainly through overhunting (Alroy 2001).

If one thinks of the first major technological revolution as the transformation of hunter-gatherer societies into pastoral, horticultural, or agricultural societies, some of the problems of modernity are already evident in incipient form. Plant and animal domestication allowed societies to adapt to the nexus of population and resource pressures, but the technological innovations also created a new scale for such problems as malnutrition, starvation, epidemic disease, social inequality, and environmental degradation. Indeed, in the anthropological literature one can find characterizations of the so-called neolithic revolution as "the worst mistake in the history of the human race" (Diamond 1995: p. 114).

A long-term evolutionary perspective offers other, more usable lessons. Johnson and Earle (1987: pp. 16–18) suggest that responses to resource depletion, population growth, or both, include warfare, technological innovation, trade, and stockpiling resources. That all four processes characterize the current world system raises the question of the adequacy of such responses, given the size of the world's population today, its rate of consumption of resources, its increasing levels of environmental degradation, and the parallels between these problems and those encountered by other civilizations. In other words, within this theoretical framework, the question of the long-term viability of the human species emerges as a central issue in a study of technology and modernity. This central issue is relatively absent from much of the history and sociology of technology, particularly the constructivist variants.

The emergence of environmental consciousness today can be viewed in ecological terms as a response to a crisis of culture and ecology, *pace* the "risk society" hypothesis (Beck 1995). Viewing the environmental crisis as an ecological problem—that is, as both a numerator problem of population size and a denominator problem of resource availability—one finds that medical science and public health measures occupy an important place in the history of the crisis. Large portions of the world's population, especially in the developed countries but increasingly in the less-developed countries, benefit by living in a postepidemiological transition world of chronic disease. (The term "epidemiological transition" can refer to various kinds of changes in health—morbidity, mortality, nutrition, etc.—that characterize modernity and modernization.) The technologies of food production (allowing nutritional improvements), public health (vaccinations, sewerage and water systems, quarantines) and, to a lesser extent, biomedical therapies (antibiotics, surgery) have played a crucial role in the high levels of global population and urbanization that are characteristic of modernity (McKeown 1976, 1979).

Ironically, the successes of the biomedical sciences, agricultural and food-processing technologies, and public health measures have also led to an impasse for biomedicine. Modernization of the disease ecology, together with alterations in food and nutrition, have meant that people are less likely to die from influenza, pneumonia, or tuberculosis than from cancer, cardiovascular disease, or diabetes. The diseases of modernity are largely of a different type; that is, they are the result of genetic predisposition interacting with noninfectious risk factors such as environmental toxins, stress, poor diet, and lifestyle.[4] Biomedicine has been much less successful at treating the "new" diseases than the older, infectious ones.

In the world of chronic disease, general ecological concerns with societal sustainability intersect with individual concerns of health and longevity. As people recognize the personal health risks of exposure to toxic substances, Beck's risk society thesis becomes both individualized and medicalized. However, the problem of health in the risk society goes well beyond concern with the negative health effects of environmental pollutants and toxicities resulting from lifestyle, such as smoking. In our interviews with leaders of the social movement for complementary and alternative medicine (CAM) for cancer treatment, "toxicity" was a central concern, not just for the etiology of diseases, but also for the therapies (Hess 1999; Wooddell and Hess 1998). In the United States, the chemotherapy industry has historical links to biological warfare; the first generation of chemotherapy drugs was developed from the biological warfare programs of World War II (Moss 1996). Over the decades, a global social movement has developed that seeks greater availability of "nontoxic" therapies for chronic diseases. In a sense CAM activists call for "greener" therapies; that is, therapies based primarily on dietary programs; supplements and herbs; nontoxic, immunity-enhancing drugs; and mind-body techniques.

The movement toward nontoxic therapies involves a transformation in the way of thinking about chronic disease—as reversible rather than degenerative. The term "chronic, degenerative disease," especially when based on genetic explanations, puts patients in a hopeless downward frame in which toxic interventions can only slow the progression of disease or mitigate its symptoms. In contrast, the CAM literature and its conferences are populated by case studies and testimonials of patients who have reversed chronic disease by using nutritional, mind-body, and other nontoxic therapies. The political implications of the testimonials are enormous: to the extent that a significant percentage of patients take the reversibility principle as a prescription for their own chronic disease, they challenge many of the high-tech therapies that assume nonreversibility (e.g., surgery, angioplasty, cytotoxic chemotherapy, gene therapies, symptom-masking drugs), as well as the industries and professions that are affiliated with those therapies. A conventional technology studies analysis might frame such conflicts as a clash of heterogeneous networks of patients, clinicians, disease knowledge claims, therapies, equipment, and insurers, but I am suggesting here that the analysis would miss the broader terrain of shifts in environmental consciousness and disease ecology that is both shaping and being shaped by the network clashes.

Furthermore, in the case of mind-body therapies, there is an additional shift of terrain from the problems of exposure and detoxification to a general critique of the lifestyle of contemporary society and culture. Often the critique centers on "stress" and the diseases generated by the lifestyle of modern society. Stress itself can become linked to toxicity. Listen, for example, to Charlotte Louise, an actress and long-term survivor of ovarian and lung cancer who became a leader in the CAM cancer therapy movement. Upon discovering that scientific medicine offered no solutions for her advanced disease, she turned to Eastern religion as a source of inspiration:

I knew that I wasn't dependent on science; I was dependent on my source of health, the divine power of the universe that I'm connected to . . . . So I started receiving clear signals about how I was supposed to be living my life, how I can get rid of toxicity, and how I've allowed it to enter my life on every level—spiritually, culturally, matrimonially—and in every relationship: my child, my mother, with my home environment, with my physical being, and my health . . . . I had to look at every way I was making my life toxic—at the emotional level, the spiritual level. That's how I did it. (Wooddell and Hess 1998: p. 172)

The passage suggests an extension of the problem of environmental risk to a generalized toxicity that is related to a modern lifestyle of rapid pace, impoverished social relations, stressful workplaces, and alienation. The passage also suggests a link between the environmental movement and alternative health and spiritual movements that secular, academic researchers and environmentalist leaders tend to underplay. In this cultural context, the spiritual becomes an antidote to the toxicities of modernity: carcinogens, chemotherapy, bad relationships, and generalized stress.

The mixture is too easily dismissed as antimodern. The embrace of a spiritual alternative does not necessarily come at the expense of rationality. Rather, the patients we interviewed seemed to hold the two together in a productive tension. One patient, a holder of an advanced degree in education, consulted with spirit guides, but checked her intuitions with medical research. As she commented, "I was trying to reconcile the educated and spiritual parts of me. I was trying to reconcile two different systems. If I made a mistake, I was putting my life on the line. This wasn't just a philosophical exercise" (Wooddell and Hess 1998: p. 214).

Rather than viewing such hybridity as antimodern or irrational, it seems better to accept it as characteristic of modernity (Harding 2000) and indeed as having a long history within western modernity (Hess 1991). There is no denial that within the countercultural formulations of the alternative spiritual, environmental, and health movements, relations between the secular, scientific strands and the religious, traditional strands can be tense. Yet, viewing the latter as antimodern, atavistic, and romantically reactionary misses the ways in which the two strands are actively engaged. Over time, nutritional science and mind-body research slowly grows around CAM, just as environmental science grows around green lifestyles and technology, bringing about the integration of the strands (see Mol, chapter 11 this volume).

More generally, the analysis of technology and modernity from the perspective of a cultural ecology of alternative health movements suggests that much more is going on in the risk society than an institutional response to hazards and risks. Rather, the issues of risk and sustainability are interwoven in the individualization of the environmental crisis in bodies and personal health histories that leads to a broader cultural questioning of the stresses of modern lifestyles. Technological alternatives such as organic food serve as crucial junctures between the alternative health and environmental movements, which suggests why the politics of food (such as genetically modified organisms or supplement regulation) can become galvanizing political issues.

However, the countermovements themselves are changing as they change understandings of technology and modernity. As the politics of alternative health, urban greening, sustainable agriculture, and other countermovements gain currency, the movements become increasingly integrated into mainstream practices. That integration process also involves selection, which in turn involves the values problem of defining legitimate grounds for selection. This leads to the second framework.

Cultural Values

As a problem of cultural values, modernity can be viewed as the continuing transformation of particularistic social relations and institutions into relatively universalistic ones. Here the focus of modernity studies is not so much industrial society with its hazards, environmental crises, and death of nature, but rather a group of interwoven institutions that includes constitutional democracy and a public sphere, market capitalism and legal universalism, nuclear families and social mobility, and religious pluralism and secular, empirical science. The institutional matrix is characterized by a continual transformation from ascribed status groups that are related hierarchically and particularistically toward a more modern institutional order and value system founded on the individual as moral actor. More than the political creature of constitutional elections or the consumer of neoclassical economies, the modern individual in this sense is any social unit (a biological individual or a corporate group) that is related to other, similar units through rules that apply, at least in theory, equally and universally to the units in the system.[5]

Although it is tempting to describe modernity in this sense as an event or series of events that occurred in northwestern Europe between 1500 and 1700, it has long been recognized that the characteristically modern matrix of values and institutions has a long history stretching back at least to the ancient world (Dumont 1986; Weber 1978). Just as the ecological framework situated the problem of modernity on a broader temporal scale, so the modernization of cultural values is better seen as a process that stretches back beyond the last half-millennium into antiquity and continues to evolve today. For example, in the case of legal universalism, which Weber (1978) traced back partly to the ritual commensality of the ancient world, modern political democracy has gradually extended the franchise from propertied white males to increasingly broader segments of society that were previously excluded on particularistic grounds, such as wealth, gender, race, ethnicity, and age. The continuing process of modernization is the product of historical struggle by excluded groups to extend the principle of equality of rights throughout their society.

Within this framework, technological change is understood less in an ecological-biological-health framework and more in terms of a legal-moral framework of individual rights and general social equality. As rights are extended, laws are made to decrease exposure to less desirable technologies and their side effects (as in not-in-my-backyard efforts to exclude industrial waste; privacy challenges to surveillance technologies; and consumer efforts for protection against unsafe devices, food, and drugs) and conversely to increase access to the more desirable technologies and technological design features. Within the medical field, access to health care, and especially its latest technologies, is a burning political issue that is characteristic of advanced or late modernity. Access to both conventional and CAM therapies tends to be framed as a discourse of rights. However, access to conventional therapies tends to emphasize the "equality" aspect of rights to material resources, whereas the politics of legalization of CAM therapy have tended to be contained in a political discourse of "medical freedom" and the patient's right to therapeutic choice. Listen, for example, to the depiction of the social movement by Michael Culbert, one of its leaders, who also flags the left-right amalgams that are characteristic of other new social movements:

What stimulated me to write my book *Vitamin B17: A Forbidden Weapon Against Cancer* (1974)—I was still a newspaper editor—was that I would go to the Berkeley municipal court to cover the Richardson trial, and there were McGovern-for-president left-wing hippies in the audience who were in favor of this John Birch doctor who had been arrested for using laetrile. This was incredible. Here was an issue that was far beyond left and right, and yet it certainly does have hotheads on both sides. The freedom-of-choice movement was a populist revolution that I participated in and helped foment. It was tremendous. We went across the country, and we never knew who was going to pop up. (Hess 1999: p. 103)

The CAM social movement in the United States has generated numerous legislative reforms at the state level, including the legalization of laetrile in twenty-four states and passage of more general "access to medical treatment" laws in many states. Those reforms were a counterpart of the more well-known reforms at the national level achieved by the AIDS movement, but by the late 1990s CAM issues were also achieving prominence at the national level.

In addition to legislative reforms, the access issue also emerged in the doctor-patient relationship, which has been transformed from a paternalistic model into more of an adult-adult partnership, thus bringing egalitarian values into the clinical encounter. The change was recognized for the chronic disease patient as early as the 1950s (Szasz and Hollender 1956), but the modernization of the doctor-patient relationship has increased in the wake of the social movements of the 1960s and 1970s. In a spillover effect from the women's health movement, women patients have tended to be the leaders in this dimension of medical modernization in the CAM field through their individualized, therapeutic programs (Wooddell and Hess 1998). Although there is always some individualization of therapy in conventional medicine (such as the tailoring of dose to age, size, and gender), CAM providers and their patients have pushed the rhetoric and practice of individualization to new levels. In the world of CAM cancer therapies, the older, Fordist therapeutic regimes are being replaced with individualized and flexible programs, that is, therapeutic packages that change over time and with the patient's health.[6] Furthermore, CAM therapies emphasize patient "adherence" rather than "compliance," that is, they accentuate the partnership between the patient and the health-care team, in which older hierarchical and paternalistic relationships are replaced with more egalitarian and individualistic ones.

The attempts to replace patented drugs with dietary regimes, high-dose vitamin supplements, herbs, mind-body techniques, and other interventions of the CAM programs also bear on a different type of equality issue in ways that are similar to "universal" design initiatives in other fields of technology.[7] The high cost of drugs for cancer, cardiovascular disease, AIDS, and other chronic diseases, together with the cost of maintaining an infrastructure to monitor the toxicities of the drugs, have made it difficult for many of the less-developed countries or even poorer strata of developed countries to gain access to such drugs. In contrast, CAM therapies are not only less toxic but also potentially less expensive, making possible greater access on a social, rather than an individual, scale.

Notwithstanding the alignments of access to CAM therapies with enhanced equality and the extension of individual rights in the medical field, the medical world has not rushed headlong into CAM therapies for chronic disease. The medical profession often employs another variant of the discourse on modernity values to block access to alternative therapies: their efficacy remains largely undocumented by conventional research methods, so why should CAM therapies be treated any differently from untested drugs? (Although the resonances with other pressing debates of late modernity around terms of "undocumented" status and "differential" treatment may sound conservative, one should keep in mind that the counterargument rests on the same set of universalistic values, only applied differently; hence its rhetorical power.)

The problem of evaluating efficacy involves another, higher-level political issue: the modernization of methodology. In clinical research today the top of the methodological hierarchy is occupied by clinical trials. In the United States a company generally cannot obtain Food and Drug Administration (FDA) approval for a proposed drug unless the substance has passed through a rigorous series of clinical trials. Clinical trials are seen as a universalistic solution to the problem of drug selection, because in theory they constitute a "level playing field" for competing drugs. In this sense clinical trials are a quintessentially modern institution; like the old figures of Justice, they are double-masked to reduce bias.

However, CAM advocates reply with their own critique, which suggests that the research methodology is not as modern as conventional medical researchers claim. In practice, the quality of information obtained from clinical trials is not substantially better than that obtained by other methods, such as retrospective cohort studies or, in some cases, even best-case series, a point that CAM researchers emphasize (Hess 1999). Yet the high cost of clinical trials means that only well-funded players (especially private capital with an interest in patentable drugs) will invest in the research. The political-economic inequalities of this method result in a problem of unfunded research on unpatentable natural substances (food supplements, herbs). Consequently, CAM researchers tend to be driven down the ladder of the methodological hierarchy to methods that are less expensive and, within the research and regulatory worlds, are considered less universalistic because of their vulnerability to biases engendered by their nonexperimental design.[8] As I move through the worlds of CAM conferences and informal discussions, I hear retorts such as: "the gold standard is well named because it takes a lot of gold to set the standard." The methods need to be modernized to take into account the new citizens of the therapeutic world: unpatented natural substances.

Legislation intended to remedy the problem of orphaned therapies represents another dimension of the modernization process as a question of health-care access. The legislation attempts to bring a greater degree of universalism into the regulatory process in light of the kinds of criticism raised here. In the United States, government research funding for CAM therapies gradually increased during the 1990s; legislation for orphaned drugs has relaxed some of the regulatory hurdles and expenses for drug approval; and dietary supplements are currently protected as food and therefore below the regulatory hurdle of drug approval. However, the marketing of supplements can only be linked to general structure and function claims (e.g., promoting prostate health) rather than disease claims (e.g., treating prostate cancer). As a result, although supplements are protected as food equivalents, they are also condemned to a second-class status with respect to drugs. Notwithstanding the charges of inequality and unfairness from both sides, the political discourse around regulatory reform in this arena takes place with reference to the value of universalism: how can fair standards be used to evaluate a technology when one form is protected by patents and supported by private capital investment, and the other is in the public domain and excluded from private interest?

To summarize, in this framework modernization largely means a universalization or democratization process. Technology constructs modernity by providing new sites for the politics of modernization, which must grapple with the paradoxes that new technologies create for existing laws (e.g., are supplements food or drugs?). Legitimate political action (which is not always the action taken) moves toward the equalization of rights of access to desirable new technologies and, conversely, avoidance for undesirable technologies or their side effects (such as pollution, safety risks, or the invasion of privacy). In the medical field, access and avoidance generally go together, because access to desirable new therapies entails avoiding undesirable, old therapies. Thus, much of health-care policy focuses on access to conventional health care as a key progressive issue.

The CAM movement throws a wrench into the politics of definitions. It challenges the desirability of the latest technology-based interventions by proposing less toxic and potentially more efficacious alternatives: chelation therapy instead of angioplasty, dietary programs and megadose supplements instead of chemotherapy, and so on. As has been recognized for AIDS patients, the clinical encounter becomes a subpolitical site (Beck 1997; Epstein 1996; Feenberg 1995), but in the case of CAM cancer therapies, patients demand access to a wide range of alternative therapies.

Arguments about how to extend the basic principles of individual rights become wrapped up in the problems of funding and resources.

For those who have no gold, the gold standard becomes a double standard. Political solutions must move between the principles of rights and the realities of resources. Here is one point of contact between the problem of cultural values and the framework of modernization as a universalization process, and the problem of political economy and the framework of modernization as capitalist expansion.

Political Economy

The political economy framework selects modern capitalism as its central object of analysis, rather than a toxic society or a family of universalizing institutions. As in the other frameworks, there is some value to situating the political economy of technology and modernity beyond the past 500 years and the birth of modern western capitalism, and even beyond a Weberian analysis of its roots in the empires and state societies of the ancient western world. (As in other sections, this argument should not be misinterpreted to claim that these premodern societies were in some sense "modern," but only to show that the problems of modernity are deeper than the history of Europe during the last half-millennium.) For example, the dominant groups of chiefdoms in early societies exhibited patterns of accumulation and conquest (Earle 1997) that, even if their economies did not have a form of monetary accounting, nevertheless bore some similarities to the patterns of production for profit and globalizing expansion that are characteristic of modern capitalism. The problem of technology and modernity therefore again needs to be framed as an intensification and scaling up of processes found in other, premodern societies. To the extent that there is a distinguishing feature for modernity in a political economy framework, it is the emergence of a rationalized, monetary form of capitalism that has "the globe as its battlefield" (Marx 1977: p. 915).

The expansion of capitalist political and economic relations is a multilayered phenomenon that includes the following: absorbing older modes of production, such as indigenous societies and peasant economies; displacing small-scale capitalism, such as family businesses and regional coops, with multinational corporations and chain stores; taking over the economic functions of neighborhoods and families, such as child care, elder care, home care, and insurance; privatizing the functions that were once located mostly in the church and state, such as education, health care, insurance, and urban services; and incorporating the work of social movements, such as corporate affirmative action programs, greening initiatives, and worker participation plans.

The common thread in capitalist expansion is the displacement of various types of community. As with "sustainability" and "equality," the term "community" is slippery and requires some qualification. A first point is that the distinction between community-oriented social relations and capitalist ones cannot be reduced to simple formulas such as gift versus commodity, monetary versus nonmonetary exchange, or altruistic versus self-interested behavior, as several social scientists have demonstrated (e.g., Appadurai 1986; Bourdieu 1977; Parry and Bloch 1989). Rather, the term "community" needs to be understood loosely as the production and exchange of goods and services based on a legitimating factor other than monetary profit, even if profitability concerns are simultaneously present. The other concerns include kin and residential obligations as well as the values that speak to humanitarian, ecological, or other types of orientation toward the general good.

The definition suggested here has the advantage of encompassing a wide range of community types. Some are relatively archaic, where membership is based on residential or kin ties; others are more modern, where membership is based more on a willingness to assume an identity or to work for a general good. Given the variety of types of community, globalization can be progressive in terms of the other frameworks. In other words, archaic communities and the lifeworld can be havens for particularistic, hierarchical relationships, so their displacement by capitalist expansion can be liberating for subaltern social groups (Fraser 1989). Likewise, closing a highly toxic family business in the wake of a less toxic multinational's market expansion may be bad for a family community or even a residential one, but it may be a step toward a more sustainable industry.

Communities in the sense used here are not simply evolutionary survivals, a *Gemeinschaft* that is giving way before the inevitable expansion of a *Gesellschaft*. Rather, new communities emerge even as old ones are displaced. A tribe may disappear, but an indigenous social movement will often spring up in its place. A church may close its doors, but a community technology center may open down the block. Even well-run large corporations produce a sense of community for employees by developing a common mission that extends beyond profitability (Collins and Porras 1994) or by encouraging friendships and informal exchanges among members of a subunit. In other words, capitalist expansion both displaces existing communities and creates the conditions for new ones.

Because archaic community structures and older forms of capitalist production die slowly, and because new forms of community emerge in their place, the result of capitalist expansion is not a wiping out of other forms of social order, but a pluralization or diversification of the other forms. As modernity in the form of capitalist expansion proceeds, layers of older social roles and technologies come to coexist alongside more recent ones that are defined by the expansion of markets. For this reason I prefer to focus the analysis on pluralistic technological "fields" rather than single technological systems such as electric power or railroads. A visual metaphor of technology in this frame is a road in a developing country where ox carts and pedestrians vie for space alongside bicycles and automobiles, with subsequent generations of transportation technology usually offering greater options of speed and scale (see Edwards, chapter 7 in this volume). Although the transformation process is dramatically evident in developing countries, even the developed centers of the world system exhibit technological pluralism. For example, within a 20-mile radius of the region where I live, there are a highway system, an airport, a train station, a seaport, a bicycle trail, sidewalks, and even a functioning lock along the old Erie Canal. New modes of transportation have displaced older ones, but with a few exceptions they have not replaced them completely, and in some American cities there has even been a return of the suppressed form (such as urban rail transportation).

The increasingly complex system formed by the historical process is not a mere syncretism of historical layers; rather, it is a selectively structured whole based on the dynamics of changing markets and emergent communities. Materially, the rich have access to most, if not all, layers of technology, whereas the poor often live in the past, or at least with past generations of a technology. Here is another point of contact with the values framework: even as access within a society becomes more or less universal for one technology or one generation of a technology (penicillin, telephones), technological innovation ensures that there are always new technological inequalities (gene therapies, digital divides). Innovation therefore casts the incessant processes of modernization as equalization of rights of access by producing a diversification of the potentially accessible. For example, whereas access to telephones may be nearly universal in some developed countries, the subsequent innovations in communication technologies (wireless phones, Internet devices) ensure that the equalization of access is a continuing process.

The diversification process leads to a major political problem that exists alongside the problems of environmental risk and social injustice raised by the other two frameworks. What are the legitimate grounds for directing a system of technological innovation? Because most technological innovation is produced or developed in the private sector, *pace* Bourdieu (1991) the system's overall development is tilted toward the profitability concerns of the dominant producers in the field. Those concerns are often in conflict with the perspectives and interests of geographically localized communities or nongovernmental organizations (NGOs) that speak for broader communities. States and international harmonization bodies, with their regulatory role, become the primary arena for sorting out the clashes of what should be produced and how it should be produced.

Again, attention to the medical field provides insights of general value to a study of technology and modernity as well as the regulatory problems that this framework helps elucidate. The complex relations among various healing systems that have different historical and ethnic backgrounds are known in the literature as "medical pluralism"; perhaps technology studies could benefit from a parallel analysis of technological pluralism. Conventional technology studies have been good at demonstrating battles between technological systems, but they have paid relatively little attention to the structures of technological pluralism that emerge from those battles.[9]

As studies of medical pluralism have progressed (e.g., Baer 1989, 1995), several major conclusions have emerged and they have some relevance for the general study of technology and modernity. First, medical pluralism is not restricted to poorer countries, where support for biomedicine is weak, or even to nonwestern countries, where cultural traditions support long-standing indigenous medical systems. Alternative health-care practitioners and other healers usually have a well-defined cultural address in a particular ethnic group or nation-state, and consequently they appeal to various levels of community loyalties. Even where the appeal of the social address is not evident—such as the recourse of European-American middle-class patients to acupuncture or Chinese herbal medicine—one finds that CAM practitioners often attract patients because they treat the whole person, rather than a body part or a disease. Office visits are frequently longer and more personalized; the practitioners return to the health-care setting the sense of community that has been lost as insurers have increasingly commodified and standardized the doctor-patient relationship. In short, partly because the alternative therapies and techniques are anchored in various types of community relationships and identities, medical pluralism is increasingly a phenomenon of all countries. More generally, the phenomenon should encourage us to look for a similar dynamic of growth in technological fields, which have been marked by an increasing diversification of both conventional technologies and various alternative designs.

Second, although CAM systems such as chiropractic (U.S. origin), Ayurveda (Indian), acupuncture (Chinese), macrobiotic (Japanese), and anthroposophy (central European) have a cultural address and a cultural appeal within their local setting and diasporas, a second dimension of medical pluralism involves the fact that CAM systems have themselves become globalized. One might conceive of the process as a countercolonial or counterhegemonic development, but from the perspective of local communities, the globalized CAM systems constitute yet another threat to the local healing traditions. When CAM systems are exported outside their locale, they become transnational communities or social movements, but their products—the therapies they offer—also tend to become commodities that compete with both biomedicine and local healing traditions in a pluralistic medical marketplace. More generally, the globalization of CAM suggests that even when one begins to discuss sustainable technologies or a universal design for alternative technologies, the new products soon become caught up in the logic of commodity exchange that will separate the products from the meanings and practices in which they were originally produced.

Third, although medical pluralism has become globalized, it is not everywhere the same. Each country has its own portfolio of biomedical and CAM therapies and practitioners based on its historical and cultural traditions. For example, chiropractic medicine is relatively large in the United States, and spiritualist and spiritist therapies are relatively small, whereas in Brazil the opposite is the case (Hess 1991). The shape of the pluralistic systems is not simply a product of the "market penetration" of global capital and biomedicine; rather, the robustness of local cultural traditions and their ecologies shape patient demand and political action for regulatory protection of the CAM systems. This point reminds us that any discussion of technology and modernity needs to take national and local variation into account, particularly when the discussion moves into policy solutions.

More generally, the existence and growth of medical pluralism points to a broader problem of the emergence of technological pluralism. The fossil fuel–based energy industry has had to come to terms with alternative energies, the transportation industry with the reemergence of bicycling or public transportation, the building industry with green design, the software industry with open source codes, and so on. Neither medical pluralism nor technological pluralism in general is well predicted by a simplistic model of political economy. Vulgar Marxism would predict the obvious: that health care is big business; that pharmaceutical companies, hospitals, insurers, and other health-care sectors constitute a major industry; and that both are globalizing and expanding into domains that were previously outside market relationships. A simplistic model of the globalization of capital would merely draw attention to the globalization of biomedicine and its hegemony, but it would have trouble theorizing the countermovements that I am pointing to here.

A good political economy model of globalization needs to take into account the resistances that emerge from various types of communities to protect practices and material entities that are threatened by the expansion of markets. Furthermore, a good model needs to take into account the attempts of globalizing capital to absorb and incorporate those resistances. Again, the expansion of biomedicine is a good case in point. The process of incorporation of alternatives is more complex than the mere extraction of local capital that occurs when, for example, medicinal herbs are taken to laboratories, analyzed for pharmaceutically active agents, and converted into drugs, usually with no royalties to the local medical tradition. Rather, biomedicine has moved more directly to incorporate whole CAM systems.

A somewhat more detailed discussion of the incorporation of CAM may provide a clearer picture of the process that the third framework brings to attention. As I have found out in my long-term fieldwork in the CAM cancer community, in the late 1990s several of the major conventional cancer hospitals in the United States opened CAM clinics in order to meet patient demand for CAM cancer therapies. Likewise, some of the major oncology practices have moved to offer "integrated" or "comprehensive" cancer care. On the one hand, the event of integration represents a victory for the social movement that called for more access to the less toxic cancer treatments associated with nutritional and mind-body therapies. Likewise, CAM providers have become increasingly mainstream as they have won licensing rights and insurance reimbursement, and with the advent of CAM clinics in conventional cancer hospitals, CAM providers are even gaining a foothold within the establishment. However, the apparent victories are also accompanied by limits on the scope of practice and status deprivation to the level of auxiliary health-care providers similar to nurses, dietitians, or physical therapists. Furthermore, the integration process selects CAM therapies that complement conventional medicine rather than provide alternatives to it; indeed, one major American cancer center now offers "CIM" therapies (complementary and integrative medicine) because it rejects "alternatives" to conventional therapies.

The colonization of a social movement that I have witnessed during the past five years is familiar to students of the other science and technology-oriented social movements (see, e.g., Mol's essay in chapter 11). Over time, grassroots activism has become increasingly institutionalized, and the social movement has fragmented as sectors have become increasingly integrated into the frameworks of former opponents. In the environmental movement, some organizations have become increasingly moderate, while the corporate sector has moved toward corporate greening initiatives (Jamison et al. 1990; Hajer 1996). In the AIDS movement, pharmaceutical companies have increasingly influenced patient advocacy organizations, which themselves have undergone a fragmented "expertification" process (Epstein 1996), and in the alternative energy movement, corporate resistance gave way to a strategy of incorporation and integration (Jørgensen and Karnøe 1995). In those and other cases, capital has played a strong hand in selecting which aspects of the social movement will grow and become prominent.

Regarding the more general problem of technology and modernity, the political economy framework focuses attention on the question of which technological systems (or in the case of CAM discussed here, which therapy-practitioner systems) will survive in the wake of innovation driven by production for profit. The dynamics of capital expansion create new products and markets that threaten the extinction of some material entities and their accompanying social roles. Either via democratic or nondemocratic means, and often after contributions from many communities, societies will decide that selected entities in the material culture and environment should exist and therefore must be protected, even if the expansion of the market would mandate their extinction.

The resulting entity, the "protected entity," is understood here to include technology as well as material and spatial culture that is protected by building codes, zoning restrictions, wilderness preserves, and animal treatment codes.[10] States and international organizations have increasingly been called upon to protect endangered entities, including technologies or desirable features of technology design, that otherwise might be swept away by the tides of technological innovation guided by the profitability concerns of global capital. Although protections may cover whole categories of entities (a wilderness preserve, a species, wind turbines, food supplements), they may also extend to design features that are protected parts of commodities. One example is the proliferation of safety regulations surrounding the design and use of consumer products, transportation vehicles, drugs, biotechnologies, workplaces, databases, guns, and food that permit or prohibit the movement of such commodities across national or regional trading boundaries. Another example is the emergence of privacy concerns around new information technologies, and the increasing demand for the protection of privacy through software designs (see Lyon, chapter 6 in this volume).

The political side of the "political economy" framework for analyzing technology and modernization draws attention to modernization as a process by which the regulatory laws of states and international organizations, together with voluntary standards set by international industrial and professional organizations, slowly redefine commodities as entities that are no longer mere products for markets. Commodities become protected entities whose existence is ensured by a code that at its best allows the perspectives of various types of communities to constrain the pure free play of market-oriented product design and innovation. In short, production for profit becomes encompassed by a broader logic of production to standards.

The commodity is therefore enmeshed in a complex, historical process, and I would suggest that the transformation of gift into commodity is not the central issue for a political economy of technology, even one of anthropological scope. Rather, regulatory law takes back some of commodity from the market by subjecting it to a double standard; not only must the commodity be profitable in the world of markets, but it must meet the legal standards of a regulatory code. Yet, regulatory law does not restore the gift to the commodity; no circle is formed. Capital reasserts itself in the battle over the structure of regulations. For example, the licensing of CAM providers may protect some of the local culture in the wake of biomedical hegemony, but such licensing also involves putting limitations on the CAM system and provider that locate it in a nondominant position within the medical field.

One might argue that globalization works against regulation, that international competitiveness drives deregulation, just as it has caused the dismantling of costly welfare states, and that the regulatory process is not as deeply interwoven in the globalization process as is suggested here. However, this argument misses the modernization process that regulatory law is itself undergoing. Increasingly, the regulations of states are being supplemented by international standard setting in processes that entail participation from NGOs and some concern with issues of general good (Feng 2002). Globalization does not imply the wholesale dismantling of regulations and standards as much as their harmonization among nation-states, and the harmonization process itself involves the complex articulations and negotiations that are suggested here. Regulation is necessary for capitalism to function, but it is also the doorway through which community can be redesigned into commodities.

Conclusions

The problem of technology and modernity as conceptualized here is not merely an analytical and descriptive one, but a deep normative question about the kind of global material-social world that should be co-constructed. The three frameworks presented here draw on different social theory traditions to direct attention to problems that require both empirical research and normative debate. The goals of sustainability, equality, and community emerge as three major criteria that provide viable points of reference for a general discussion of technological and social redesign (see Feenberg 1995; Fischer 1995; Sclove 1995; Van der Ryn and Cowan 1996; Lerner 1997; Rothschild 1999; and Schot, chapter 9, this volume). However, the goals bump up against each other and provide reference points for a triangulation of criticism. For example, communities can be full of particularistic and antiegalitarian social relationships, or they may have unsustainable ecological practices. Likewise, greening initiatives can be economically costly in ways that threaten communities or enhance inequality. Concerns with democracy, equality, and human rights can be discussed in a language of the individual that ignores concerns of community or sustainability. Consequently, the three goals provide checks on each other for a political discussion that must be anchored in specific cases.

In many if not all the technological fields, one can locate a set of complementary and alternative technologies, a CAT sector that is similar to the CAM sector described here for the case of medical pluralism. In the transportation field, there are bicycles, greenways, and public transportation systems; in the energy and chemistry field, renewable energies and alternatives to chlorine-based chemicals; in the waste-processing field, biological sewage treatment and recycling programs; in the agricultural field, organic farming and multicropping; in the computer field, privacy software and open-source systems; in the architecture and urban design field, feminist, community-oriented, and green design; and so on. Often, but not always, the alternatives can be constructed in ways that do not put the normative criteria in a zero-sum relationship. Yet even when that is achieved, the alternatives remain alternatives because they are not as viable from the perspective of the market. Consequently, the state and, increasingly, nongovernmental organizations, are needed to intervene and guarantee the existence of alternatives through regulations and standards.

When social movements mobilize to reconstitute complementary and alternative technologies as protected entities, the success of such political action usually occurs at a cost. A selection process operates on both the technologies and the movement organizations so that the complementary technologies are favored over the alternatives, just as the accommodationist organizations are favored over more radical voices. Integration leads to division as social movements are captured, old friendships and the sense of movement community are shattered, and manifestos are translated into partial policy victories. I have watched the process occur to some degree in the CAM cancer therapy movement in the United States during the 1990s. Yet, recognition of the reality of partial integration through incorporation should not lead to the paralysis of inaction. Instead, recognition merely highlights the process by which a new generation of social movements must be continually created within a new technological field with new contours of conventional and complementary and alternative technologies. In some cases and on some grounds there is progress.

Notes

1. This definition would require splitting off other types of instrumental social action, such as psychotechnologies or social technologies. The definition was developed in part in conversations with Torin Monahan, a doctoral student at Rensselaer who is working on a practices-oriented approach to technology (Monahan 2000). Some of the ideas presented here are discussed more completely in my electronic volume, *Selecting Technology, Science and Medicine: Alternative Pathways in Globalization*, Volume 1, at <http://home.earthlink.net/~davidhesshomepage>.

2. The research also includes a book of interviews with women leaders of the complementary and alternative cancer therapy movement in the United States coauthored with Margaret Wooddell (Wooddell and Hess 1998). For a quantitative documentation of the extent of CAM in the United States, see Eisenberg et al. (1998). I borrow the term "field" from Bourdieu (1991), without necessarily accepting other aspects of his framework, such as the near absence of a political analysis of technological design.

3. I use the term "cultural ecology" loosely to refer to a variety of programs that can be distinguished more properly as cultural ecology, historical ecology, political ecology, and the new ecology (Biersack 1999).

4. The possibility that apparently noninfectious chronic diseases may turn out to be infectious has become more evident since the revision of the etiology of gastric ulcers in the early 1990s. On the infectious tradition for the treatment of cancer, see Hess (1997).

5. The formulation in this paragraph draws on the social theory research tradition that includes DaMatta (1991), Dumont (1986), Parsons and Shils (1951), and Weber (1978), as well as Habermas (1989) and his critics (e.g., Fraser 1989: chap. 6).

6. See Martin (1994) for a more general discussion of flexibility in the economy and the health field.

7. "Universal" design is never completely universal, in the sense of being applicable to everyone, but the principle is to redesign technology and material culture so that they are accessible to a wider number of users. Examples include easy-grip tools and buildings with ramp access rather than steps. Material culture maintains hierarchical social distinctions (e.g., older people with arthritis, people in wheelchairs), and universal design is intended to mitigate those distinctions by making one design that is applicable to different social categories.

8. The problem is further complicated by the fact that some of the features of the gold standard of clinical research design have built-in biases in favor of conventional, pill-oriented medicine. For example, it is difficult if not impossible to provide double blinds and placebo controls for dietary programs. The more one looks at the design problems for clinical trials of CAM therapies, the lumpier the image of a "level playing field" becomes.

9. As Baer (1989, 1995) and others have recognized, the term "pluralism" suggests an equality of actors that is misleading; rather, the structure of the diversity of medical fields is hegemonic, and biomedicine is the dominant healing system in almost every society in the world.

10. This approach differs somewhat from the European actor-network theory (Callon 1995), from which I borrow the term "entity," in that I would maintain as desirable the normative distinction between humans and things (see Pickering 1992). The law distinguishes between the rights of humans and the protections of things, but increasingly it must grapple with the conflict between the two goods.

11

The Environmental Transformation of the Modern Order

Arthur P. J. Mol

For decades, environmentalists and their theoretical interpreters had a rather clear and undisputed position toward modernity and the project of modernization. Just 20 years ago the Dutch environmental sociologist Egbert Tellegen (1983) identified the common denominator of environmental movements around the world as their antimodern ideology. Environmentalists of the time, with their many distinct theories and practices, and widely varying tactics, shared an antimodern attitude. Whether they were small-is-beautiful adherents, Club-of-Rome critics, neo-Malthusians, or neo-Marxists, these environmental movements seemed united in attacking the basic institutions of modernity, such as capitalism, industrialism, modern science and technology, and the bureaucratic nation-state. In the past two decades, however, the attitudes of environmentalists toward modernity and modernization have changed dramatically. The landscape of "green" positions and ideologies toward modernity has become far more complex, ranging from demodernizers or antimodernists, through various kinds of modernists (including neo-Marxists) to postmodernists. If anything, we can conclude that compared with the 1970s and 1980s, environmentalists have become more modernist or at least less hostile toward modernity.

During the past two decades as well, social scientists and social theorists have identified the environment as one of the "battlegrounds" for understanding the changing character of modernity. While for a long time environmental studies flourished only at the margins of many social science disciplines, such major figures in sociology as Anthony Giddens, Zygmunt Bauman, and Ulrich Beck have recently focused on environmental issues. A similar upsurge of academic activity can be seen in environmental history and environmental philosophy. This upsurge of interest was of course partly inspired by the reappearance of environmental issues on the international public and political agendas in the late 1980s and early 1990s. In addition, it has become clear that responses to environmental concerns, at many levels, have begun to change the basic institutions of modern society.

This chapter deals with this shifting relation between modernity and environment. More precisely, it explores how environmental considerations and interests are contributing to the transformation of modernity. I start with a brief overview of the major schools of thought in academic environment and modernity studies. Then I elaborate one specific perspective, ecological modernization, which spotlights the social transformation processes and dynamics concerning environmental questions. Next, I use this perspective in showing how environmental considerations are reshaping the business strategies of chemical producers and consumers. Finally, I examine sectoral and national variations in the environmental transformation of the modern order.

Modernity and the Environment: An Overview

Scholars in environment and modernity studies can be grouped into four schools of thought: neo-Marxists who especially criticize the capitalist ordering of the modern economy but not necessarily modernity itself; scholars who are rather critical toward modernity and modernization processes (demodernization or counterproductivity adherents); scholars who argue that modernity has been changed beyond recognition (postmodernists); and scholars who stress the significant changes of modernity's institutional order (reflexive modernization theorists).

Neo-Marxism as Modernization

In the 1970s, neo-Marxist studies of the modern capitalist economy were particularly influential in bringing to light the origins and logic of the environmental crises. Focusing attention on the internal economic contradictions of capitalism, neo-Marxist environmental sociologists such as Ted Benton, Peter Dickens, Allan Schnaiberg, and James O'Connor analyzed the end of the capitalist economic order, as it would jeopardize the resource base of the production and consumption treadmill. These scholars combined the idea of aggressive global expansion of the capitalist economy with the continuing and intensifying (global) environmental crisis to formulate a hypothesis about the "second contradiction of capitalism": the economic growth and expansion inherent in the global capitalist economy will run up against environmental boundaries that will in the end upend and transform the global capitalist economic order beyond recognition.

In their analyses of the modern environmental crises, neo-Marxists were keen to focus on the capitalist economy rather than on modernity as a whole. In contrast to their critical views on the capitalist market economy, these neo-Marxists maintained that the modern bureaucratic state, modern science and technology, and modern norm and value systems were important elements of a sustainable society—only under different (noncapitalist) relations of production. In this sense these neo-Marxist environmental sociologists were modernists.

Yet even among neo-Marxists today, there persists disagreement about the environmental consequences of (global) capitalism and the repercussions of the environmental crisis on global capitalism. A leading American neo-Marxist, James O'Connor (1998: p. 235), recently concluded that, "a systematic answer to the question, 'Is an ecologically sustainable capitalism possible?' is, 'Not unless and until capital changes its face in ways that would make it unrecognizable to bankers, money managers, venture capitalists, and CEOs looking at themselves in the mirror today.'" Peter Dickens, a renowned European neo-Marxist, has a more balanced assessment (1998: p. 191): "According to this second contradiction argument, nature will continue to wreak 'revenge' on society as a result of capitalism. Several related questions remain, however. First, will capitalism be able to restructure itself once more, this time in the form of what has been called, 'ecological modernization'?" Leff takes the discussion of ecological modernization one step further. From a neo-Marxist perspective, he initially resists simply incorporating environmental concerns into global capitalist development (through standard economic means such as the internalization of externalities), but finally reaches the conclusion that an environmentally sound development is not "totally incompatible with capitalist production" (Leff 1995: p. 126).

Demodernization and Antimodernization Perspectives

Scholars adopting demodernization and antimodernization perspectives, often building on neo-Marxist analyses, also focus on contradictions in the capitalist economic system. If these demodernization scholars depart from neo-Marxist perspectives, it is because they claim that neo-Marxist analyses are incomplete. A group of counterproductivity theorists have criticized neo-Marxist analyses from a "radical" demodernization perspective (Spaargaren and Mol 1992; Mol 1995). These authors include Murray Bookchin, Ivan Illich, the later André Gorz, the earlier Rudolf Bahro, Otto Ullrich, Wolfgang Sachs, and Hans Achterhuis, and their ideas have resonated throughout the environmental movement from the 1970s to today. Otto Ullrich (1979), for example, in his book *Welt-niveau*, criticized Marxists for their preoccupation with the social relations of production, and their corresponding inattention to the forces of production. In Ullrich's view, the analysis of environmental crises ought to incorporate the "myth of the great machine" embodied in the organization of the industrial system, to understand why the effects of the system of production are contradictory to the goals for which it was designed. The industrial system is minutely administered, Ullrich argued, in an ever more centralized, hierarchical way, which reflects the imperatives of the technical systems that are omnipresent in the system of production, but that are no longer adapted to the demands of humans and nature.

The solutions that demodernization or counterproductivity theorists advocated did not emerge from an analysis of existing tendencies in contemporary society. Most scholars in this tradition agreed that we were and still are moving further into modernity, creating catastrophic side effects. The core of the demodernization ideas focused rather strongly on the normative and prescriptive analyses of the changes and transformations necessary to maintain society's resource base. What the normative stances of demodernization theorists have in common with environmentalists in the modern traditions (discussed later) is their call for upgrading environmental criteria and introducing environmental perspectives and rationalities in designing future institutions and social practices. It is exactly against this idea of a new central, leading principle that *post*modernists argue.

Postmodern Critiques and Perspectives

According to postmodernists, if "sustainability" is taking such a central position in diagnoses of the present and prescriptions for the future, there is a new "grand narrative" in the making. When formulated in this way (de Ruiter 1988), it becomes clear why postmodern authors are among the fiercest critics of modernist approaches to environmental problems. They see many schemes for dealing with environmental problems, as remnants of the old modernization theories that dominated the 1950s and 1960s and as an extension of the much troubled Enlightenment. Postmoderns have directly challenged the knowledge claims that are the foundation of ecological transformations. Postmodern critiques are in some respects even more radical than those of counterproductivity theorists because they flatly deny that sustainability criteria could or should be developed in any way. A recent, rather radical exponent of this position, Blühdorn (re)starts the debate on what exactly is the ecological problem, and ends up with the conclusion that environmental problems are no longer there. "To the extent that we manage to get used to (naturalize) the non-availability of universally valid normative standards, the ecological problem . . . simply dissolves" (Blühdorn 2000: p. 217). Large segments of contemporary society no longer see environmental change as problematic, or at least not in any universal way. According to postmodernists, this diversity of environmental-problem definitions radically devalues any ecological critique of modern developments, even though few members of contemporary, postmodern societies fully acknowledge this consequence.

These radical postmodernists want to hammer home the point that the distinction between society and its natural environment is always a time- and space-bound "social construction." No distinction can be made between more or less objective, true, or widely held intersubjective understandings of reality, including the understandings of the environment. According to postmodern thinking, every grand narrative can and should be deconstructed and shown to be arbitrary.

Yet, the rather imprecise and loose use of the label "postmodern" frustrates any thorough evaluation of the postmodern tradition. For example, Zygmunt Bauman (1993: pp. 186–222) considers himself a postmodernist, although his definition of environmental problems and his elaborations of desirable solutions resemble deindustrialization and demodernization ideas, rather than the postmodernism of Blühdorn. Bauman shares with both the radical postmoderns and the de- or antimodernists a strong rejection of modernity and modernization as relevant categories for environmental reform. Not surprisingly, Bauman also strongly criticizes reflexive modernization, especially its aim to "save" modernity.

Reflexive Modernization

If Ulrich Beck did not invent the concept of reflexive modernization, he certainly brought it to the center of present-day social theory with his book *Risikogesellschaft* (Beck 1986). According to Beck, reflexive modernization entails the "self-confrontation" of modern society with the negative consequences of modernization, among which is the environmental crisis. While the distribution of goods and prosperity (and conflicts about them) is a crucial factor in the constitution of industrial society during high modernity, with the transition to reflexive modernity it is conflicts over risks that dominate. Risks become a dominant feature of everyday life, causing paralyzing feelings of anxiety among large groups of individuals. And the risks produced by modern institutions strike these very institutions like a boomerang; social conflicts about environmental and technological risks are in essence conflicts about the social and economic consequences of risk *management*, and can thus threaten the responsible modern institutions: the state, science and technology, and the market economy.

Anthony Giddens unmistakably feels an affinity with Beck's work (Giddens 1990, 1994a). He parallels Beck to a considerable extent in emphasizing the changing "risk profile" of modern society, in which scientific and technological developments have reduced many premodern risks such as famine and natural disasters, but at the same time have increased new types of ecological risks. However, Giddens balances Beck's apocalyptic risk society scenario by emphasizing the transformations of social institutions in order to deal with these new risks. These institutional transformations are the central focus of ecological modernization.

Ecological modernization theorists identify the institutions of modernity, not only as the main causes of environmental problems but also as the principal instruments of ecological reform. At the same time, these institutions are themselves transformed through the process of ecological restructuring. Economic institutions such as the commodity and labor markets, regulatory institutions such as the state, and even science and technology are transformed in that they take on characteristics that diverge from their productivity-oriented predecessors. The constant influx of new information about the ecological consequences of social practices and institutional arrangements results in a continual redirection of the core institutions of modernity. In this sense these institutions have lost their "simple modernization" character and are open for continual restructuring and redefinition according to environment-inspired requirements. Ecological modernization can thus be interpreted as the reflective reorganization of industrial society's institutions to cope with the ecological crisis. It is open to empirical investigation whether this ongoing institutional restructuring and these institutional learning processes can overcome the self-destructive tendencies of industrial society (Beck 1986, 1994). Similarly, it is an open question to what extent modern institutions will be transformed.

Although there exists a certain tension between the more apocalyptic undertones of Beck's risk society and the gradualist perspective of ecological modernization (Mol and Spaargaren 1993), the two views do not fundamentally contradict each other as some have argued (e.g., Blowers 1997; Buttel 2000). Both strains of reflexive modernization—in sharp contrast to proponents of de- and postmodernization—share the perspective that all ways out of the ecological crisis will lead further into modernity.

Ecological Modernization: How the Environment Moves into and Transforms the Modernization Process

In broad agreement with reflexive modernization, the ecological modernization perspective analyzes the transformation of modernity as a result of the growing importance of environmental considerations and interests in society. This section surveys ecological modernization and locates this perspective in relation to reflexive modernization.

Ecological Modernization Theory

The basic premise in ecological modernization theory is the centripetal movement of ecological interests, ideas, and considerations in social practices and institutional developments, which results in the constant ecological restructuring of modern society. Ecological restructuring refers to the ecology-inspired and environment-induced processes of transformation and reform of the central institutions and social practices of modern society. Institutional restructuring should, of course, not be interpreted as a new phenomenon in modern societies, but rather as a continuous process that has accelerated in the phase of reflexive modernity. According to ecological modernization scholars, the present phase (roughly since the 1980s) is distinctive because of the centrality of environmental considerations in these institutional transformations.

Ecological modernization theorists echo a Weberian view in drawing attention to the growing autonomy of an ecological sphere and a growing independence of ecological rationality in relation to other spheres and rationalities (Mol 1995, 1996; Spaargaren 1997). In the domains of policies and ideologies, some notable environment-informed changes took place beginning in the 1970s. Most environmental ministries and departments, as well as many environmental laws and environmental planning, date from that era. While a separate "green" ideology—manifested in environmental nongovernment organizations (NGOs) and environmental periodicals—started to emerge in the 1970s, in the 1980s this ideology became more and more independent from—and could no longer be interpreted in terms of—the old political ideologies of socialism, liberalism, and conservatism (Paehlke 1989; Giddens 1994b).

The crucial transformation, which makes the notion of growing autonomy of the ecological sphere and rationality especially relevant, is of even more recent origin. In the 1990s, the ecological sphere and ecological rationality grew increasingly independent from the economic sphere and economic rationality, the bedrock as it were of classic modernization.

The consequence will be that slowly but steadily economic processes of production and consumption will be and indeed are increasingly designed, organized, analyzed, and judged from an economic *and* an ecological point of view. From the late 1980s onward, institutional changes have started to appear in the economic domain of production and consumption (as discussed later). The claim that we should analyze these transformations as *institutional* changes recognizes their semipermanent character. Although the process of ecology-induced transformation should not be interpreted as linear and irreversible (as was common in the modernization theories in the 1950s and 1960s), the changes have some permanency and are difficult to reverse.

Ecological Transformation Processes: Core Features

Most studies adopting an ecological modernization framework focus empirically on environment-induced transformations in modern social practices and institutions. The core features of such transformations— including the main dynamics, actors, and mechanisms—can be described by five heuristics. Taken together, these core features distinguish ecological modernization from neo-Marxist, demodernization, and postmodern ideas.

• Science and technology become contributors to environmental reform. First, science and technology are not only judged for their role in causing environmental problems but also are valued for their actual and potential role in curing and preventing them. Second, conventional curative and repair options (such as "end-of-pipe" technologies) are replaced by more preventive sociotechnological approaches that incorporate environmental considerations from the design stage onward. Finally, despite a growing uncertainty with regard to scientific and expert knowledge concerning environmental problems, there is continued appreciation of the contributions of science and technology to environmental reform.

• Economic and market dynamics and economic agents gain in importance. Producers, customers, consumers, credit institutions, insurance companies, the utility sector, and business associations increasingly turn into social carriers of ecological restructuring, innovation, and reform (in addition to state agencies and new social movements; cf. Mol and Spaargaren 2000).

• The modern "environmental state" (Mol and Buttel 2002) is transformed. First, there is a trend toward decentralized, flexible, and consensual styles of national governance at the expense of top-down hierarchical command-and-control regulation, a trend sometimes referred to as "political modernization" (Jänicke 1993). Second, there is greater involvement of nonstate actors in the conventional tasks of the nation-state, including privatization, conflict resolution by business-environmental NGO coalitions, and the emergence of "subpolitics" (Beck 1994). Finally, there is an emerging role for international and supranational institutions that to some extent undermines the sovereign role of the nation-state in environmental reform.

• New positions, roles, and ideologies for environmental movements emerge in the processes of ecological transformation. Instead of positioning themselves on the periphery or even outside the central decision-making institutions, environmental movements become increasingly involved in decision-making processes within the state and to a lesser extent the market. This is accompanied by a bipolar or dualistic strategy of cooperation and conflict, and the resulting internal debates and tensions (Mol 2000).

• There are changing discourses. New discursive practices and new ideologies emerge in political and societal arenas, where neither the fundamental counterpositioning of economic and environmental interests nor a total disregard for the importance of environmental considerations are accepted any longer as legitimate positions (Hajer 1995). Intergenerational solidarity in preserving the sustenance base emerges as the undisputed core and common principle.

These five heuristics, which together describe ecological modernization, can be used in analyzing and describing specific sectors, such as chemical production and consumption in Europe. Some scholars and political agents also apply these heuristics as normative paths for change, using them to construct a desirable route to a sustainable future. In the next section I focus especially on the analytical and descriptive (rather than normative) qualities of ecological modernization.

From Theory to Practice: Transformations in Chemical Production and Consumption in Europe

Although the origins of the chemical industry can be traced back to the sixteenth century, it expanded significantly in Europe during the industrial revolution in the nineteenth century. While France had been a major producer of chemicals in the late eighteenth century, Great Britain and later Germany took over in the nineteenth century. Today, the United States, Germany, the United Kingdom, Japan, Italy, Switzerland, and the Netherlands are usually mentioned among the top chemical-producing countries. Although developments in industrial nations were far from homogeneous, both spatially and temporally, most contemporary industrial countries have acquired a chemical industry of a more or less similar structure (if not size). Consumption of chemicals, chemical products, and goods containing significant amounts of chemicals is spread worldwide. The chemical industry and its products have been and still are notorious for their damage to the environment. Since its early stage, chemical production has caused severe environmental deterioration and led to large public protests. Environmental movements have recently targeted chemical products such as pesticides, coloring agents, polyvinyl chloride (PVC), chlorofluorocarbons, and organic solvents, to name a few.

Only from the 1980s onward can one really speak sensibly of the ecological restructuring of chemical production and products. Even so, this reform process did not reduce anxieties about chemical dangers and risks among significant segments of the population. I first look at the scope of this environmental reform process in western Europe and then analyze the main dynamics behind these transformations.

Ecological Reform: Quality and Degree

In the past 15 years, important changes have occurred in individual chemical companies and at the level of the chemical sector. The majority of western European chemical companies have established environmental management systems that are coordinated by in-house environmental, health, and safety officers and departments, although this is true to a lesser extent in the smaller chemical industries in Europe (Franke and

Wätzold 1995). Company strategies frequently include monitoring and management of the in- and outflow of materials and energy, alongside more traditional strategic concerns such as monitoring and management of financial (capital) and human resources. New instruments such as annual environmental reports, environmental certification systems, and environmental audits have become common. In the Netherlands, for instance, 119 out of the 143 chemical firms produced an annual environmental report for 1999. The same number of companies (119) had an environmental management system, but only 43 (36 percent) of these were certified according to International Standards Organization (ISO) 14000, the European Environmental Management and Audit System (EMAS) guidelines, or British Standard 7750 (FO Industrie 2000). Similar developments can be identified in other western European countries. Companies have appointed special environmental officials to translate general environmental requirements—often set by government agencies—into operating specifications and criteria.

Company expenditures on environmental measures and investments have increased during the past decade, both in absolute and relative terms. Company expenditures on environmental measures, which typically were 10 percent of total annual investment in the early 1990s, are about 15 percent at present and are expected to increase to 20 percent in the coming decade (Commission of the European Communities 1993, 1997). In addition, research and development resources have been reoriented toward the environment. In the pesticides industry, R&D resources spent on the environment have skyrocketed with the development and introduction of new products. The expansion has been considerable in other chemical sectors, too. Although definitions of environment-oriented R&D vary, most authors and most chemical firms claim that between 30 and 80 percent of company R&D costs are related to the environment (Mol 1995).

Ecological reform can be seen not only in these investment activities but also in internal company decision making. The development and introduction of new products that do not have clear environmental benefits, managers of chemical companies indicate, will be vetoed in the internal decision-making process because the commercial risks are too high. It is now standard practice to conduct ex ante ecological evaluations of new products (sometimes via life-cycle analysis) and environmental audits of production sites. These exercises can result in modifications in the kinds of raw materials used and the design of new production processes. In addition, chemical industries have engaged in new activities. For instance, polymer producers have investigated new recycling technologies for plastics; many have acquired a majority share in plastics recycling companies. These technical, economic, and organizational changes at the company level clearly do not consist of merely tinkering with an existing development path. They should rather be interpreted as the precursors of a broader industrywide transformation. ·

Viewed from the aggregated sectoral level, the environment has become an increasing factor in the competition among chemical companies. For example, low organic solvent paints (including water-based paints, high solids, and radiation-cured systems) have successfully challenged the market for traditional organic solvent paints. While small niche-market firms initiated the production of low organic solvent paints, all the major European paint companies have by now complemented their conventional paints with the new products or switched to these new paint systems. This reform enabled the Dutch government to ban organic solvent-based paints from the professional markets. Some small traditional paint companies lacked the resources and expertise to develop such more ecologically sound paint systems, and some of them were taken over or even collapsed. Producers of PVC plastics have lost market share to producers of polypropylene and polyethylene (PP and PE). The unsatisfactory environmental performance of PVC, in the view of influential sectors of society, is the main cause of this shift in market shares, especially in Germany, the Netherlands, and Denmark.

The environmental initiatives of governments have added entirely new dimensions to chemical-sector competition. Recycling requirements affect the product development and polymer choice of plastics manufacturers as well as industrial end users such as the automobile industry. Recycling requirements also led to the emergence of fixed contracts between polymer producers, industrial end users, and recycling companies, changing the industry's structure and limiting free competition. The mandatory registration of pesticides and especially the related costly R&D

on environmental effects resulted in an acceleration in (de)merging and joint ventures among pesticide industries in the 1980s (Mol 1995). One of the consequences is that so-called active ingredient production has almost disappeared from the Netherlands (while it has become concentrated in France, Germany, and the United Kingdom).

Besides these new frontiers of competition, cooperation within the chemical industry has been augmented in environmental matters. Industry or trade associations, both at the national and the European Union (EU) level, have stepped up their environmental activities and often doubled their staff to do so. The industry's negotiations with regulatory agencies are often coordinated by these so-called branch associations, which also handle public relations and communications with other interest groups and the wider public. The Responsible Care program—coordinated by the Council of European Federations of the Chemical Industry, known also as CEFIC from the acronyn in French (CEFIC 1999)—is among the best known of these communication programs. In addition, branch organizations have begun to engage in the translation of regulatory requirements down to the level of individual companies, to some extent evolving into a kind of neocorporatist organization in environmental politics.

Last but not least, decreases in emissions and wastes, and the reuse and recycling of waste, should be seen as indicators of environmental reform. But, in the best tradition of the disenchantment of science, often it is not easy to obtain reliable data for the European chemical industry or for national chemical industries (for some examples, see CEFIC 1999; FO Industrie 2000; European Environmental Agency 1998). Most data show decreasing emissions for most substances throughout the 1990s, although in a few cases growing production volumes offset decreasing emissions per unit of output (e.g., greenhouse gases).

Transformation Processes: Actors and Dynamics
This ecological restructuring of the chemical industry can be understood as indicating the growing importance of ecological factors and arguments in industrial development in relation to economic ones, although the latter of course will remain dominant for some time. The chemical industry has institutionalized this increasing importance of ecology

through various mechanisms, dynamics, and actors (Mol 1995; Paquiet et al. 1996).

Within the market for chemical products, the environment has become a relatively independent factor that can no longer be controlled by economic factors. Consumers of chemical products articulate demands from both economic and ecological points of view; conventional economic and quality criteria have been extended to environmental standards. Consumer organizations are including environmental criteria in their testing and evaluation of product quality. Customers not only ask for environmentally sound products but also are starting to demand environmentally sound chemical production processes in the form of certified environmental management and audit schemes and environmental product specifications. Companies are responding to these new dimensions of consumer and customer demand with new marketing strategies, new product information standards, and changing advertisement designs.

The environment has also exerted an influence on financial markets. Insurance companies increasingly carry out an environmental audit before they insure chemical industries. Indeed, international insurance companies are among the main defenders of the Kyoto Protocol. In some cases, financial organizations such as banks make investment loans conditional on an environmental evaluation. However, chemical producers should not be seen as purely reactive actors, confronted with an "ecologized" market demand, for they have partly created these new demands. Specialized chemical producers have identified many niche markets for environmentally sensitive products, while large transnational chemical companies see environmental specifications as an area of competition. Incidentally, employees within chemical industries play a significant role in initiating and implementing these ecological transformations (Baylis et al. 1998a; Wingelaar and Mol 1997).

Besides these economic factors, governmental measures, public pressures articulated by NGOs, and international developments are also shaping the pace of environmental transformations. Governmental interventions in chemical production and products have a dual aspect. At times, authorities still follow the traditional line of command-and-control (regulation and enforcement), while sometimes more communicative and cooperative strategies have emerged. The latter negotiations often involve long-term agreements with the chemical sector on overall environmental goals, taking into account the sector's (technological) knowledge, environmental information, and preferences on time paths and (technological) measures. The move to a larger degree of flexibility and self-regulation in environmental policy seems to work especially well for the large chemical complexes that have well-organized internal environmental monitoring and management systems, and where government agencies do not have sufficient knowledge, monitoring devices, and manpower for direct regulation. Liability policies have reinforced this cooperative strategy, stimulating some chemical companies to use "white lists" (instead of black lists) for chemical substances allowed in their products. The division between the two modes of intervention differs from country to country, depending on policy style and political culture (see Franke and Wätzold 1995).

A central characteristic of contemporary ecological reform is that the quest for environmental improvements does not have to be continuously enforced by the state, since environmental concerns have become institutionalized (to some extent) in economic practices, as attested by the examples of insurance companies, consumer demand, and liability policy. This institutionalization would become even greater if the most powerful mechanism in capitalist market institutions—prices—was used on a larger scale. Until now, price differences reflecting ecological standards have been introduced by regulatory organizations (for instance by means of different value-added tax percentages, taxes, or deposit systems), to different extents in the various EU countries (see Ekins and Speck 2000). Nevertheless concerns about "competitiveness" have largely exempted the most heavily polluting sectors from these new price signals (e.g., on energy use or CO_2 emissions), and economic incentives at the EU level are mired in political debate. The major chemical producers have so far resisted major tax reforms.

Despite the improved ecological performance of the chemical industry, and its continuing institutional transformations, the industry still generates powerful feelings of insecurity and anxiety in the public. Recent polls by both independent scientific institutes and chemical associations indicate that the public remains wary of chemicals and the chemical industry because of their environmental risks. The generally reassuring messages coming from risk assessments, life-cycle analysis, and scientific-technological control and management of the chemical industry's expert systems are challenged time and again by counterexpertise as well as chemical accidents. And these challenges have moved up to a global scale. While in the 1950s and 1960s chemical risks were primarily of local origin, since the 1980s they have become global through the pervasiveness of far-flung food and commodity chains (including pesticides), international transportation of bulk chemicals, and global ecological interdependencies (such as depletion of the ozone layer).

The adherents of Ulrich Beck's risk society theory may rightly conclude that these confrontations with chemicals and chemical production, in almost every aspect of daily life, have not resulted in an unquestioning, basic trust in the chemical industry. Still, one searches in vain to locate any massive movement away from a lifestyle dependent on chemicals or to find signs of fundamental distrust in the scientific foundations underlying the chemical industry. Protests in the 1970s against the plasticized "throwaway society," accompanied by calls for the dismantling of chemical production, contrast markedly with today's scientifically informed counterexpertise and scientific controversies on specific product and processing alternatives (for instance, on PVC, see Tukker 1999; Bras-Klapwijk 1999). And contemporary environmental NGOs strongly support a sustainable chemical industry instead of requesting the dismantling of chemical production. Perhaps only in the natural food sector do we see serious initiatives to abolish chemicals (mainly pesticides and chemical fertilizers).

Despite the many signs of change, the ecological transformation of the chemical industry is only beginning. The chemical industry is still challenged by critics of its ecological performance, and we are nowhere near a sustainable chemical industry, as most data on emissions, environmental quality, and accidents show. The point is that first, transformation processes in the chemical industry are to a significant extent informed by environmental considerations and second, this ecological transformation is a process involving (and transforming) the institutions of modernity. In that sense the ecological restructuring of the chemical industry resembles what has been labeled the "modernization of modernity" (Beck 1986).

Other, once-promising alternatives seem to have come to a dead end, as will become evident in the next section.

Soft Chemistry: A Stagnating Alternative for Restructuring

The most clearly defined alternative to an ecological modernization of the chemical industry is the idea of soft chemistry. Soft chemistry (*sanfte chemie*) resembles Lovins' soft energy path and Ullrich's alternative for *sackgasse* (dead end) technology (Lovins 1977; Ullrich 1979: pp. 149ff). Soft chemistry moves away from some of the central characteristics of modern technological systems and revitalizes the environmental concepts that were prominent in the early 1970s.

According to von Gleich (1988, 1991), one of the founders and interpreters of the soft chemistry paradigm, three criteria distinguish "soft" from "hard" chemical science and technology. First, soft chemical technology intervenes only superficially, less profoundly, in chemical structures. The level or degree of intervention (*Eingriffstiefe*) of hard chemical technology has had three consequences: increasing the power of humans over nature, increasing the potentials of risk because of extended time-space dimensions and irreversibility, and increasing the gap between the scope of our knowledge of nature and the scope of our intervention in nature. The fact that the intervention level of soft chemistry is less deep means that soft chemical technologies, in von Gleich's view, retain a use-dependent neutrality: their negative consequences are not inherently related to the technology itself, but rather to its application.

Second, soft chemical technology is distinguished by its instrumental character (*Werkzeugcharakter*), i.e., the possibility for production-level workers (the primary producers) to use and control the properties of the natural resources used in production. Whereas hard chemical technology requires standardized and uniform natural resources and Fordist production processes in which primary producers and natural resources are adapted to production technology, soft chemical technology is adapted to the properties of the natural resources as they are found. Finally, soft chemical technology makes use of the coproductivity (*Mitproduktivität*) of nature, incorporating biological and ecological processes as an integral part of chemical production. Soft chemistry thus departs from a modernist, instrumental view of nature as an element to be controlled by or excluded from industrial processes.

Despite its obvious appeal to advocates of alternative technologies, the idea of soft chemistry has found few applications. The production of so-called natural paints was the most promising soft chemical technology, but even during the wave of environmental consciousness in the late 1980s and early 1990s the market for natural paints did not rise above a 1 percent share of the European market. Even in Germany and the Netherlands, where state programs aimed to improve the environmental performance of conventional paint systems, government authorities have been hesitant to support natural paints, partly because of their inferior product quality and their poor environmental performance, but also because government authorities wish to maintain good relations with the regular chemical paint industry. Moreover, the mainstream environmental movements in Germany and the Netherlands have founded their ideology, not on soft chemistry, but on the environmental modernization of the chemical industry. Natural paints have found only meager backing from environmental organizations and have even been fiercely criticized by these organizations. In other chemical sectors and products, such soft chemistry plays an even more limited role. Soft chemistry, then, as a "way out" of the environmental crisis or as an alternative to ecological modernization, seems to possess little descriptive power or normative value.

Environmental Transformation of Modernity: Sectoral and Regional Variations

My analysis of ecological modernization in European Union countries raises at least two questions of representativeness. Can we analyze or expect similar dynamics and transformation processes in other (industrial) sectors in Europe? And to what extent do these ecological modernization processes differ among countries and regions? This section tries to give a preliminary answer to these questions.

Sectoral Variations
In evaluating distinct industrial sectors in one country, I focus on the characteristics of the sector (e.g., economic structure and organization)

instead of making qualified estimates as to which sector(s) might have ecological restructuring processes comparable to that of the chemical industry. Three sectoral dimensions are relevant.

First, we must consider the contribution of the sector under analysis to (national) environmental problems. Of course the sector's objective contribution to environmental problems is relevant, but even more important is the perception of its contribution held by key collective actors: government agencies, environmental organizations, the media, consumers, and citizens. A sector perceived by these social actors to be an environmental problem inevitably faces greater pressures to alter its production processes and products. Taking this into account, it is no surprise that the chemical sector, facing intense pressure, has instituted the most far-reaching ecological modernization.

Second, distinct sectors differ fundamentally in their structural qualities, including such features as production processes and organization, economic concentration, and vertical integration. Keith Pavitt's (1984) well-known four-part typology of industrial sectors might be useful in indicating how different structural qualities of industrial sectors either promote or interfere with ecological restructuring:

- In *supplier-dominated* sectors such as textiles, clothing, leather, printing and publishing, and wood products, innovations are mainly triggered by the diffusion of capital goods and innovative intermediate products, while R&D expenditures as such are limited. Suppliers dominate the development of these sectors to a considerable extent. Endogenous changes relate to incremental improvements in equipment, procedures, and organization. The firms are typically not very large.

- *Specialized suppliers*, such as mechanical and instrument engineering firms, mainly aim at product innovations that enter other sectors in the form of capital inputs. These firms tend to be small and knowledge intensive and maintain close relations with their customers. Product innovations are carefully coordinated with these customers, while fundamental process innovations are relatively scarce.

- *Scale-intensive* sectors generally involve complex systems in which economies of scale are significant (in R&D, design, production, or distribution). Companies tend to be large; they tend to integrate vertically;

and they typically produce their own production technologies. Significant financial and personnel resources are dedicated to innovation (both product and process). This category includes transport equipment, electronic durables, metal manufacturing, food products, and glass production.

• *Science-based* sectors, such as the chemical, pharmaceutical, electronics, and bioengineering industries, are characterized by large R&D departments with large R&D resources. Their innovative products enter a large number of sectors as intermediate inputs or capital. The companies tend to be large, with the exception of new "Schumpeterian" or entrepreneurial ventures (e.g., bioengineering).

Third, the international orientation of the sector is highly relevant. Changes in an internationally oriented sector such as the chemical industry will be triggered to a considerable extent by international competition and policy processes, while the international context is obviously less relevant for more nationally oriented industrial sectors. Individual chemical corporations, in striking contrast to companies in most other industrial sectors, are international political actors in their own right, with direct contacts, for instance, to European Union institutions at the highest levels.

Empirical research on ecological modernization processes in different industrial sectors is just beginning. In the United Kingdom, Baylis and his colleagues (1998a,b) found that the chemical, electrical, and electronics industries had most thoroughly embraced environmental transformations, while mechanical engineering was a clear laggard and the food sector was intermediate. The food sector's concern for food safety actually interfered with environmental protection, with major efforts in product innovation but few efforts in reshaping production processes to reduce emissions. Baylis and his colleagues concluded that the driving forces for environmental change in these sectors are conventional regulation, pressures from customers and consumers (especially in the food sector), and wider public concern. In addition, they found that large companies, and sectors with many large companies, are the leaders in environment-informed transformations. Similar differences among sectors were reported by Boons et al. (2000) in a wide-ranging survey of all

Dutch studies in environment-induced transformations. This study found that the kind of ecological modernization as well as the mechanisms, dynamics, and actors involved in these transformations varied among industrial sectors and industries.

Globalization and Regional Styles

Earlier I outlined the principal ecological modernization heuristics, which are clearly visible in the western European countries. To what extent can these heuristics shed light on environmental transformation processes outside western Europe?

One would expect significant differences in ecological modernization outside Europe, owing to distinct national and regional institutions. These institutional differences include profound variations in state–market relations, ranging from "developmental" states to "predatory" states, or from a "Rheinländisch" model to an Anglo-Saxon model (Evans 1995; Staute 1997). In addition there are distinct national policy styles (Richardson 1982; Vogel 1986; van Waarden 1995); varying regimes of accumulation (Lipietz 1987); and different national systems of innovation, with their nation-specific network of institutions that initiate, import, modify, and diffuse new technologies (Nelson and Rosenberg 1993; Edquist 1997). Furthermore, we may need to be alert for distinct "national characters" (Cohen 2000). Variations in any of these aspects will give ecological modernization a specific national or regional flavor.

At the same time, these national and regional variations may be offset or even homogenized by the processes of globalization. To the extent that globalization processes continue in the domains of economy (via global markets and transnational corporations), of policy and governance (by transnational institutions and multilateral agreements), and of culture (by norm and value formation in a global civil society), there will be institutional pressures that might make ecological modernization similar across countries or regions. Garcia-Johnson's (2000) recent study on the export of environmentalism by U.S. chemical companies to Brazil and Mexico demonstrates how global economic dynamics can produce similarity in environmental reforms. Globalization processes mean that the dynamics of ecological modernization (e.g., shifting technological paths, new forms of environmental governance, internalization of external effects, new environmental roles of market actors) are increasingly looking similar in very different modern societies. Even so, the heuristics that govern environmental reform will always be codetermined by national and regional characteristics and institutions: one might well speak of environmental "glocalization."

Studying and defining regional (or national) variations or styles of ecological modernization seems to me a promising middle course that will enable us to see how far the commonalities of ecological modernization reach, where the specifics of the regional variation start, and how the commonalities and variations change across time.

12

Technology, Modernity, and Development: Creating Social Capabilities in a POLIS

Haider A. Khan

He puts his engine [a watch] to our ears, which made an incessant noise like that of a water-mill; and we conjecture it is either some unknown animal, or the god that he worships; but we are more inclined to the latter opinion.—Jonathan Swift, *Gulliver's Travels* (1726)

We Westerners are absolutely different from others!—such is the moderns' victory cry, or protracted lament. They do not claim merely that they differ from others as the Sioux differ from the Algonquines, or the Baoules from the Lapps, but that they differ radically, absolutely, to the extent that Westerners can be lined up on one side and all the cultures on the other, since the latter all have in common the fact that they are precisely cultures among others. In Westerners' eyes the West, and the west alone, is not a culture, not merely a culture.—Bruno Latour, *We Have Never Been Modern* (1993)

E. M. Forster, in his novel *Passage to India* observes that the "restfulness of an Indian gesture reveals a civilization which the West can disturb but will never acquire" (Forster 1924: pp. 251–252). In this brief passage Forster sets up the dichotomies between East and West in terms that can reveal in their ambivalence a deconstructive gesture, perhaps unknown to its individual "western" author, that has the potential to interrogate the key terms of this essay. For technology, modernity, development—all three of these terms—are to be written under such a deconstructive gesture. The tension that is inherent whenever such theoretical terms are used with intended empirical correlates comes quickly to the fore under such writing *sous rature*, which is also a way of putting these terms inside question marks.

Such is the intention of this essay.[1] Yet this venture is fraught with the perils of relativizing these terms, erasing their meaning, and perhaps ultimately pointing to nihilism. How does one give the study of technology, modernity, and development an "empirical turn" under such circumstances? Reflection on this question soon leads to another. How does one offer a theory of technology, modernity, and development so that correct empirical applications are indeed within reach? Therefore, the first task is to take seriously the tensions within the theoretical terms that lead to a real threat of their dissolution (destruction, in Heidegger's terms). An analysis of these tensions and a way out of them would seem to be the minimum theoretical conditions for an empirical turn to be meaningful. Such an analysis also carries the potential for exposing the contradictions inherent in modernization theories and the conventional dichotomies encapsulated by the simple oppositional formula of an East–West dichotomy.

Accordingly, I begin with some analysis of the theoretical connections among technology, modernity, and development in a nonwestern context. The discussion here is intended to suggest some methodological aspects of connecting theories of modernity with empirical approaches in the context of technology and development. Of particular significance are the modern and postmodern aspects of technological development in the newly industrialized economies (or NIEs). I try to draw some lessons from my own studies of the Taiwanese innovation system. The argument presented here will entail some suggestions for future directions of empirical investigations in other nonwestern societies and some warnings about possible pitfalls in this type of work. Throughout the essay, the idea of technological systems as social, economic, and political constructions that are historically path dependent plays a crucial role.[2]

Technology, (Post) Modernity, and Development: The Western–Nonwestern Distinction

If the very nature of technology is put under scrutiny and thus problematized, we end up with what Hughes (1987: p. 51) calls "messy, complex, problem-solving components." Any such system gives rise to problems ("reverse salients" in Hughes's terminology) that require further negotiations. Pinch and Bijker (1987) underline the historicity and social construction of the very idea of the "safety bicycle" in the nineteenth century during a protracted process of problem formulation, stabilization, and (social) closure. Thus, they point out:

The "invention" of the safety bicycle was not an isolated event (1884), but a nineteen-year process (1879–98). For example, at the beginning of this period the relevant groups did not see the "safety bicycle" but a wide range of bi- and tricycles—and, among these, a rather ugly crocodilelike bicycle with a relatively low front wheel and rear chain drive. By the end of the period, the phrase "safety bicycle" denoted a low-wheeled bicycle with rear chain drive, diamond frame, and air tires. As a result of the stabilization of the artifact after 1898, one did not need to specify these details: They were taken for granted as the essential "ingredients" of the safety bicycle. (Pinch and Bijker 1987: p. 39)

If technology as a theoretical term and its empirical correlates are thus shown to be socially contested and constructed over significant time intervals, the connections between technology and modernity are twice problematized. In the first place, the ensemble of attitudes and institutions that are assumed to be coterminous with the idea of modernity are themselves in flux and need to be described as a system in motion. In the second place, technology, as a crucial dynamic component of modernity, is also destabilized and is itself destabilizing; it can find stabilization only through a historically contingent interplay of social forces. In this context, what the economic historians like Paul David (1985) or theorists such as Brian Arthur (1994) have called "path dependence" needs clear articulation and focus in the social construction of temporal sequences of technological systems and subsystems. These sequences can now be seen as elements chosen from a large set of intertemporally connected technologies. Indeed, one can redefine the traditional choice set of microeconomic theory from one that deals only with an object of individual choice to the social choice of technology in historically conditioned and socially contingent circumstances. This redefinition of choice, as we will see, can have rather profound theoretical and empirical consequences.

Recognition of the problem of path dependence as a problem of sequencing (and subsequencing) of historically and socially contingent technological paths at once creates an opportunity and a seeming impasse. Focusing on the contingency aspect gives a strong justification for specific, empirically grounded projects in technology studies. However, if contingency is all there is, then there seems to be no way out of pure (Humean) empiricism. Avoiding such an empiricist (as opposed to merely empirical) turn requires an epistemological grounding beyond that of sense data. However, this is precisely where the dangers of a subjectivist neo-Kantianism or dogmatically objectivist realism become the greatest. This is also the point where debates on modernity and postmodernity often lose their way in a nihilistic type of relativism (see Lyotard 1984a, 1988, 1993; Derrida 1981, 1988; Rorty 1989).[3]

One way out of the impasse is to take the idea of freedom as a key feature of technology, modernity, and development (Sen 1992, 1999; Khan 1998). As we will see, we are by no means free to characterize freedom any way we like. However, for the moment I want to take freedom as a primitive term and examine the claim that it is a key ingredient of all three of my principal terms. For this purpose, and for this purpose only, it is sufficient to think of freedom as an extension of the scope of action for the individual, society, or nature.

Limiting ourselves thus to such a thought experiment with freedom, we can think of technology as extending our scope of action over space and time. Such an extension is institutionalized in the history of development in the West through a coherent set of social, economic, and political institutions and articulation of ideologies of modernity. The contested rise of modernity in technology politics (Schot, Chapter 9, this volume) describes the complex forms such articulations and practices can take. Development, in this (western) sense, then, means the extension of similar types of technological progress embodied in similar types of institutions and expressed in similar ideational forms, for example, the modernization ideology.

The extensive debates on modernization, westernization, and progress have put these ideas through much critical sifting. Very little that is intellectually coherent remains after the colonizing (sometimes racist), imperialist, and patronizing shibboleths are laid to rest. In fact, the very dichotomy of East versus West seems to be a peculiar western hegemonic construction that remained uncontested only as long as the incoherence of modernity itself remained unrecognized.[4] One way to read (sympathetically) the modern versus postmodern debate and the jihad versus McWorld type of polarization (Barber 1995; Khan 1996, 1998) is to see in these concepts a visceral response to modernity both in the West and in the East.

Yet what is often lost in the intense heat of such debates is the intuition that the notions of technology, modernity, and development all have to do with enhancing a complex sort of freedom. It is only through clarification of the meaning of freedom and its connection with the other terms that we can hope, if at all, to avoid the destructive (again in Heidegger's sense) implications of postmodern gestures. To avoid the simplistic slogans of modernization versus antimodernization, westernization versus nativism or development versus nondevelopment, we must deconstruct in order to reconstruct the meaning of freedom. Again, the entire literature of postmodern and poststructuralist questioning is valuable precisely for this reason. However, we need to find a way out of simple and misleading nihilism. We need to fight our way out of a complacent irony into the "real"[5] world of uncertainty and confusion. I want to show that this movement in thought brings us face to face with technology as creativity. I also want to draw attention to the close links between technology as creative activity and human freedom as social capabilities (Sen 1992, 1999; Khan 1998; Levine 1997). Finally, I wish to explore the connections between this approach and reflexivity as a socially embedded relation. In the next section I critically discuss the relationship among technology, growth, and development in the context of reflexive modernity and reflexive development.

Technology, Modernity, and Development: Refractive Reflexivities

Ulrich Beck's (1992) contrast of a "simple modernity" with reflexive modernity has opened a wide area of reflection on matters central to the concern of this chapter. If advanced capitalist societies (presumably they are more developed) have entered a stage that Beck calls the "risk society," are the less advanced to be characterized as "scarcity societies" with simple modernity as the object of social, political, and economic construction? Raising the question in this way allows one to reflect on the nature of the construction of a plurality of modernities. It may turn out that simple versus reflexive modernity or simple versus reflexive development are intellectual dichotomies that are too neat to sustain in the face of global complexities.

As Lash et al. (1996: pp. 6–7) point out: "*Risk Society* itself had raised worries for many readers that Beck seemed to be offering a vision of a kind of hyper-Enlightenment, where individuals and institutions were becoming increasingly able consciously to reflect on the premises of their own and others' commitments and knowledge claims." However, Beck et al. (1994) acknowledged the possibility of an automatic and blind reflex. For Beck, risk societies are led by the riskiness of large-scale chemical, thermonuclear, and other technologies to an abyss beyond calculability.

If reflexive modernity is limited in this way to a nonpredictive society where any convergence on a semblance of a solution is contingent upon processes that are only partially visible, much less under anyone's control, then reflexive development cannot be any more than a partial attempt to manage the problems of development. Thus the whole idea of developing into modernity through technical progress is made opaque. This opacity is not natural, but rather is political, social, and cultural. In fact, it can be said to be an integral part of the modernist project. There is a large irony involved here; the modernist project that emphasizes self-awareness and reflexivity is largely unaware of the limits of such reflection and the continuous production of opacity in the social, political, and ideational realms. This is a tendency that is so pervasive that perhaps we need a name for it. Since one of the key aspects of the phenomenon I have in mind is an endemic distortion, it may be appropriate to call this a "refractive" reflexivity.

Under conditions of refractive reflexivity, our partial knowledge and further reflections do not necessarily converge to produce a given, "optimal" solution in the economist's sense. Rather, the institutional setting in which the discourse is carried out determines to a large extent the limits of our reflections. Wynne's example of the struggles between the sheep farmers in the north of England and the scientists from various agencies shows empirically how refractive reflexivity is indeed a question of power and not merely a debate on epistemological matters (Wynne 1996). Dependence on the power of experts leads to a "constructed" insecurity and anxiety. Lack of democracy heightens such dependence.[6] In his discussion of the contested construction of modernity in technology politics, Schot (Chapter 9, this volume) mentions the strikes at the Rotterdam harbor against grain elevators, which were unforeseen by either the capitalists or the socialists; each group was in its own way under the thrall of modernist rationality.

To clarify further, conceptually refractive reflexivity can be seen at two distinct but related levels. First, at the level of the individual, it is the limitedness of rational calculation that appears to be primary. In this sense, refractive reflexivity points to the bounded nature of individual rationality. Miscalculation, limited computability, uncertainty, and unintended consequences are endemic. At best, reflexivity marks an incomplete rationality at the individual level.

The second, social aspect of reflexive rationality is perhaps even more important. Here we have the social embeddedness of virtually all human institutions, including the reflexive institutions of modernity. Take for example, the universities—which are perhaps reflexive institutions par excellence. Yet in the history of the modern universities, unforeseen developments arising from deeper social and economic forces have shaped the discourses and practices more than the conscious strategic plans that the administrators produce with such readiness and regularity. In the economic sphere, the sharp debates on macroeconomic policies, technology, and industrial and trade policies result in partial advances at times. However, the complex domain of socially embedded institutions makes the economic policies less than optimal and subject to revisions (which themselves are subject to revision) without ever converging on an optimal path in even a limiting sense. Political conflicts of the sort discussed by Schot and others in this volume are inevitably a part of this complex, nonconvergent process.

Postulating refractive reflexivity as a dominant condition of modernity and development can help us grasp why such fear and anxiety are part of these constructs. It can also help us understand why "scientific" surveys of risk can exacerbate this anxiety. Unlike Latour (1993), who claims that "we have never been modern," I observe that the refractive nature of our reflexivity shows that modernity can never escape its own complexities and contradictions. There is no epistemic safehouse of otherness to be found in a premodern or undeveloped state of affairs.[7] Technological development is also a refractive development. It enhances as much as it distorts our freedom to become fully human, as the next section shows.[8]

Like Beck, then, I accept that the critical issue is "that industrial society sees itself as risk society and how it criticises and reforms itself"

(Beck 1996: p. 34). But the refractive operations of the social body and the unevenness of the distribution of power are precisely at stake here. What is implicated through the globalization of the risk society even when scarcity persists both as an ideological construct and a real redistributive issue could be described as the limits of a (fractured) global society's capacity for self-correction. In a fractured globalization (Kumssa and Khan 1996) there is both globalization and regionalization of risk. By exporting risk to the South through relocation of dangerous industries, the wealthy "risk" societies may make poor countries into ecological waste dumps. At the same time, there may also be some transfer of risk from the South to the North in the form of communicable and infectious diseases.

All these interactions between risk societies and the rest of the world under the aegis of a new liberal international economic order also create risk awareness and conflicts among different groups. Hence the transnationalization of a critique of science, technology, and corporate policies is countered by partial cooptation and intensification of efforts to offer technocratic solutions. Even nongovernment organizations (NGOs) are not immune from this counter-countermovement on the part of modernizing forces.

Thus we have to confront a situation in the field of development that can not be simply characterized as a modernization and economic growth perspective to be opposed by a social transformation and multidimensional perspective. The problem is deeper than the simple dichotomies (economic versus noneconomic, one versus many, or modernization versus alternative development). Beck's insight that the critique itself must be democratised and "a critical theory of society is replaced by a theory of *societal self-critique*" applies here exactly (Beck 1996, p. 33). In the never-ending process of questioning one's proneness to refraction, conscious reflection clearly plays a key role. But so do qualities such as empathy, compassion, and more than mere intellectually consistent regard for others by self-regarding individuals.

If the argument about refractive reflection and the need for bringing both intellectual and emotional resources to our collective rescue is correct, then freedom for the individual and for society involves a new recognition of necessity: a necessity for self-critical and other-regarding reflection as well as the limits of such instrumentally rational reflection becomes apparent. In contrast to Hegel's classical phrase (interpreted by Engels), "freedom is the recognition of necessity," necessity here points to the need for going beyond instrumental reason to call upon all of our humanness and creative power so that new institutions of solidarity and freedom can be built globally. Only from such a perspective of freedom can we empirically assess the nature of technological systems and their development in nonwestern parts of the world.

Freedom as Social Capability and Development as Freedom

Amartya Sen, the recent Nobel laureate in economics, has done much to broaden and deepen the discussion of freedom in development economics. Sen's initial project was to offer a critique of utilitarianism and social choice theories based on utilitarian approaches. Eventually the critique led to a complete rejection of utilitarianism in favor of a framework of positive freedom called "capabilities" (Sen 1992, 1999). Aristotelian philosophers such as Martha Nussbaum have pointed out the ambivalence in Sen's initial formulation. In Nussbaum's view, an Aristotelian approach based on a concrete concept of a "good life" is a better foundation for the capabilities approach than the classical liberal view of individual goods being determined by individuals' subjective preferences.

Khan (1998) attempts to establish a dialectical relationship between the individual and the (ethical) community and thus bring out the fully social nature of capabilities. While the determination of capabilities can be social, their concrete manifestation is only possible through individuals. Individuals are the bearers of capabilities. Calling capabilities "social" merely draws attention to the fact that the freedom of individuals to lead a certain type of life is always constructed and constrained by their social context.

At this point, a partial list of the most important capabilities may make matters concrete for the reader. In table 12.1 (compiled by David Crocker 1995), "N" and "S" refer to Nussbaum and Sen. Both Sen and Nussbaum agree that these capabilities are distinct and of central importance. One cannot easily trade off one dimension of capability against

Table 12.1
Social Capabilities

---

1. Capabilities in relation to mortality
   N and S: Being to live to the end of a complete human life, so far as is possible
   N: Being able to be courageous

2. Bodily capabilities
   N and S: Being able to have good health
   N and S: Being able to be adequately nourished
   N and S: Being able to have adequate shelter
   N: Being able to have opportunities for sexual satisfaction
   N and S: Being able to move about from place to place

3. Pleasure
   N and S: Being able to avoid unnecessary and non-useful pain and to have pleasurable experiences

4. Cognitive virtues
   N: Being able to use the five senses
   N: Being able to imagine
   N: Being able to think and reason
   N and S: Being acceptably well informed

5. Compassion
   N: Being able to have attachments to things and persons outside ourselves
   N: Being able to love, grieve, to feel longing and gratitude

6. Virtue of practical reason (agency)
   N: Being able to form a conception of the good
   S: Capability to choose: ability to form goals, commitments, values
   N and S: Being able to engage in critical reflection about the planning of one's own life

7. Community: Being able to live for and with others, to recognize and show concern for other human beings, to engage in various forms of familial and social interaction
   N: Being capable of friendship
   S: Being able to visit and entertain friends
   S: Being able to participate in the community
   N: Being able to participate politically and being capable of justice

8. Ecological virtue
   N: Being able to live with concern for and in relation to animals, plants and the world of nature

9. Leisure
   N: Being able to laugh, to play, to enjoy recreational activities

10. Separateness
    N: Being able to live one's own life and nobody else's
    N: Being able to live in one's very own surroundings and context

Table 12.1
(Continued)

11. Self-respect S: Capability to have self-respect S: Capability of appearing in public without shame
12. Human flourishing N: Capability to live a rich and fully human life, up to the limit permitted by natural possibilities S: Ability to achieve valuable functionings

Source: Crocker (1995).
N, Martha Nussbaum; S, Amartya Sen.

another. At most, one can do so in a very limited way. They cannot be reduced to a common measure such as utility.

As Crocker points out, "capability ethic" has implications for freedom, rights, and justice going far beyond simple distribution of income considerations. If one accepts the capability approach as a serious foundation for human development (see Sen 1992, 1999; Khan 1996), then it follows that going beyond distributive justice is necessary for a complete evaluation of the impact of economic policies.

The social capability approach outlined here emphasizes the positive freedom to choose a good life. Given the limitations of space, it is impossible to elaborate on all the aspects of both well-being and agency freedoms that such a concept of development must encompass. I briefly touch upon the area most relevant to the theme of this chapter: the freedom of difference and diversity in developing societies that are attempting to modernize.

The postmodern turn has correctly focused our attention on these two aspects of our planetary civilization in the age of high technology— especially transportation, information, and communication technologies. The goal of modernization theories in the past was to emphasize a certain kind of uniformity that might obtain with economic growth and technical progress: free markets, formal democracy, high technology, westernization, and related values. The recent revival of these ideas in the international organizations and in western academia has less to do with their intellectual merits than with the collapse of any coherent alternative vision of development. The relativism (and in extreme cases nihilism) fostered by the postmodern turn has not prevented, and cannot prevent, this narrow modernism from being the only game in town.

In contrast, the emphasis on freedom in a socially determinate way that follows from the capability approach is indeed a viable alternative to an intellectually discredited and narrow form of modernism. By celebrating differences of race, gender, and ethnicity, and emphasizing the underlying unity of our need to develop the above-mentioned functionings, the capability approach can offer a clear-sighted alternative to both an absolutist modernism and an indiscriminately relativist postmodernism. Fostering technology systems congruent with these goals would seem to be especially relevant for such an approach to development. A critical look at technology systems in the newly industrialized economies such as South Korea and Taiwan can help ground the discussion in concrete, theory-based empirical research. Before such an examination can be carried out, however, we need to address an important puzzle regarding the capabilities approach as an appropriate evaluative framework.

A Non-Essentialist View of Capabilities: A Network Approach to Technology and Capabilities

The foregoing account of development as the process of humans flourishing through the positive freedom of capabilities can help to overcome the nihilism of the postmodern turn. However, does it not carry some danger—even with the qualifier "social" tacked before capabilities—of an essentialist bias? The question is a serious one because as Feenberg (1999a: p. 15) reminds us, an essentialist view of technology "interprets a historically specific phenomenon in terms of a transhistorical conceptual construction."[9] If this is true of technology, it can apply in a similar way to a historically formulated view of capabilities as well.

Therefore it is important to add that capabilities are fully social only when they can be viewed in a concrete, historical context. Viewed in this manner, they appear, not as some unchanging human nature or need, but as an evolving, socially mediated activity that makes sense of what human needs really are in the context of the actual evolution of human societies and artifacts.

The use of the word "artifacts" is not accidental. Humans make artifacts and are in turn transformed by them. This is not technological determinism. Rather, to put it in Latour's terms, there is a need to go beyond purification—the forcible separation of nature and society—and acknowledge the hybridity that modernity produces (Latour 1993). The complex production of hybridity is both an integral part of modernity and a problem that it generates. Going beyond this into a "parliament of things" acknowledges the network of quasi-objects that exists as the very condition of our own existence and discourse.

Thus, taking the capabilities perspective beyond a humanist interpretation will involve integrating a nonessentialist, network approach to technology with a view of humans as one significant form of life among others. At the same time, human social and political activities—interpreted historically and in a field of social networks and forces—must be given their appropriate recognition. Without such recognition, the very meaning of the world of humans and technology is in danger of serious distortion. In his critique of Heidegger's "seen-from-above" view of modern technology, Feenberg expresses this point quite forcefully:

From the standpoint of the ordinary human being—and even system managers and philosophers are ordinary human beings in their spare time—networks are lived worlds in which humans and things participate through disclosive practices. This lifeworld of technology is the *place of meaning* in modern societies. Our fate is worked out here as surely as on Heidegger's forest paths. (Feenberg 1999a: p. 197, emphasis added)[10]

If technologies and human beings thus form the same network, space is opened from below for a genuine critique of technology, modernity, and development. I would argue that this is in fact where the empirical turn beyond the East–West dichotomy should really take us. In the next section I offer a critical empirical view of national innovation systems (NIS) in developing countries. Most likely, what the developing countries need—to anticipate a little—are not systems of innovation from above in order to catch up with the West, but the creation of a Latourian parliament of things. But first it is necessary to clarify the relationship between technology and democratic freedom in the context of development.

When the process of development is characterized as the creation of social capabilities, the issue of positive freedom comes to the foreground.

Political freedom, that is, the freedom of citizenship, is inextricably connected with any reasonable list of positive freedoms. In order for citizenship to be meaningful, citizens must be able to participate fully and democratically in the affairs of society and the state. I have used the term "deep democracy" to indicate the necessary capabilities—economic, social, and cultural—that must exist for citizenship to make any sense at all.[11] In a similar way Feenberg (1999a) stresses the importance of agency.

In this context, claims Feenberg (1999a, p. 101), "the fundamental problem of democracy today is quite simply the survival of agency in this increasingly technocratic universe."

Despite occasional resistence the design of technical institutions disqualifies modern men and women for meaningful political participation. The division of labor becomes the model for the division of society into rulers and ruled. As in the factory or hospital or school, urban centers, media, even unions are reconstructed around the paradigm of technical administration. Expertise legitimates power in society at large, and 'citizenship' consists in the recognition of its claims and conscientious performance in mindless subordinate roles. (Feenberg 1999a: p. 101)

Feenberg analyzes this translation of the efficiency of a technocratic system into legitimacy via the delegation theory of Latour. In this theory, norms are delegated to technical devices. Even Latour's simple example of a device for automatically closing the door embodies such a norm. Much more than simple conventions are at stake when we move from a door-closing device to the organization of technical-social life in areas such as education, scientific training and practice, medicine and health-care delivery systems, or public administration. The rise of technocracy and an elitist, hierarchical order is ultimately tied to a certain antidemocratic conception of development through technical progress (Khan 1997, 1998).

In using the theoretical framework developed earlier to understand the creation of technical modernity, two approaches—both seemingly empirical—are contrasted. The first, a national innovation system, approaches the problem of technology and development from above. Hence it remains trapped in a technological determinism. The second, alternative approach emphasizes the role of democratic struggles in choice, development, and design of technology. Clearly, agency, conceived in a nonessential, heterogeneous way—particularly the agency of the nonelite, ordinary people—plays a key role in this alternative empirical approach to technology in developing societies. This approach, called a "positive feedback loop innovation structure" (POLIS) is cognizant of the complex interactions among technology, economy, and polity. Ultimately it emphasizes the teleological desideratum of equalizing social capabilities as the end of development. Given this end, technology is much more than an instrumental means. Depending on how the above relations are conceived, institutional structures can be judged as promoting more or less freedom in concrete historical contexts.

Empirical Approaches to Technology, Modernity, and Development: A Critique of National Innovation Systems

An appropriate example—one might even be tempted to say, an exemplary one—of the multiple contradictions between technology systems in a modernizing, development context and democratic norms of freedom is the idea and practice of national innovation systems. The concept of an NIS, like many other concepts in the field of the economics of innovation, was originally proposed for analyzing the advanced industrial countries (Freeman 1987; Nelson and Rosenberg 1993; Lundvall 1992). As a systems-oriented, holistic way of thinking about technological change, it has undoubted strengths. By identifying links between R&D, development of human resources, formal education and training, and innovating firms, an NIS presents an analytical schema for relating a cross-cutting array of activities that lead to a dynamic, innovative economy. The proponents of this approach also advocate an evolutionary as opposed to a mechanistic approach (based on a classic physics-type study of equilibria) for studying the economics of innovation.

Given the obviously sincere and serious intentions of the theorists of NIS, and the intellectual break with neoclassical economics, the study of NIS held promise of providing a retrospective understanding of economic history and a prospective, prescriptive approach to help countries innovate. Nowhere was this promise more eagerly believed than in the developing countries. No one was more excited by the prospects of NIS than the avid modernizers in their governments, universities, and international organizations and think tanks. I have documented in great detail elsewhere (Khan 1997, 1998) the reach and sweep of NIS in newly industrializing countries such as South Korea and Taiwan.

However, so far the thinking about an NIS and its connections to modernity and development has been entirely technocratic. The argument always proceeds in terms of the function of technologies and their role in increasing per capita gross domestic product in the most efficient manner. The intense and inconclusive debate raging with respect to whether East Asia has really grown because of a simple accumulation of labor and capital or because of a productivity increase through genuine technical progress and learning neatly illustrates this technocratic bias. Neither side is willing to step beyond the economic inputs and outputs, production functions, and technology as a black box. It is, of course, important to know whether learning has taken place in, for instance, textiles or electronics sectors. But there is no recognition of the point made by Feenberg and others, namely that "design ... incorporates broader assumptions about social values" (Feenberg 1999a: p. 86).

This "cultural horizon" of an NIS, which legitimately can be said to constitute a hermeneutic, interpretive dimension, should offer some interpretive flexibility. A recent paper by Murata (1999) illustrates the relevance and importance of such interpretive flexibility by simple but elegant examples such as street speed bumps (to slow traffic) and attaching a car key to the driver so it is not left in the car in a fit of forgetfulness. When an underdeveloped economy accepts an NIS whose components come from abroad, a societywide hermeneutic process is unleashed. Yet this is where interpretive flexibility is frequently thwarted by the closure undemocratically imposed on the rest of the population by the technocratic elite and their modernizing allies from the West.

Such premature closures can certainly produce success stories in modernist technological terms. In Taiwan, for example, the NIS has succeeded to the extent that it has been able to capture worldwide market shares in several high-technology areas. The Taiwanese manufacturers' swift capture of the lion's share of worldwide information-technology hardware markets is nothing short of amazing. In most relevant product categories, Taiwan has more than 50 percent of the market share. In some categories such as scanners, it has almost cornered the whole market. In many other high-technology areas also, companies based in Hsinchu Science Park have been quite successful. Yet this very success in exports may have forced the Taiwanese companies to seek a closure that largely excludes their domestic constituencies.[12] Only the preferences of the technical, business, and bureaucratic elites are reflected in the design and development of technology in the Taiwanese NIS. A more detailed empirical analysis can substantiate this criticism.

The key conceptual term in my critique of the NIS is the idea of a POLIS. A POLIS can be seen as both a critique and an extension of an NIS. Like an NIS, a POLIS also emphasizes the salience of institutional structures, both economic and noneconomic, in creating positive feedback loops in technical progress and productivity increases. However, going further, a POLIS connects such technical progress as may occur to the normative issues of enhancing freedom in all spheres—economic, political, and cultural. Using the terminology introduced earlier, we can say that a POLIS enhances both economic productivity and social capabilities.

Taiwan: Building a POLIS?

In this subsection the theoretical model developed earlier informs an analysis of a leading East Asian "miracle" country: Taiwan. The history of development in Taiwan shows a greater reliance on direct foreign investment, more direct government ownership of enterprises, and a greater role for small and medium enterprises in the manufacturing sector than the other large East Asian "miracle" economy, South Korea.

The early development policy in Taiwan was aimed at increasing agricultural output, developing an infrastructure, and promoting light manufacturing industries. Import substitution was pursued until the mid-1960s. U.S. foreign aid played a crucial role in financing imports and in early capital formation. Even though the theoretical thrust of aid was to help the country modernize, a curious silence pervaded the technical analyses when it came to the structures of authority. In fact, quite often antidemocratic structures were strengthened by such aid.

Taiwan's switch to a regime of export promotion took place in the mid-1960s, as in South Korea. Initially, the government backed exports of the light manufacturing industries, such as textiles and consumer electronics. At the same time, Taiwan pursued a long-term strategy of building a more complex industrial structure that included steel, petrochemicals, machine tools, and electronic equipment.

The new outward-looking strategy was accompanied by a series of financial and fiscal measures to facilitate export financing and to help establish export processing zones. From the beginning, Taiwan made a special effort to promote high-technology sectors through publicly funded research laboratories. Later, an industrial park at Hsinchu was created specifically for high-technology industries.

In the wake of the 1973 oil crisis, the government introduced a policy of major infrastructure projects and subsequently promoted the capital goods-producing sectors. As a result, Taiwan broadened its export base to include machinery and related equipment. The second oil shock led to substantial changes in Taiwan's industrial policies. The country's overcapacity and the lack of competitiveness in a number of firms were addressed by a strategy of scaling down industrialization plans. Strategically selected firms, however, still received special grants and loans. Foreign investment in capital-intensive sectors was encouraged to further effect a transfer of technology and knowledge.

A new orientation in the 1980s emphasized high-technology and skill-intensive activities. Specifically, three areas—information, electronics, and machinery—were identified as strategic. Products targeted for special treatment included precision instruments, machine tools, videocassette recorders, telecommunications equipment, and computers.

In spite of its openness, flexibility, and strategic vision, the Taiwanese economy has yet to create a well-balanced POLIS. The predominance of small firms is a handicap where high-tech ventures require large R&D expenditures. The strategic complement of R&D—skilled human components—may also create a bottleneck in some sectors. More important, a hierarchical, authoritarian managerial and financial control structure may prevent a democratizing move toward equalizing capabilities. Both within the enterprises and at the macroeconomic level, the task of making power responsible has been very difficult. Thus, whether Taiwan has succeeded in creating a POLIS is not a trivial question. However, there is one particular sector—electronics—in which Taiwan has achieved a mature capability to innovate. A discussion of the electronics sector can serve as a prelude to a discussion of an economy-wide capability to innovate.[13] Even here, a detailed empirical investigation will expose crucial areas of difficulty in making innovation and control genuinely democratic.

The Electronics Sector in Taiwan   From humble beginnings in the 1950s, when Taiwan first started producing transistor radios, the electronics sector has grown to include many advanced products. Among them are the various components of personal computers, advanced workstations, and other microelectronic products. Companies such as Tatung and ACER have sales exceeding U.S.$1 billion. A number of small firms such as Sampo Corporation and United Microelectronic Corporation have shown tremendous growth in recent years. The share of foreign-owned firms declined during the 1980s and 1990s. However, even now foreign-owned firms account for more than 25 percent of the electronics industry's output. Small- and medium-sized firms (defined as firms with fewer than 300 employees) dominate the industry. This means that innovation in Taiwan, unlike South Korea, occurs in relatively small firms.

Table 12.2 shows the plans for the electronics industry for the year 2004. This can be compared and contrasted with the situation in 1990. In 1990, nearly U.S.$6 billion of total computer production was exported, with information products leading the way. Of this, 40 percent

Table 12.2
Electronics and Information Technology, Production Values, and Forecasts (U.S.$ billions)

	Output 1990	Forecast 2004	Average annual growth (%)
Information products	6.9	34.0	15.1
Automation	2.8	12.0	13.5
Consumer electronics	2.3	6.5	7.0
Telecommunications	1.9	10.2	16.0
Semiconductors	1.5	8.0	14.8
Total	15.4	70.7	

Source: Hobday 1995: p. 100; 2004 estimates by the present author.

went to North America and 41 percent to Europe. Japan imported only 2 percent of the computer exports, but Asia-Pacific accounted for about 14 percent.

Although the takeoff in the electronics sector appears to be a market phenomenon, government policies played a key role. In May 1979, the Executive Yuan presented the Science and Technology Development Program, which identified information technology systems as an area of emphasis for future R&D. The idea for an institute for information industry also emerged during this period.

The ministry of economic affairs moved quickly. In July 1979, the implementation plan for computer technology was contracted out to the Industrial Technology Research Institute. The Council for Economic Planning and Development prepared a 10-year plan, 1980–89, which provided targets for R&D expenditures and human capital supply. The Electronics Research Services Organization took charge of coordinating the transfer of technology from foreign companies. These responses were technocratic and frankly authoritarian. No democratic pretenses were expected or offered.

By all indicators, the ambitious plans succeeded for the most part. Many new companies, such as the success story Datatech, were started in the 1980s. By the 1990s, Taiwanese firms were among the world's innovative designers of PCs, electronic notebooks, and circuit boards. During these years Taiwan also surpassed Great Britain to become the world's fifth largest producer of semiconductors.

Under an overall imitative strategy (Chiang 1990), Taiwan decided to follow the leaders in already established technologies and to compete by cutting costs through production efficiencies. The government has taken the responsibility for acquiring technology from abroad. It has also fostered advanced research. The government-supported research institutes, utilizing skilled scientists and engineers, conduct the research and the results are then transferred to the private sector. Furthermore, economic incentives are provided to the strategic sectors. In terms of complementary acquisition of human capital, many Taiwanese went abroad to acquire advanced education and skills in science and technology. A number of local employees were also trained in the foreign multinationals where they were employed as engineers, technicians, and managers.

Lucrative financial incentives were offered to attract skilled Taiwanese living abroad.

As Hobday (1995) points out, there are at least five types of strategic firms in the electronics industry. These are foreign corporations and joint ventures, the major local manufacturing groups, high-technology startup firms, government-sponsored ventures, and the traditional small and medium enterprises that cluster together in special market niches. Strategic interactions among these actors resulted in the industry's rapid growth and expansion as a whole, even as some individual firms declined. There is an almost classic Schumpeterian "creative destruction" scenario. It is also classically undemocratic—a phenomenon not noticed by technocratic analysts such as Hobday.

Hobday (1995) has discussed the role of the major private manufacturing groups and government-sponsored startups in Taiwan. The following brief discussion highlights the actions of these diverse economic agents in creating the conditions for an NIS (but not a POLIS) within the electronics sector, and through its linkages, in the broader economy.

The Electronics Sector: Firms   The progress of the industrial group Tatung, according to Hobday, is representative of the entire electronics industry in Taiwan. In the 1970s, electronics became the industrial group's largest operation. The electronics maker began to produce black-and-white televisions by 1964, videocassette recorders by 1982, and 14-inch color monitors for computers by the early 1990s (see table 12.3). The company currently produces a range of household electronics and electric goods in its manufacturing plants around the world.

Tatung, like the typical South Korean *chaebol* (South Korean corporate groups), first gained its manufacturing knowledge through technical cooperation deals. By investing capital in joint venture projects with foreign companies, the Tatung group participated in licensing agreements while learning technological skills through "original equipment manufacturing" (OEM) deals. Tatung absorbed and adapted foreign technology, learning to modify, reengineer, and redesign consumer goods to fit customer needs. While initially production involved little R&D, by 1990 the group employed more than 500 R&D staff. However, the job of this staff was mainly in advanced engineering rather than "blue sky"

Table 12.3
Tatung's Progress in Electronics

Product	Introduction date
Black-and-white televisions	1964
Color televisions	1969
Black-and-white television picture tubes	1980
Videocassette recorders	1982
High-resolution color television picture tubes	1982
Personal computers	Mid-1980s
Hard disk drives	Mid-1980s
Television chips/Application Specific Integrated Circuits (ASIC)	Late 1980s
Sun workstation "clones"	1989
Fourteen-inch color monitors	1991

(basic and theoretical) research. Finally, by the mid-1980s Tatung was transferring its production technologies to its subsidiaries in East Asian countries that offered lower production costs.

ACER is representative of the high-technology startup companies that began to appear in Taiwan in the late 1970s and early 1980s. For years, ACER relied on product innovation and original equipment manufacturing (OEM) with experience gained by individuals who had worked overseas in U.S. firms or universities (see table 12.4). Many of the other recent startups, like ACER, have used OEM to some extent, and most were unknown outside of Asia despite brand name sales.

ACER, according to many observers, exemplifies the strengths and weaknesses of Taiwan's high-technology startups. ACER started with only eleven engineers in 1976; its total sales reached some U.S.$1.4 billion by 1993. ACER led the local computer industry in the 1980s, with 60 percent of sales being name brand through "own-brand manufacture" (OBM). In this decade the company began to distribute directly to customers abroad to challenge other brand leaders and move beyond OEM. However, the company retreated from this forward strategy after heavy losses between 1990 and 1993.

This discussion suggests the uncertain position of companies like ACER. On the positive side, these companies were able to benefit

Table 12.4
ACER: Behind-the-Frontier Innovations toward an NIS

Year	Innovation
1984	Developed its own version of the 4-bit microcomputer (later followed by 8-bit, 16-bit, and 32-bit personal computers (PCs))
1986	Launched the world's second 32-bit PC, after Compaq but ahead of IBM
1988	Began developing supercomputer technology using the Unix operation system
1989	Produced its own semiconductor Application Specific Integrated Circuits (ASIC) to compete with IBM's PS/2 technology
1991	Formed a joint company with Texas Instruments (and the Taiwanese government) to make dynamic random access memory chips (DRAMs) in Taiwan
1992	Formed alliances with Daimler Benz and Smith Corona to develop specialist microelectronics technology
1993	Produced a novel PC using a reduced instruction-set (RISC) chip running Microsoft's Windows NT operating system
1993	Licensed its own U.S.-patented chip technology to Intel (in return for royalties)
1993	Received royalties from National Semiconductor, Texas Instruments, Unisys, NEC, and others for licensing its PC chipset designs

tremendously from the improving technological infrastructure and established market channels; they were able to bypass the "consumer" electronics phase of the 1970s and to enter the market at a higher technology level; and they have benefited greatly from managers and engineers educated abroad. On the other hand, these companies have encountered many difficulties as latecomers. ACER sustained heavy losses in own-brand sales. This forced the company to retreat to its earlier OEM strategy, once again making ACER dependent on the global leaders of core technologies. Unless and until these latecomers develop in-house technologies, they will be unable to compete with the global leaders on an equal basis.

The final group to be discussed here consists of the government-sponsored startups. Table 12.5 shows the companies working at the government-developed Hsinchu facility and their relationship with

Table 12.5
High-technology Startups in Hsinchu Science-Based Industrial Park (1980s)

Firm	Start date	Sector	Sources of senior staff, technology, and training
Microelectronics Technology Inc.	1983	Telecom	Hewlett-Packard, Harris, TRW
United Fiber Optic Communications Inc.	1986	Telecom	Sumitomo, Philips, AT&T, STC (UK)
TECOM	1980	Telecom	Bell Labs, IBM
Macronix	1989	Semiconductors	Intel, VLSI-Tech
Winbond Electronics Corp.	1987	Semiconductors	RCA, Hewlett-Packard
Taiwan Semiconductor Manufacturing Corp.	1987	Semiconductor foundry	Harris, Burrows, RCS, Philips, IBM

Source: Hobday 1995: p. 118.

international companies. With these special startups, the government has taken a "hands on" approach, offering direct and indirect assistance, including tax incentives and loans, and the use of science park facilities at Hsinchu to entice overseas Taiwanese to return to Taiwan. In one case, Microelectronics Technology Inc., a telecommunications equipment maker, the government was greatly responsible for initiating this firm. In another instance, the government arranged for technology transfers for Winbond Electronics Corporation. Winbond's founder and eventually many of its employees came from the Industrial Technology Research Institute, a state-controlled organization that trained engineers in advanced semiconductors. With government-sponsored technology transfers, Winbond was able to compete not only locally but internationally as well. However, problems with shortages in investment capital, poor brand name recognition, and uncertain distribution arrangements kept the company dependent on international leaders for technological innovation and capital goods.

United Fiber Optic Communications Inc. (UFOC), despite an auspicious start, faced many of the same problems of other latecoming startup companies in Taiwan. The government, specifically the Ministry of Economic Affairs' Industrial Development Bureau felt that Taiwan needed an indigenous fiber optic producer. This ministry called together the four largest copper producers within Taiwan and the local telecommunications operator to form a joint venture company, UFOC. The new venture sought licensing agreements with four other international companies, finally deciding on AT&T. Faced with the difficult choice of continuing to purchase its know-how from international competitors or investing heavily in its own in-house technology, these companies have typically relied on the former for continued learning and technology. This suggests some of the difficulties of latecomers in overcoming the OEM path to further development (Hobday 1995). The underlying problem, from the point of view of creating a POLIS, is that neither the state policies nor the private enterprises attempt to directly address the question of creating social capabilities. It is as if the battle for economic gains has crowded out all other considerations. Economic models, no less than technological systems, are also path dependent.

As scholars of technology have pointed out, initial disputes and controversies about technologies and their characteristics are "closed" by making one configuration the privileged one (Rip and Kemp 1998), or using Kuhn's later terminology, an exemplar. The exemplar then defines the boundaries of discourse, establishing the standard way of seeing both problems and solutions. This paradigmatic artifact and the associated procedures establish a "technological frame" (Bijker et al. 1987: pp. 167–187). The world of technology and people are, to a significant degree, perceived only within this frame. The faltering attempt to build a POLIS in Taiwan shows how an elite-based model of an NIS has served as a systemic exemplar.[14] One might speak of a "development frame."

As I have argued elsewhere, in the case of the so-called developing countries, the debate on what development frame to choose was closed very early on (Khan 1997, 1998). After World War II, the two dominant paradigms of development—western capitalism and Soviet-style socialism—both advocated large-scale, heavy industry. The role of technical elites was paramount in either case. It was only through the "deviations" of Chinese socialism in the countryside in the 1950s and 1960s, and the revolt against technology in the West in the late 1960s, that technocracy came to be questioned. Yet the seeming triumph of capitalism globally in the past two decades, and the imposition of a neoliberal order through the structural adjustment programs, narrowed the debate once again to state versus market, technological learning versus factor accumulation, and other oppositional terms.

What needs to be done in the way of posing a theoretical challenge is to bring to the fore the normative issues connected with freedom as social capabilities. In Taiwan, the NIS has apparently succeeded. However, the normative issues are still very much contestable areas of discourse, as indeed are the technologies and practices themselves. As Taiwan matures as a polity and society, such contests are likely to become more visible. The refractive reflexivities of modernity will manifest themselves (as they already have to some extent in the sphere of ecology) through a complex set of social, economic, and political struggles that cannot be predicted in advance.

It is in this context that I have proposed replacing the idea of the national innovation systems with a new concept that recognizes the connections, which are often suppressed or ignored, between technology on the one hand, and the culture and politics of modernity on the other. Coining a new abbreviation, POLIS, for the positive feedback loop innovation structure,[15] I wish to draw attention precisely to the political and cultural aspects of an NIS. Normativity of social life and struggles for freedom are paramount aspects of this complex concept. Furthermore, replacing the word "system" with "structure" flags the contradictory elements within the "innovation systems" and the society where these are to be implanted. There are many concrete aspects of the NIS that appear in a different light when we think of them as part of a POLIS. Two examples will suffice.

First, the NIS in the developed countries embody assumptions regarding citizen's rights, environmental regulations, and the needs of at least the higher categories of workers (for instance, the so-called knowledge workers).[16] By contrast, the NIS as they exist in developing countries would often exploit child workers and women, and turn a blind eye to environmental degradation and violations of citizen's rights. When these are pointed out, the response—not too infrequently—is that these are the necessary prices to pay for development and modernity. Conceptualizing the innovation process as a POLIS, on the other hand, immediately draws attention to the lack of congruence between technology and social capabilities, including the suppression of democratic freedoms. Future empirical work along these lines in actual development processes can reveal these contradictions and perhaps suggest various democratic ways of resolving them, at least partially.

My second example has to do with information technology as a component of an NIS and a POLIS. The standard NIS approach is to see information technology as the harbinger of a new era in a globalized economy. If this is so, information technology will certainly result in a new technological regime, as Rip and Kemp (1998) have defined it.[17] Again, since such regimes make up "the totality of technology" and prestructure the "the kind of problem-solving activities that engineers are likely to do" there is a huge component of path dependence at issue. Without quite recognizing it, we may well be choosing the contours—the structures that enable and constrain—of our future society.

If information technology will result in a new technological regime in this sense in developing societies, some socially relevant questions must be asked. A perspective of a POLIS leads to such a set of critical questions. For example, what are the social values at stake here? Are we going to emphasize efficiency in hierarchically organized production as the prime value, or will we think of citizenship, social communication, and creation of a public sphere as equally important? Who will define the "technical code"? How will these codes be institutionalized? How will information technology be codified in the developing societies when the codification is already under question in the West? Will the progressives, including scientists, engineers, students, intellectuals, and ordinary people, in these "modernizing" societies join with the critical-minded progressives from the modern West? Or will they simply follow the "imperatives" of the computer, software, and telecommunications companies and their own modernizing impulses? Or will they turn their back completely on modernity, counterculture fashion?

These are complex questions that force us to confront a complex reality. Will the Latourian "parliament of things" arrive in both East and West, thus erasing one of the invidious distinctions between these two equally imaginary (in the Lacanian sense) entities, or will the status quo continue? It can, of course, get much worse than that. Positive feedback loops accentuate precisely and remorselessly the initial differences between the advanced and the backward regions unless countervailing action is taken. Perhaps a new internationalism from below will recognize and strengthen the actor network that can achieve a reflexive modernity (which, of course is also refractive at the same time) with a progressive technological structure leading toward increasing at least some of our salient social capabilities. However, at this time, it is not clear what particular social and political conditions can make such internationalism from below a real historical prospect.

Conclusions

The social and political failures of "successful" information-technology and other high-technology firms in Taiwan and elsewhere in developing countries provide empirical data that need to be taken seriously in science and technology research. As long as one focuses on narrow economic costs and benefits, tidy indicators of success and failure can be constructed. Part of the point of this essay has been to warn the readers against such narrow interpretations of successes and failures.

Broadening our criteria, however, means questioning modernity and development in the specific contexts of technology policies. A critique of national innovation systems is an example of such a contextual approach. Contrasting an NIS with a POLIS reveals the technocratic bias and nondemocratic framing of technologies, even in technologically modern and economically successful developing countries. This is a far from accidental, though by no means inevitable, result. It is rooted in the historical development of imperialism, and the attendant international division of labor. Ironically, achieving technologically based modernization, viewed through the uncritical lens of an NIS, is usually misconstrued as the inevitable necessity of constructing an NIS in a world that is really the result of a series of concrete historical contingencies. Clearly, this epistemological gesture cannot envision a process of development where technology can be designed and controlled through a deep democratic process.

An economic (and perhaps even technological) determinist position argues that the poor countries must first grow rich by adapting an elite-defined NIS and other policies for economic growth. Only later, when the country is more affluent, can the people afford luxuries such as democratic freedoms and ecological consciousness. This position ignores both the real historical democratic tradition and ecological awareness in indigenous peoples' cultures because its modernist bias and determinism will not allow such "anomalies" to enter into the modernization paradigm. Yet, as Latour has so acutely observed, the current collective global situation will not allow such easy recipes for success. Attitudes and practices must change, in the East as well as in the West. Ironically, it may be more difficult, as the empirical study of Taiwan here illustrates, to recover and extend democratic freedoms and transform the NIS into a POLIS when too much economic "development" has already taken place. Only a series of further negotiations within the economy, civil society, and state—the outcomes of which are far from transparent—can determine whether a move from an NIS to a POLIS can be made by the newly industrialized economies. This future, though far from completely open, is not simply one inscribed by a closed national system of innovation.

Notes

1. I would like to thank Karin Hillen and Gyeong Jei Lee for excellent research assistance. Pat Baysa also provided valuable assistance. Comments from David Hess, Michiel Korthals, and other workshop participants—Thomas Hughes, Arie Rip, Tom Misa, and Philip Brey in particular—were very helpful in preparing the final version. All remaining errors are my own.

2. This idea is elaborated on later; here it can be thought of as somewhat akin to "the seamless webs" described by Bijker et al. (1987: pp. 9–15), or more particularly, of Callon in the same book. It should be clear, however, that my epistemology and ontology are firmly nonrelativistic, yet postmodern.

3. In Khan (1998) I have tried to move the modern versus postmodern debate beyond the rather sterile terminological controversies about high, late, advanced, neo (and other) types of modernity. Reflexive modernity (Beck 1992; Bourdieu and Wacquant 1992; Beck et al. 1994; Giddens 1991) is another fruitful point of entry into a similar set of issues.

4. On this see the very illuminating *Orientalism* by Edward Said (1995); see also Hay (1970).

5. Even Derrida (1988: p. 137) has been moved to remark: "A few moments ago, I insisted on writing, at least in quotation marks, the strange and trivial

formula, 'real-history-of-the-world', in order to mark clearly that the concept of text or of context which guides me embraces and does not exclude the world, reality, history. Once again ... as I understand it (and I have explained why), The text is not a book. It is not confined in a volume itself confined to a library. It does not suspend reference-to history, to reality, to being, and especially not to the other since to say of history, of the world, of reality, that they always appear in an experience, hence in a movement of interpretation which contextualizes them according to a network of differences and hence of referral to the other, is surely to recall that alterity (difference) is irreducible. Difference is a reference and vice versa."

6. It is important to keep in mind here the distinction between "formal" and "deep" democracy (Khan 1998).

7. This is one of the important points made by Latour (1993). See especially the chapter on revolution and his discussion of the principle of symmetry generalized.

8. It should be clear to the reader that I do not object to "collectives" as ensembles of human and nonhuman agents or even "actants" as explanatory categories. However, the issue of becoming human remains salient. I do not think that Latour's antihumanist position would reject this. However, to the extent that certain antihumanist positions do reject the importance of "becoming a free human being," I am willing to part company with them without getting back into the fold of classical humanism.

9. Feenberg shows that many thinkers who try to think of technology critically may nevertheless fall prey to this tendency. His list includes Heidegger, Borgmann, and Habermas, among others.

10. In the first sentence Feenberg (1999: p. 194) is referring to the power of disclosure (*Erschlossenheit*) in Heidegger.

11. See Khan (1998), chapter 6 and appendix 6.2 for a discussion of the cluster conditions for deep democracy (see also Gilbert 1990).

12. Of course, it could be argued that to the extent that the closures abroad embody progressive social values, such export dependence is a good thing. There are several problems with this argument, however. First, the closures abroad may not be that progressive. Second, even if they were, there is still the question of agency of the domestic producers, designers, and users. The extent to which this agency problem is solved is vital to the assessment of specific technologies as well as the national innovation system (NIS) of which these are a part.

13. Of course, it is not being claimed that having an apparently self-sustaining innovation structure in one sector is sufficient for a POLIS. For this we must examine the economywide links.

14. In a recent paper Rip and van der Meulen (1996) argue that research systems also shift over time. In their view, research systems are moving from a modern to a postmodern framework, with a potential for less steering and more aggregation. Unfortunately, it would seem that the theorists and policymakers in

the less-developed countries are still in the thrall of a modernist NIS. The Taiwanese case is an all too clear and disturbing example.

15. It is important to realize that being nationwide is not a necessary condition for a POLIS. It could very well be regional, or even confined to a city. For a beautiful example of a citywide POLIS in Boston, see Hughes (1998). At the other extreme, a POLIS could in principle be supranational.

16. For example, Feenberg (1999a: pp. 90–91) discusses reflexive design and his own experience in studying groupware.

17. Rip and Kemp define "regime" as follows: "The whole complex of scientific knowledge, engineering practices, production process technologies, product characteristics, skills and procedures, and institutions and infrastructures that make up the totality of technology. A technological regime is thus the technology-specific context of a technology which prestructures the kind of problem-solving activities that engineers are likely to do, a structure that both enables and constrains certain changes." (Rip and Kemp 1998: p. 340)

13

Modernity and Technology—An Afterword

Arie Rip

Modernity and technology are too important to study in isolation, as Tom Misa indicates in a proposition in his chapter. This implies a further proposition about technology and modernity being interconnected. It is this idea of interconnectedness that led us to deplore the "great divide" between detailed technology studies, with their claim of situated developments and contingency, and abstract or theoretical discussions of modernity. We exaggerated a bit in order to make a point and set up a twin argument for an empirical turn in modernity studies and for recognizing broader structures and long-term dynamics in technology studies. Conjuring up a field of technology and modernity studies in this way was made easier because the authors had already been looking for bridges across this great divide before, and they could build on the work of colleagues and discussions with them at the November 1999 workshop at the University of Twente. In other words, we did not start from zero. Yet the divide between technology studies and modernity studies remains difficult to bridge. There are methodological challenges, often summarized as the contrast between micro (or local) and macro (or global) levels of analysis. There are also substantial issues about the nature of modernity (and of technology, for that matter) and about the different perspectives that can be brought to bear, especially when further diagnosis is required that concerns openings for change and desirable directions.

Tom Misa's introduction outlined a program but also left room for the other authors to analyze the tensions and offer their own approaches and insights. It is fitting to look back, at the end of this volume, and ask how far we have come. In this way we continue the conversation about modernity and technology among the authors, and now also include the readers of this volume.

The conversation is about methods and approaches (of modernity studies and technology studies, and their *rapprochement*) and about substance, namely, concerns about our world with its modernist projections, its technological achievements, and its vulnerabilities. Thus, while the conversation begins with the conviction that academic reflection can contribute to real-world issues, it is not "just" an academic discussion. The chapters in this volume amply testify to real-world issues when they discuss infrastructures, surveillance, the environment and the chemical industry, and national innovation systems. There is also a concern with the dominance of modernist regimes and what, rightly or wrongly, they exclude; and thus with the possibility of lateral views, or ruptures, as these occur or are sought after. In this way, reflections may create openings for transformation.

Going on from there, one might try to identify concrete possibilities for change and to justify such attempts. There is a risk of reification because such justifications must be a platform for action. Recognizing their constructed character is necessary but may run another risk when contingency is emphasized and agency becomes irrelevant. The idea of co-construction emphasized in this book transcends contingency, but does not lead to simple suggestions for individual agency. The ambivalence can be addressed by what Barbara Marshall in her chapter calls "strategic essentialism"; she refers specifically to feminist theory and feminist practice, but the approach is general.

In this afterword I touch on these issues. My interest is not only in showing what we have learned, but also in identifying what remains to be taken up. I start with the methodological issue of how one can "see" co-construction at work, or the global in the local.

Methodological Issues

The chapters of this book offer windows on the modern world and its technologies. Through such windows we "see" something. Think of how anecdotes and examples draw our attention when they let us recognize something that strikes us as important and relevant. This is how

Misa presents examples of modern and postmodern technologies. More than just slick corporate packaging is at play in the contrast between the IBM Museum, an exhibition on computers in the modern world by IBM, and Sony World, a similar exhibition by Sony. Both were staged in New York; IBM's exhibition exemplified the hierarchical and functional mode, Sony's the fluid and imagery mode. The contrast signaled modern versus postmodern (whatever that may be), with overtones of America being prisoner of its earlier successes and Japan moving quicker and more playfully (at least in consumer products). The contrast between the two exhibitions—and their link with corporate culture and corporate images—functions as a window on the modern world and carries a certain immediacy. Don Slater, in his chapter, adds the idea of a "crystallizing example" that clinches earlier groping toward understanding.

How can such examples and their attendant analysis be *windows*? The local and specific practices allow us a view on what is of wider significance. Our view is of the global as it appears in the local and is refracted by it (in turn, the global structures the local). Windows on the world (as offered by analysts) reveal our intimations about the world. Something we knew, perhaps, but could not articulate. An example gives us a sense of recognition and helps us (analysts, readers) to articulate our intimations. Obviously, there are risks to the analyst: what are the grounds for recognizing one structure or trend rather than another? There are ways to handle this problem, such as triangulation or reflective equilibrium. What remains is the immediateness of the example and how it is structured. This derives from the story it tells. In a story, the global can be incorporated and made explicit by zooming in on a word or a phrase—say, Japan versus America.

In Junichi Murata's chapter, we hear about domestic industrial expositions in Japan in the Meiji era, the fifth such exposition in 1903 drawing more than four million visitors to Osaka. With such popular interest in modern technologies, it becomes understandable that the introduction of trains pulled by steam locomotives was a running showcase: people could see the modern western world "through" a train. These windows on modernity were, of course, vastly popular in the West itself. Johan Schot reminds us of the deliberate technological framing of modernity at the 1939 New York World's Fair. And for that matter, for years Amsterdam's Schiphol Airport had greater revenues from (nonflying) visitors paying for a glimpse of modernity than from airplanes actually flying.

This notion of a "window" is similar to Dorothy Smith's argument about the situated nature of knowledge and how it can be unfolded to show the "apparatus" involved in the background—even in the everyday-life case of walking your dog in the neighborhood (Smith 1987, 1990). Phrased in this way, it is a purely methodological point about local and global. There is also a substantive aspect, however. If we use the right windows, we can "see" something interesting and important about technology and modernity that we had not seen before. Schot takes this approach, highlighting slices of development over time, with recent changes in modern technology and modern politics becoming salient. In addition, the "global" is not just a methodological category, but also a force for better or for worse, as is very clear in the chapters by David Hess, Arthur Mol, and Haider Khan.

Such windows also work by surprising us. We see things we had not imagined, but now that we are told about them, our vision is expanded. Don Slater shows how Trinidadians in their use of the Internet take up modernity enthusiastically, as a way to reinforce and expand an identity from the periphery. In a study of telecommunication technology and modernity in Indonesia, we identified a dual dynamic: a strong push from the state to create a national identity in the Indonesian archipelago, through information and communication technologies, and a heterogeneous, bottom-up dynamic of creating Internet access and exchange driven by engineers and other users, and now including Indonesian-style Internet cafes. It is interesting that the metaphor of guerilla tactics was used, which in Indonesia has nationalistic overtones (independence having been fought for and achieved through such tactics). Yet now such high-tech guerilla tactics have helped to undermine President Suharto's New Order and its reference to high-tech modernization (Barker et al. 2001). Through such an analysis, we can see—in action, as it were—the co-construction of modernity and technology.

Closer by, literally just outside our homes, we can find surprises as well. What about the morality of sidewalks? They are part of a functional separation between the different modes of using a street as a public space. This became important in the early twentieth century with the multiplication of vehicles using the street (bicycles and motor cars, carts and horse-drawn carriages, as well as various trams). Allocating parts of the street to different kinds of users disciplined each category of users (socioculturally and materially) and created a way to optimize streets and their use. Over the decades of the twentieth century, this disciplining led at first to encouraging motor car traffic—an icon of modernity—and more recently, to attempts to limit motor cars' freedom. The infrastructure of city streets (another example for the analysis in Paul Edwards' chapter) was actively co-constructed by engineers and city planners, on the drawing boards and in response to actual patterns of use. Cross-profiles of city streets (presenting the multidimensionality of electricity cables, telephone wires, sewers, gas lines) in relation to their various users became a planning and construction tool to master this multidimensional complexity (Disco et al. 2002). Technological complexity increased, but there is a continuity with the earlier hygienic movement and modernist city planning—think of the Italian Futurists' multilevel transport systems in their city planning schemes before World War I. In contrast to utopian schemes, problems needed to be solved on location, and the messiness had to be confronted time and again. Cities were shot through with complexity—postmodern?—before they were rationalized, and their modernist reconstruction cannot completely contain their basic heterogeneity.

In the infrastructure of cities, then, we "see" how ideals and structures of modernity interact with local practices and evolving technologies. The tangle that results gets tied up with material "knots" of cross-profiles and their planning, with institutionalized disciplining of behavior and interaction, and with the professionals who claim expertise over the construction and reconstruction of cities. Clearly, co-construction can be traced; it can work as a methodological point of departure. As Philip Brey argues in his chapter, and the chapters in parts II and III show by example, there is interaction between different levels and scales. In particular, there is the historical phenomenon of the emergence of institutions and institutionalized activities between the local and the global, between the micro and the macro, that mediate the interactions. Anthony Giddens' mechanisms—money, timetables, and expert systems—that disembed social life depend on such mediation for their effectiveness. Thus Misa's (1994) argument for a middle-level methodology becomes even more pressing. There is an emerging and by now well-articulated intermediate layer in modern societies that carries the work of co-construction. Its study must be one of the preferred "windows."

There is a final methodological point to be made: looking out of a window, one positions oneself as an observer. The framing involved in using a window to look at the world is not my main concern here, even if framing is an important phenomenon in social life. Framing a problem in a certain way (modernist or otherwise) and being able to get that frame accepted allows certain solutions to be more successful than others. The struggles analyzed in the chapters of part III show many examples.

What I want to comment on is how the question of agency has been forced to the background by the concern to show the co-construction of modernity and technology. Thus the modernist view of agency as purposeful action leading to the achievement (or not) of an intended goal was not thematized and compared with other views in which agency is more broadly seen as making a difference (see Law 1994). Andrew Feenberg's observation in his chapter is particularly illuminating: "Human beings and their technologies are involved in a co-construction without origin." Agency can then be no more than modulation of such processes informed by an understanding of their dynamics. However, knowing how such tangles get tied up, and how mediators become established, one might want to anticipate how and where to act. Some "knots," and some mediating institutions, are better places than others. Arriving at them is a sort of bootstrap operation. We can then ask what productive bootstrap operations might look like, and whether there might be productive arrangements generally. In the end, this would lead to an interest in the "constitution" of a late-modern technological society, and attempts to improve it.

Haider Khan in his chapter actually proposes a new arrangement, a POLIS, as part of such a constitution. Because his is an explicit attempt at constitution building, as it were, it can in principle be evaluated for its possibilities and limitations. A POLIS should overcome limitations of national innovation systems by introducing feedback loops and more bottom-up learning—a bootstrap operation. It is not yet integrated with the references in his chapter to differentials in power structures and to making a difference, and this highlights a challenge to the co-construction approach. Co-construction suggests activities that lead to outcomes, but it is not quite clear *whose* activities and *which* outcomes. The critical tradition would introduce the reference to power structures; Langdon Winner's criticism of recent technology studies is a clear example (see Winner 2001). In addition to the obvious reply that power structures are constructed as well, I would argue that it is important to understand the limitations of modernist views of agency and the fact that critical action can easily fall into the trap of alternative modernism. In our work on "constructive technology assessment" (see Rip et al. 1995; Schot, chapter 9, this volume) we had to address this issue. We have come up with notions such as modulating processes (of the co-evolution of technology and society), based on an understanding of how prospective structures are projected as promising options and, to some extent, made true (Kemp et al. 2001; van Lente and Rip 1998).

A Late-Modern Technological World

Windows on the world give us partial views, but these will add up to an amalgam, as David Hess calls it in his chapter, or a kaleidoscopic closure, an intriguing phrase coined by Murray (1997) when she analyzed interacting narratives made possible through the Internet. What kind of world becomes visible in this way? And what sort of history of technology and modernity can be articulated?

Paul Edwards, in the opening paragraphs of his chapter, suggests a history of successive backgrounding and naturalization of technologies as invisible infrastructures that are assumed to function smoothly and serve their purpose. A lot of ingenuity and care is invested in not disappointing this assumption. Engineers in particular feel this ethics of care for the world of artifacts. It is part of their mandate, as it were, with another part deriving from their working toward technological progress, and being allowed to do so relatively autonomously.

The engineers' thoroughgoing modernism—"we can do great things, if you let us"—is very visible in the late nineteenth and early twentieth centuries. It is the same period in which many social groups called for emancipation and anarchists played havoc with the existing order. Electricity started to transform the sociotechnical landscape. Airplanes captured the imagination, with dreams of aviation elevating humankind—it was even called "the winged gospel" (Corn 1983). Investments in civil aviation continued even if they did not turn a profit until after World War II. By now, aviation has become part of the infrastructure of modern society. Its technical nature and functioning have become invisible for the general public and politicians. With the naturalization of the air travel infrastructure came an inevitable lack of attention to its vulnerabilities—with the attendant surprise on September 11, 2001, that terrorists could turn this modern infrastructure against its own projections of security.

There are other storylines about the co-construction of modernity and technology, but all of them appear to have an ironic twist of earlier successes creating problems. Is this a defining characteristic of the co-construction of technology and modernity in the late twentieth century? Think of plastics and other new materials being hailed as the key to the future and later condemned as unsafe and a threat to the environment. Or of surveillance, an age-old technology materializing in panoptic arrangements and then partially dematerializing again through information and communication technologies—even while cameras and other physical devices and hybrid items like barcodes remain of central importance. David Lyon in his chapter suggests that technological dependence coupled with consumerism (both features of modernity) are together changing modernity out of recognition, with surveillance of various kinds getting a central position in this new or postmodernity.

In health care, the co-construction with modernity is particular striking, with the hygienist and eugenic movements of the nineteenth and early twentieth century, the strong development of sanitation and better housing, new medical drugs and therapies, and the experience of the national-socialistic Third Reich—in its own way strongly modernist. After World War II, eugenics was transformed into genetic counseling, with a strong individualistic thrust. Individual autonomy in health care, especially in clinical genetics, is now a sacred principle—while tensions appear because of advances in genetics and the wider introduction of preventive medicine.

Just as we see technologies in the plural, there might be modernities in the plural. Why do we speak of technology and of modernity as being of one kind? The chapters in this book show their variety, and many of the authors insist on the limitations of abstract notions of technology (or for that matter, of modernity). Still, the abstract notion of technology is widely available as one of the modernist keywords that emerged as forceful in the course of the nineteenth century—hence, a Massachusetts Institute of *Technology*. What is needed is an analysis of how and why such abstract notions emerge and can be forceful, not just with modernity theorists, but also with various actors.

To indicate the abstract, iconic character of technology in a text, one can write it in capitals: TECHNOLOGY. Indeed, the transformation of technology as the term was used in the early nineteenth century into the iconic TECHNOLOGY of the twentieth century can be traced historically; this is a research project that is long overdue. Actors rarely use the term "modernity," but they often appeal to the "modern" and "being modern." And this has inclusion and exclusion effects (just as declaring a "modern period" in history includes "us" and excludes "them," as Tom Misa phrases it). Views of technological progress and its modernizing role imply collusion between TECHNOLOGY and MODERNITY. This continues to the present, even if we are sadder and wiser.

In Don Slater's chapter, the rhetorical connections between INTERNET and MODERNITY are particularly striking; David Lyon sees SURVEILLANCE as a continuation of MODERNITY; while David Hess shows how alternative therapies become accepted as "complementary" to modern medicine. (SUSTAINABILITY is yet another such icon.)

In these alliances and sometimes battles of abstract notions— ideographs, as van Lente (1993) calls them—one sees the creation of protected spaces for further development (as with technological progress in most of the twentieth century) as well as the creation of specific modernities linked to the particular combination that is dominant. And as happened with TECHNOLOGY and MODERNITY in the late twentieth century, when one is criticized, they both suffer. This can be a stimulus for change, as in the case of ecological modernization in the chemical industry, which Arthur Mol in his chapter offers as an argument in the analysts' discussions of MODERNITY.

Under these iconic and thus forceful terms, a variety of developments occur that are shaped by the icons. What is more, the constellation of icons shifts because of such processes and their outcomes. As it turns out, the chapters take a particular cross section through this multilevel and kaleidoscopic co-construction process. They tend to discuss technologies as they appear, more or less ready-made, in our societies. Tom Misa's phrase about the "infrastructure of daily life, choreographing the members of modern societies" characterizes this tendency beautifully, and Paul Edwards adds to this when he emphasizes "fluency" in infrastructures.

Yet, on balance, the chapters in this volume say too little about the contextual dynamics of *new* technologies and their embedding in society. Our conversation on technology and modernity needs to be widened to include evolutionary economics and recent innovation studies (see Rip and Kemp 1998), which are mentioned only in passing. There is reference to innovation systems, as when Haidar Khan discusses Taiwan. Johan Schot, in his final proposal addresses the dynamics of development, but remains programmatic. Arthur Mol assumes that there are such dynamics (in the chemical industry).

My reason for highlighting this limitation to our book is that we may well be experiencing a new wave of engineers' (and politicians') technological modernism. This can be seen in the push for biotechnology, in spite of its being contested; the massive R&D investment in the life sciences; and not least the promises made for genomics whether "green" (agricultural) or "red" (medical). True, these are heavily hyped visions, but they capture the imagination. Why is it that they capture our imagination in the late-modern world? I see a new version of technological modernism, linked to the search for, and promise of, upstream solutions for downstream problems. It is a further variant on the modernist idea of action as emanating from a source, and leading to effects because of the nature of this source, rather than through processes of co-construction.

In genomics, this view is particularly clear. It was pushed strongly because of its scientific promise, and taken up by policy makers to show that they could support important research. Heads of state in the United States and the United Kingdom wanted to announce the unraveling of the human genome. Collusion of scientists and policy makers in priority programs, each for their own reasons, is not unusual. But here there is a specific dream: if you know the genetic map and how it relates to functions and dysfunctions, you can purposefully manipulate a gene and successfully prevent an illness. This is the age-old modernist dream: causal links from an agent source to desired effects. Moreover, genomics is linked to another modernist feature, individualization, with the promise of gene passports allowing individualized therapy. Symbolic aspects are not far behind: "The genome is viewed as the core of our nature, determining both our individuality and our species identity. [It is] the true essence of human nature, with external influences considered as accidental events" (Mauron 2001: p. 831).

Ironically, the Human Genome Project itself has produced a setback to this modernist program: there are far fewer human genes than expected (about one-third of the original estimates). This implies that genes are polyvalent; not so much their mere existence but rather the *regulation* of their expression becomes important. Some genomics researchers want to keep up the front of the original promises. But for human diagnosis and therapy, and for producing better plant (and perhaps animal) varieties, the message is clear: there will be no linear relationship between genetic information, intervention, and impact.

A new wave of technological modernism is surely building in nanotechnology, which is projected as transforming the economy of the twenty-first century. Nanotechnology is defined in part by its scale, but also by the ability to manipulate matter on the nano scale and to produce materials and devices with properties that will "make a difference" at the phenomenal level. There is debate about how far we can go in this direction (Smalley 2001; Drexler 2001), but the direction itself is clear: by manipulating the minute and by creating devices, we can set in motion a chain of effects that will make a difference "downstream." There are indeed promising devices, but so far the successful ones, such as sensors and a "lab-on-a-chip," are devices for analysis, not for creating a difference.

There is a storyline of (modernist) promise and (subsequent) disappointment here. The actual co-construction processes will determine the specific outcomes and thus the new technological modernity. That is why it is important to understand the contextual dynamics of the development of new technologies and their actual embedding in society.

New Modernities

There are new technologies, but also new modernities. Simplifying a more complex process, one might argue that in the West the environmental movement introduced pluralism with respect to the alliance between technology and modernity. Multiculturalism is becoming a further challenge in the West, and globally as well, in a postcolonial world where American dominance is not only criticized but can also be undermined, at least for a time, as September 11th has shown. Adopting the U.S. State Department's view for just a moment, very modern means (flying modernist jets into modernist skyscrapers) were used to further antimodern ends (ridding the East of the West). There are multiple oversimplifications here, in particular the equation of "the" West with "the" modern (Harootunian 2000a,b), but the point is that there are cracks in the modernist alliance with technology.

North–South differences introduce a further tension. Within the South itself, this tension is played out between technoeconomic ambitions that almost of necessity reflect western-modernist approaches, and postcolonial resistance to such western dominance. In a country like South Africa, both tendencies are visible in the ruling African National Congress. Mutual "othering" occurs, with Said's (1995) diagnosis of orientalism in the West being counterposed by an African Renaissance movement following the same pattern.

For new technology, "othering" appears to occur equally in the West, North, South, and East. It appears in different forms, but all in response to the danger of an existing order being broken by the novel entrant. Modernity supports novelty, but on its own terms, and will push aside the novelties and variety that do not conform to these terms. In this sense, Arthur Mol's intriguing phrase "the modernization of modernity" (*pace* Beck) may not be able to accommodate plurality and multiculturalism.

Co-construction is a broader concept, but its recognition of multiple agency is not enough to create concerted effort in desirable directions. As David Hess would put it, a multiplicity of modernities also drains modernity of its emancipatory potential, undermining the grounds for normative critique. On the other hand, as Don Slater emphasizes, modernity is under construction (always, but definitely now), and emancipation might be a contingent effect.

Does co-construction, then, lead us to an impasse? Perhaps, but let us go back to the actors, using a "window on the world." The first window is about the Royal Dutch-Shell Company building scenarios of a future world. One of their methodological prescriptions is to identify key developments that can be assumed to be present in all scenarios, and then introduce variety. Such key developments are called TINAs (there is no alternative). In the Shell scenarios, TINA is about information and communication technologies and their effects. Is this a projection of technological determinism? Or a realistic diagnosis of how the co-evolution of technology and society turns out to work in our world?

Such scenarios are "theories in practice," and make a normative point by accepting a particular development as TINA. This becomes clear when one uses a window on another part of the modern world, where agricultural science (and to a lesser extent medical science) has to come to terms with local specificities and individual particularities. The modernist approach is to position the local and individual as specifications of global regularities and laws. Knowing the composition of the soil, or the gene makeup of the individual, one can prescribe what is necessary. This modernist approach runs into difficulties. In agriculture, the history of the (living) soil, and thus local knowledge, turns out to be important. In the medical sector, one sees a proliferation of alternative and complementary medicine, as David Hess outlines.

There need not be a contrast with the irresistible development and use of information and communication technologies. Agriculture and medicine can and do profit from the new possibilities. The tension that remains is between the disciplining requirements of information and communication technologies (a code is necessary to transmit) and the local idiosyncracies. In other words, there are elements of governance embedded in the various technologies, which in practice add up to hybrid and possibly conflicting governance.

The question of governance might lead to a fifth proposal, which would complete Tom Misa's introductory chapter. One element must be about the importance of spaces rather than action. Spaces offer opportunities for agency and for a variety of agents (depending on the nature of the space). Zygmunt Bauman has an intriguing phrase about living in the cracks of modernity (see Law 1994). Such cracks should become spaces; and agency in those spaces might shift the modernist structures a bit.

A second element starts with the recognition that co-construction happens anyway and adds that actors will anticipate and reflect and act strategically all the time. In and through their interactions, there is an emergent design effort in the small and in the large. Such design efforts should not be modernist, starting with a concept or prototype that must then be implemented. As designers, in particular in information and communication technology, but also in other technologies and projects, have been learning—often the hard way—interactive design that includes input from projected users is very important.

These two elements do not add up to a fifth proposal that can be formulated in one sentence. But such a proposal must be about creating space and making sure that something productive is done in that space. One cannot and should not define beforehand what is to "count" as productive. Descriptively, one could say that modernities and technologies co-evolve, interacting on the basis of relative autonomy. The key question is what this co-evolution will lead to and whether one can use an understanding of the dynamics and patterns as products of co-evolution to shape one's own and perhaps others' actions oriented toward an unknown future.

The fifth proposal might be the affirmative version of this question: it *is* possible to find patterns in co-evolution and show that they will return. But since these patterns are co-constructed, the actors involved must be addressed and mobilized to improve anticipation, reflection, and learning.

References

Abbate, Janet. 1999. *Inventing the Internet*. Cambridge, Mass.: MIT Press.

Abrams, Philip. 1982. *Historical Sociology*. Shepton Mallet, UK: Open Books.

Achterhuis, Hans, ed. 1997. *Van Stoommachine tot Cyborg*. Amsterdam: Ambo (Translated into English as *American Philosophy of Technology: The Empirical Turn*).

Achterhuis, Hans, ed. 2001. *American Philosophy of Technology: The Empirical Turn*. Bloomington, Ind.: Indiana University Press.

Adas, Michael. 1989. *Machines as the Measure of Men: Science, Technology, and Ideologies of Western Dominance*. Ithaca, N.Y.: Cornell University Press.

Adler, Paul S. 1990. "Marx, Machines, and Skill." *Technology and Culture* 31: 780–812.

Adorno, T. et al. 1976. *The Positivist Dispute in German Sociology*. Translated by G. Adey and D. Frisby. London: Heinemann.

Akrich, Madeleine. 1992. "The De-Scription of Technical Objects," in W. Bijker and J. Law, eds. *Shaping Technology/Building Society: Studies in Sociotechnical Change*. Cambridge, Mass.: MIT Press, pp. 205–224.

Akrich, Madeleine. 1995. "User Representations: Practices, Methods and Sociology," in A. Rip, T. J. Misa, and J. Schot, eds. *Managing Technology in Society: The Approach of Constructive Technology Assessment*. London: Pinter, pp. 167–184.

Albrow, Martin. 1996. *The Global Age*. Stanford, Calif.: Stanford University Press.

Alder, Ken. 1997. *Engineering the Revolution: Arms and Enlightenment in France, 1763–1815*. Princeton, N.J.: Princeton University Press.

Allen, Michael Thad. 2001. "Modernity, the Holocaust, and Machines without History," in Michael Thad Allen and Gabrielle Hecht, eds., *Technologies of Power*. Cambridge, Mass.: MIT Press, pp. 175–214.

Alroy, John. 2001. "A Multispecies Overkill Simulation of the End-Pleistocene Megafauna Extinction." *Science* 292 (5523): 1892–1896.

Anbeek, Ton. 1994. *Geschiedenis van de Nederlandse Literatuur 1885–1985.* Amsterdam: Arbeiderpers, 2nd ed.

Antonio, Robert. 2001. *Marx and Modernity: Key Readings and Commentary.* Oxford/Cambridge: Blackwell.

Appadurai, Arjun. 1986. *The Social Life of Things.* Cambridge: Cambridge University Press.

Arquilla, John, and David F. Ronfeldt, eds. 1997. *In Athena's Camp: Preparing for Conflict in the Information Age.* Santa Monica, Calif.: RAND Corporation.

Arthur, Brian. 1994. *Increasing Returns and Path Dependence in the Economy.* Ann Arbor: University of Michigan Press.

Aytac, I. A., J. B. McKinlay, and R. J. Krane. 1999. "The Likely Worldwide Increase in Erectile Dysfunction between 1995 and 2025 and some Possible Policy Consequences." *British Journal of Urology* 84: 50–56.

Baer, Hans. 1989. "The American Dominative Medical System as a Reflection of Social Relations in the Larger Society." *Social Science and Medicine* 28(11): 1103–1112.

Baer, Hans. 1995. "Medical Pluralism in the United States." *Medical Anthropology Quarterly* 9(4): 493–502.

Bakardjieva, Maria, and Andrew Feenberg. "Community Technology and Democratic Rationalization." *The Information Society* (forthcoming Spring 2002).

Ball, Kirstie. 2000. "Situating Surveillance: Representation, Meaning, Movement, and Manipulation." Paper presented at the Surveillance and Society conference, Hull University. Available at <business.bham.ac.UK/business/papers/easst.htm> (18 Nov. 2001).

Ball, Kirstie, and David Wilson. 2000. "Power, Control, and Computer-Based Performance Monitoring: Repertoires, Subjectivities, and Resistance." *Organization Studies* 21 (3): 539–565.

Banham, Reyner. 1986. *A Concrete Atlantis: U.S. Industrial Building and European Modern Architecture, 1900–1925.* Cambridge, Mass.: MIT Press.

Baran, Paul, et al. 1964. "On Distributed Communications: RAND Memorandum Series." Santa Monica, Calif.: RAND Corporation.

Barber, Benjamin. 1995. *Jihad vs. McWorld.* New York: Random House.

Barker, Joshua, Nico Schulte Noordholt, and Arie Rip. 2001. *The Societal Construction of Technology in Indonesia.* Enschede, Netherlands: University of Twente. Final Report to the Royal Netherlands Academy of Sciences.

Barns, Ian. 1999. "Technology and Citizenship," in Alan Petersen, Ian Barns, Janice Dudley, and Patricia Harris, eds. *Poststructuralism, Citizenship, and Social Policy.* London/New York: Routledge, pp. 154–198.

Barrett, M. 1992. "Words and Things: Materialism and Method in Contemporary Feminist Analysis," in M. Barrett and A. Phillips, eds. *Destabilizing Theory.* Cambridge: Polity.

Baudrillard, Jean. 1995. *Simulacra and Simulation*. Translated by S. Fraser. Ann Arbor: University of Michigan Press.

Bauer, Martin, ed. 1995. *Resistance to New Technology: Nuclear Power, Information Technology, Biotechnology*. Cambridge: Cambridge University Press.

Bauman, Zygmunt. 1992. *Intimations of Postmodernity*. London/New York: Routledge.

Bauman, Zygmunt. 1993. *Postmodern Ethics*. Oxford: Blackwell.

Baxter, James Phinney. 1948. *Scientists Against Time*. Boston: Little, Brown.

Baylis, R., L. Connell, and A. Flynn. 1998a. "Sector Variation and Ecological Modernization: Towards an Analysis at the Level of the Firm." *Business Strategy and the Environment* 7 (3): 150–161.

Baylis, R., L. Connell, and A. Flynn. 1998b. "Company Size, Environmental Regulation and Ecological Modernization: Further Analysis at the Level of the Firm." *Business Strategy and the Environment* 7 (5): 285–296.

Beck, Adrian, and Andrew Willis. 1995. *Crime and Security: Managing the Risk to Safe Shopping*. London: Perpetuity Press.

Beck, Ulrich. 1986. *Risikogesellschaft: Auf dem Weg in eine andere Moderne*. Frankfurt/Main: Suhrkamp Verlag.

Beck, Ulrich. 1992. *Risk Society: Towards a New Modernity*. London: Sage.

Beck, Ulrich. 1994. "The Reinvention of Politics: Towards a Theory of Reflexive Modernisation," in Ulrich Beck, Anthony Giddens and Scott Lash, *Reflexive Modernization: Politics, Tradition and Aesthetics in the Modern Social Order*. Cambridge: Polity; Stanford, Calif.: Stanford University Press, pp. 1–55.

Beck, Ulrich. 1995. *Ecological Politics in an Age of Risk*. Cambridge: Polity.

Beck, Ulrich. 1996. "Risk Society and the Provident State," in Scott Lash, Bronislaw Szerszynski, and Brian Wynne, eds. *Risk, Environment and Modernity: Towards a New Ecology*. London: Sage, pp. 27–43.

Beck, Ulrich. 1997. *The Reinvention of Politics*. Cambridge: Polity.

Beck, Ulrich, Anthony Giddens, and Scott Lash. 1994. *Reflexive Modernization: Politics, Tradition and Aesthetics in the Modern Social Order*. Cambridge: Polity; Stanford, Calif.: Stanford University Press.

Bell, Daniel. 1976. *The Coming of Post-Industrial Society: A Venture in Social Forecasting*. Harmondsworth, UK: Penguin, Peregrine Books.

Beniger, James R. 1989. *The Control Revolution: Technological and Economic Origins of the Information Society*. Cambridge, Mass.: Harvard University Press.

Berg, Maxine. 1980. *The Machinery Question and the Making of Political Economy 1815–1848*. Cambridge: Cambridge University Press.

Berg, Maxine. 1985. *The Age of Manufactures: Industry, Innovation, and Work in Britain, 1700–1820*. New York: Oxford University Press.

Berman, Marshall. 1982. *All That is Solid Melts into Air: The Experience of Modernity*. New York: Simon and Schuster. Paperback reprinted 1988.

Berners-Lee, Tim, and Robert Cailliau. 1990. "WorldWideWeb: Proposal for a Hyper Text Project." CERN (European Laboratory for Particle Physics), Geneva. <www.w3.org/proposal.html> (last accessed April 15, 2002).

Best, Steven, and Douglas Kellner. 1991. *Postmodern Theory: Critical Interrogations*. Houndmills and London: Macmillan.

Biersack, Aletta. 1999. "Introduction: From the 'New Ecology' to the New Ecologies." *American Anthropologist* 101(1): 5–18.

Bijker, Wiebe. 1992. "The Social Construction of Fluorescent Lighting, Or How an Artifact was Invented in Its Diffusion Stage," in Wiebe Bijker and John Law, eds. *Shaping Technology/Building Society: Studies in Sociotechnical Change*. Cambridge, Mass.: MIT Press, pp. 75–102.

Bijker, Wiebe. 1993. "Do not Despair: There is Life after Constructivism." *Science, Technology & Human Values* 18: 113–138.

Bijker, Wiebe. 1995a. *Of Bicycles, Bakelites, and Bulbs*. Cambridge, Mass.: MIT Press.

Bijker, Wiebe. 1995b. "Sociohistorical Technology Studies," in S. Jasanoff, G. Markle, J. Peterson, and T. Pinch, eds. *Handbook of Science and Technology Studies*. London: Sage, pp. 229–256.

Bijker, Wiebe, and John Law, eds. 1992. *Shaping Technology/Building Society: Studies in Sociotechnical Change*. Cambridge, Mass.: MIT Press.

Bijker, Wiebe, Trevor Pinch, and Thomas Hughes, eds. 1987. *The Social Construction of Technological Systems: New Directions in the Sociology and History of Technology*. Cambridge, Mass.: MIT Press.

Blaszczyk, Regina. 2000. *Imagining Consumers: Design and Innovation from Wedgwood to Corning*. Baltimore, Md.: Johns Hopkins University Press.

Blaug, R. 1997. "Between Fear and Disappointment: Critical, Empirical and Political Use of Habermas." *Political Studies* XLV: 100–117.

Blechmann, M., ed. 1999. *Revolutionary Romanticism*. San Francisco: City Lights Books.

Blondheim, Menahem. 1994. *News over the Wires: The Telegraph and the Flow of Public Information in America, 1844–1897*. Cambridge, Mass.: Harvard University Press.

Bloor, David 1976. *Knowledge and Social Imagery*. London: Routledge and Kegan Paul; Chicago: University of Chicago Press. Reprinted 1991.

Blowers, A. 1997. "Environmental Policy: Ecological Modernization and the Risk Society?" *Urban Studies* 34 (5–6): 845–871.

Blühdorn, I. 2000. "Ecological Modernisation and Post-Ecologist Politics," in G. Spaargaren, A. P. J. Mol, and F. Buttel, eds. *Environment and Global Modernity*. London: Sage, pp. 209–228.

Bogard, William. 1996. *The Simulation of Surveillance*. Cambridge/New York: Cambridge University Press.

Bologh, Roslyn. 1990. *Love or Greatness? Max Weber and Masculine Theorizing*. London: Unwin Hyman.

Boons, F., L. Baas, J. J. Bouma, A. Groen, and K. Le Blansch. 2000. *The Changing Nature of Business: Institutionalization of Green Organisational Routines in the Netherlands 1986–1995*. Utrecht: International Books.

Borg, Kevin. 1999. "The 'Chauffeur Problem' in the Early Auto Era: Structuration Theory and the Users of Technology." *Technology and Culture* 40: 797–832.

Borgmann, Albert. 1984. *Technology and the Character of Contemporary Life*. Chicago: University of Chicago Press.

Borgmann, Albert. 1992. *Crossing the Postmodern Divide*. Chicago: University of Chicago Press.

Borning, Alan. 1987. "Computer System Reliability and Nuclear War." *Communications of the ACM* 30: 112–131.

Bourdieu, Pierre. 1977. *Outline of a Theory of Practice*. Cambridge: Cambridge University Press.

Bourdieu, Pierre. 1991. *Language and Symbolic Power*. Cambridge, Mass.: Harvard University Press.

Bourdieu, Pierre, and Loïc J. D. Wacquant. 1992. *An Invitation to Reflexive Sociology*. Chicago: University of Chicago Press.

Bouwens, A. M. C. M., and M. L. J. Dierikx. 1996. *Tachtig jaar Schiphol, Op de Drempel van de Lucht*. Den Haag: SDU.

Bowker, Geoffrey C., and Susan Leigh Star. 1999. *Sorting Things Out: Classification and its Consequences*. Cambridge, Mass.: MIT Press.

Boyer, Christine. 1996. *Cybercities: Visual Perception in the Age of Electronic Communication*. New York: Princeton Architectural Press.

Boyne, Roy. 2000. "Post-Panopticism." *Economy and Society* 29 (2): 285–307.

Boyne, Roy, and Ali Rattansi, eds. 1990. *Postmodernism and Society*. London: Macmillan.

Bracken, Paul. 1983. *The Command and Control of Nuclear Forces*. New Haven, Conn.: Yale University Press.

Bras-Klapwijk, R. M. 1999. "Adjusting Life Cycle Assessment Methodology for Use in Public Discourse." Ph.D. dissertation, Technical University of Delft, the Netherlands.

Braun, Ingo, and Bernward Joerges, eds. 1994. *Technik ohne Grenzen*. Frankfurt/Main: Suhrkamp.

Bray, Francesca. 1997. *Technology and Gender: Fabrics of Power in Late Imperial China*. Berkeley: University of California Press.

Brey, Philip. 1998. "Space-Shaping Technologies and the Geographical Disembedding of Place," in A. Light and B. Smith, eds. *Philosophy & Geography*. vol. 3, *Philosophies of Place*. New York/London: Rowman & Littlefield, pp. 239–263.

Brooks, Michael W. 1997. *Subway City: Riding the Trains, Reading New York*. New Brunswick, N. J.: Rutgers University Press.

Buttel, F. 2000. "Ecological Modernization as Social Theory." *Geoforum* 31(1): 57–66.

Calhoun, Craig. 1994. "The Infrastructure of Modernity: Indirect Relationships, Information Technology, and Social Integration," in Hans Haferkamp and Neil J. Smelser, eds. *Social Change and Modernity*. Berkeley: University of California Press, pp. 205–236.

Calhoun, Craig. 1995. *Critical Social Theory*. Oxford: Blackwell.

Calhoun, Craig. 1998. "Explanation in Historical Sociology: Narrative, General Theory and Historically Specific Theory." *American Journal of Sociology* 104(3): 846–871.

Callon, Michel. 1987. "Society in the Making: The Study of Technology as a Tool for Sociological Analysis," in W. Bijker, T. Pinch, and T. Hughes, eds. *The Social Construction of Technological Systems: New Directions in the Sociology and History of Technology*. Cambridge, Mass.: MIT Press, pp. 83–103.

Callon, Michel. 1995. "Four Models of the Dynamics of Science," in S. Jasanoff, G. Markle, J. Peterson, and T. Pinch, eds. *Handbook of Science and Technology Studies*. London: Sage, pp. 29–63.

Callon, Michel, and Bruno Latour. 1981. "Unscrewing the Big Leviathan: How Actors Macro-structure Reality and How Sociologists Help Them to Do So," in Karin D. Knorr-Cetina and Aaron V. Cicourel, eds. *Advances in Social Theory and Methodology: Toward an Integration of Micro- and Macro-sociologies*. Boston: Routledge & Kegan Paul, pp. 277–303.

Callon, Michel, and Bruno Latour. 1992. "Don't Throw the Baby Out with the Bath School! A Reply to Collins and Yearley," in Andrew Pickering, ed., *Science as Practice and Culture*. Chicago: University of Chicago Press, pp. 343–368.

Callon, Michel, John Law, and Arie Rip, eds. 1986. *Mapping the Dynamics of Science and Technology*. London: Macmillan.

Campbell-Kelly, Martin, and William Aspray. 1996. *Computer: A History of the Information Machine*. New York: Basic Books.

Castells, Manuel. 1989. *The Informational City: Information Technology, Economic Restructuring, and the Urban Regional Process*. Cambridge: Blackwell.

Castells, Manuel. 1996. *The Rise of the Network Society. Information Age*. vol. 1 Oxford/Cambridge: Blackwell.

Castells, Manuel. 1997. *The Power of Identity. Information Age*. vol. 2 Oxford/Cambridge: Blackwell.

Castells, Manuel. 1998. *End of Millennium. Information Age*. vol. 3 Oxford/ Cambridge: Blackwell.

CEFIC. 1999. *CEFIC Responsible Care 1999*. Brussels: Council of European Federations of the Chemical Industry.

Chandler, Alfred D. 1977. *The Visible Hand: The Managerial Revolution in American Business*. Cambridge, Mass.: Belknap Press.

Charney, Leo. 1998. *Empty Moments: Cinema, Modernity, and Drift*. Durham, N.C.: Duke University Press.

Charney, Leo, and Vanessa R. Schwartz, eds. 1995. *Cinema and the Invention of Modern Life*. Berkeley: University of California Press.

Chiang, J. T. 1990. "Management of National Technology Program in a Newly Industrializing Country: Taiwan." *Technovation* 10 (8): 531–554.

Christ, G. J. 1998. "The Control of Corporal Smooth Muscle Tone, the Coordination of Penile Erection and the Etiology of Erectile Dysfunction: The Devil Is in the Details." *Journal of Sex Education and Therapy* 23(3): 187–193.

Clark, Kim B. 1985. "The Interaction of Design Hierarchies and Market Concepts in Technological Evolution." *Research Policy* 14: 235–251.

Clarke, Adele E. 1998. *Disciplining Reproduction: Modernity, American Life Sciences, and "the Problems of Sex."* Berkeley: University of California Press.

Clarke, Roger. 1988. "Information Technology and Dataveillance." *Communications of the ACM* 31, 5 (May): 498–512.

Clunas, Craig. 1999. "Modernity Global and Local: Consumption and the Rise of the West." *American Historical Review* 104: 1497–1511.

Cockburn, Cynthia, and Susan Ormrod. 1993. *Gender and Technology in the Making*. London: Sage.

Cohen, Adam. 2000. "Spies Among Us" *Time* (June 26): 7–15 (technology section).

Cohen, J. L., and A. Arato. 1992. *Civil Society and Political Theory*. Cambridge, Mass.: MIT Press.

Cohen, M. J. 2000. "Ecological Modernisation, Environmental Knowledge and National Character: A Preliminary Analysis of the Netherlands." *Environmental Politics* 9(1):77–105.

Cohen, Stanley. 1985. *Visions of Social Control*. Oxford/Cambridge: Blackwell.

Collins, Harry, and Trevor Pinch. 1998. *The Golem at Large: What You Should Know About Technology*. Cambridge: Cambridge University Press.

Collins, James, and Jerry Porras. 1994. *Built to Last: Successful Habits of Visionary Companies*. New York: Harper & Row.

Commission of the European Communities. 1993. *Panorama of EU Industry 1993*. Luxembourg: Office for Official Publications of the European Communities.

Commission of the European Communities. 1997. *Panorama of EU Industry 1997*. Luxembourg: Office for Official Publications of the European Communities, vol. 1.

Constant, Edward. 2000. "Recursive Practice and the Evolution of Technological Knowledge," in John Ziman ed., *Technological Innovation as an Evolutionary Process*. Cambridge University Press, pp. 219–233.

Corn, Joseph J. 1983. *The Winged Gospel: America's Romance with Aviation, 1900–1950*. New York: Oxford University Press.

Cortada, James. 1993. *The Computer in the United States: From Laboratory to Market, 1930 to 1960*. Armonk, N.Y.: M. E. Sharpe.

Cortada, James. 1996. *Information Technology as Business History*. Westport, Conn.: Greenwood.

Crocker, David A. 1995. "Functioning and Capability: The Foundation of Sen's and Nussbaum's Development Ethic," in Martha Nussbaum and Jonathan Glover, eds., *Women, Culture and Development*. New York: Oxford University Press, pp. 153–198.

Crossley, Nick. 1997. "Corporeality and Communicative Action: Embodying the Renewal of Critical Theory." *Body and Society* 3(1): 17–46.

Culbert, Michael. 1974. *Vitamin B17: A Forbidden Weapon Against Cancer*. New Rochelle, N.Y.: Arlington House.

Cutcliffe, Stephen H. 2000. *Ideas, Machines and Values*. Savage, Md.: Rowman & Littlefield.

Cutcliffe, Stephen H., and Carl Mitcham, eds. 2001. *Visions of STS: Counterpoints in Science, Technology, and Society Studies*. Albany: State University of New York Press.

Cutcliffe, Stephen H., and Robert C. Post, eds. 1989. *In Context: History and the History of Technology*. Bethlehem, Pa.: Lehigh University Press.

DaMatta, Roberto. 1991. *Carnivals, Rogues, and Heroes*. Notre Dame, Ind.: Notre Dame University Press.

Dandeker, Christopher. 1990. *Surveillance, Power, and Modernity*. Cambridge: Polity.

David, Paul. 1985. "Clio and the Economics of QWERTY." *American Economic Review* 75 (2): 332–337.

Delanty, G. 1997. *Social Science: Beyond Constructivism and Relativism*. Buckingham, UK.: Open University Press.

Delanty, G. 1999. *Social Theory in a Changing World: Conceptions of Modernity*. Cambridge: Polity.

Derrida, J. 1981. *Positions*. Chicago: University of Chicago Press.

Derrida, J. 1988. *Limited Inc*. Evanston, Ill.: Northwestern University Press.

Dessauer, F. 1958. *Streit um Technik*. Frankfurt/Main: Josef Knecht.

Diamond, Jared. 1995. "The Worst Mistake in the History of the Human Race," in Jean-Luc Chodkiewicz, ed. *Peoples of the Past and Present*. New York: Harcourt Brace, pp. 113–117.

Dickens, P. 1998. "Beyond Sociology: Marxism and the Environment," in M. Redclift and G. Woodgate, eds. *The International Handbook of Environmental Sociology*. Cheltenham/Northampton, UK: Edward Elgar, pp. 179–192.

Dierkes, Meinolf, Andreas Knie, and Peter Wagner. 1990. "Engineers, Intellectuals and the State." *Industrial Crisis Quarterly* 4: 155–174.

Disco, Cornelis, Hans Buiter, and Adrienne van den Bogaard. 2002. *Stad en Stedelijke Technologie*. vol. 6, *History of Technology in the 20th Century*. Zutphen: Walburg.

Dohrn-van Rossum, Gerhard. 1996. *History of the Hour: Clocks and Modern Temporal Orders*. Translated by Thomas Dunlap. Chicago: University of Chicago Press.

Dosi, Giovani, Chris Freeman, Richard Nelson, G. Silverberg, and L. Soete, eds. 1988. *Technical Change and Economic Theory*. London: Pinter.

Douglas, Susan J. 1987. *Inventing American Broadcasting, 1899–1922*. Baltimore, Md.: Johns Hopkins University Press.

Douglas, Susan J. 1995. *Where the Girls Are: Growing Up Female With the Mass Media*. New York: Times Books.

Douglas, Susan J. 1999. *Listening In: Radio and the American Imagination*. New York: Times Books.

Drew, Brian, ed. 1998. *Viagra and the Quest for Potency*. Sarasota: Health Publishers.

Drexler, K. Eric. 2001. "Machine-Phase Nanotechnology." *Scientific American* 285 (September): 74–75. Available at <www.sciam.com/nanotech/> (9 Dec. 2001).

Driel, Hugo van, and Ferry de Goey. 2000. *Rotterdam: Cargo Handling Technology 1870–2000*. Zutphen/Eindhoven, Netherlands: Walburg.

Driel, Hugo van, and Johan Schot. 2001. "Regime-shifts in de Rotterdamse haven." *NEHA Jaarboek*: 286–318.

Driver, Felix, and David Gilbert, eds. 1999. *Imperial Cities: Landscape, Display and Identity*. Manchester, UK and New York: Manchester University Press.

Dryzek, J. S. 1995. "Critical Theory as a Research Program," in S. K. White, ed. *The Cambridge Companion to Habermas*. Cambridge: Cambridge University Press, pp. 97–119.

Du Mont, J., and D. Parnis. 2000. "Sexual Assault and Legal Resolution: Querying the Medical Collection of Forensic Evidence." *Medicine and Law* 19(4): 779–792.

Dumont, Louis. 1986. *Essays on Individualism*. Chicago: University of Chicago Press.

Duncombe, Stephen. 1997. "I've Seen the Future—and It's a Sony!" in Thomas Frank and Matt Weiland, eds. *Commodify Your Dissent*. New York: W. W. Norton, pp. 99–111.

Earle, Timothy. 1997. *How Chiefs Come to Power*. Stanford, Calif.: Stanford University Press.

Edgerton, David. 1998. "De L'Innovation aux Usages: Dix thèses eclectiques sur l'histoire des techniques." *Annals HSS* (July–October) nos. 4–5: 815–837. Also published in *History and Technology*.

Edquist, Charles, ed. 1997. *Systems of Innovation*. London: Pinter.

Edwards, Paul N. 1996. *The Closed World: Computers and the Politics of Discourse in Cold War America*. Cambridge, Mass.: MIT Press.

Edwards, Paul N. 1998a. "Virtual Machines, Virtual Infrastructures: The New Historiography of Information Technology." *Isis* 89 (1): 93–99.

Edwards, Paul N. 1998b. "Y2K: Millennial Reflections on Computers as Infrastructure." *History & Technology* 15: 7–29.

Eisenberg, David, Roger Davis, Susan Ettner, Scott Appel, Sonja Wilkey, Maria Van Rompay, et al. 1998. "Trends in Alternative Medicine Use in the U.S., 1990–1997." *Journal of the American Medical Association* 280(18): 1569–1575.

Eisenstein, Elizabeth L. 1983. *The Printing Revolution in Early Modern Europe*. Cambridge: Cambridge University Press.

Ekins, P., and S. Speck. 2000. "Proposals of Environmental Fiscal Reforms and the Obstacles to their Implementation." *Journal of Environmental Policy and Planning* 2 (2): 93–114.

Elam, M. 1994. "Anti Anticonstructivism or Laying the Fears of a Langdon Winner to Rest." *Science, Technology & Human Values* 19: 101–106.

Ellis, Stephen. 1999. "Beware What You Leave on the Net: Someone's Watching." *The Australian* (October 14): 30.

Engels, Friedrich. 1970. *The Housing Question*. Moscow: Progress.

Engerman, David C. 2000. "Modernization from the Other Shore: American Observers and the Costs of Soviet Economic Development." *American Historical Review* 105: 383–416.

Epstein, Steve. 1996. *Impure Science*. Berkeley: University of California Press.

Ericson, Richard, and Kevin Haggerty. 1997. *Policing the Risk Society*. Toronto: University of Toronto Press.

European Environmental Agency. 1998. *Europe's Environment: The Second Assessment*. New York/Amsterdam: Elsevier.

Evans, P. 1995. *Embedded Autonomy: States and Industrial Transformation*. Princeton, N.J.: Princeton University Press.

Eyerman, Ron, and Andrew Jamison. 1991. *Social Movements: A Cognitive Approach*. Cambridge: Polity.

Featherstone, Mike. 1991. *Consumer Culture and Postmodernism.* London: Sage.

Feenberg, Andrew. 1986. *Lukács, Marx, and the Sources of Critical Theory.* New York: Oxford University Press.

Feenberg, Andrew. 1991. *Critical Theory of Technology.* Oxford University Press.

Feenberg, Andrew. 1992. "Subversive Rationalization: Technology, Power and Democracy." *Inquiry* 35 (3–4): 301–322.

Feenberg, Andrew. 1995. *Alternative Modernity: The Technical Turn in Philosophy and Social Theory.* Los Angeles/Berkeley: University of California Press.

Feenberg, Andrew. 1999a. *Questioning Technology.* London/New York: Routledge.

Feenberg, Andrew. 1999b. "Reflections on the Distance Learning Controversy." *Canadian Journal of Communication* 24 (3): 337–348.

Feenberg, Andrew. 1999c. "Whither Educational Technology?" *Peer Review* (Summer): 4–7.

Feenberg, Andrew. 2000a. "From Essentialism to Constructivism: Philosophy of Technology at the Crossroads," in E. Higgs, A. Light, and D. Strong, eds. *Technology and the Good Life?* Chicago: University of Chicago Press, pp. 294–315.

Feenberg, Andrew. 2000b. "Technology in a Global World." Paper presented at APA Eastern Division meeting in New York.

Feenberg, Andrew. 2002. *Transforming Technology: A Critical Theory Revised.* New York: Oxford University Press.

Felski, Rita. 1989. "Feminism, Postmodernism and the Critique of Modernity." *Cultural Critique* 13: 33–56.

Felski, Rita. 1995. *The Gender of Modernity.* Cambridge, Mass.: Harvard University Press.

Feng, Patrick. 2002. "Global Standards, Local Cultures," Ph.D. dissertation. Science and Technology Studies Department, Rensselaer Polytechnic Institute, Troy, N.Y.

Ferguson, Eugene. 1992. *Engineering and the Mind's Eye.* Cambridge, Mass.: MIT Press.

Ferguson, Kathy. 1993. *The Man Question: Visions of Subjectivity in Feminist Theory.* Berkeley: University of California Press.

Feynman, Richard. 1988. *"What Do You Care What Other People Think?"* New York: W. W. Norton.

Fischer, Claude S. 1992. *America Calling: A Social History of the Telephone to 1940.* Berkeley: University of California Press.

Fischer, Frank. 1995. *Evaluating Public Policy.* Chicago: Nelson-Hall.

FO Industrie. 2000. *Uitvoering intentieverklaring Chemische industrie: Jaarrapportage 1999.* Den Haag: Facilitaire Organisatie Industrie.

Forster, E. M. 1924. *A passage to India*. New York: Grosset. Reprinted in 1952.

Foucault, Michel. 1977. *Discipline and Punish: The Birth of the Prison*. Translated by A. Sheridan. New York: Pantheon Books. (Paperback edition by Vintage 1979.)

Foucault, Michel. 2000. *Power: The Essential Works of Foucault, 1954–1984*. vol. 3. New York: New Press.

Fox, Robert, ed. 1999. *Technological Change: Methods and Themes in the History of Technology*. Amsterdam: Harwood.

Franke, J. F., and F. Wätzold. 1995. *The Chemical Industry and the Eco-Management and Audit Scheme*. Berlin: Technische Universität Berlin (Institut für Volkswirtschaftslehre).

Fraser, Nancy. 1989. *Unruly Practices: Power, Discourse and Gender in Contemporary Social Theory*. Cambridge: Polity.

Freeman, C. 1987. *Technology and Economic Performance: Lessons from Japan*. London: Pinter.

Friedlander, Amy. 1995a. *Emerging Infrastructure: The Growth of Railroads*. Reston, Va.: Corporation for National Research Initiatives.

Friedlander, Amy. 1995b. *Natural Monopoly and Universal Service: Telephones and Telegraphs in the U.S. Communications Infrastructure, 1837–1940*. Reston, Va.: Corporation for National Research Initiatives.

Friedlander, Amy. 1996. *Power and Light: Electricity in the U.S. Energy Infrastructure, 1870–1940*. Reston, Va.: Corporation for National Research Initiatives.

Frisby, David. 1986. *Fragments of Modernity*. Cambridge, Mass.: MIT Press.

Fuller, Steve. 1997. *Science*. Minneapolis: University of Minnesota Press.

Fuller, Steve. 2000. *Thomas Kuhn: A Philosophical History for Our Times*. Chicago: University of Chicago.

Garcia-Johnson, R. 2000. *Exporting Environmentalism: U.S. Multinational Chemical Corporations in Brazil and Mexico*. Cambridge, Mass.: MIT Press.

Giddens, Anthony. 1973. *Capitalism and Modern Social Theory: An Analysis of the Writings of Marx, Durkheim and Weber*. Cambridge: Cambridge University Press.

Giddens, Anthony. 1979. *Central Problems in Social Theory: Action, Structure, and Contradiction in Social Analysis*. London: Macmillan.

Giddens, Anthony. 1981. "Agency, Institution, and Time-space Analysis," in K. Knorr-Cetina and A. V. Cicourel, eds. *Advances in Social Theory and Methodology: Toward an Integration of Micro- and Macro-sociologies*. Boston: Routledge & Kegan Paul, pp. 161–174.

Giddens, Anthony. 1984. *The Constitution of Society*. Berkeley: University of California Press.

Giddens, Anthony. 1985. *The Nation-State and Violence*. Cambridge: Polity.

Giddens, Anthony. 1990. *The Consequences of Modernity*. Stanford, Calif.: Stanford University Press.

Giddens, Anthony. 1991. *Modernity and Self-Identity: Self and Society in the Late Modern Age*. Cambridge: Polity; Stanford, Calif.: Stanford University Press.

Giddens, Anthony. 1994a. "Living in a Post-Traditional Society," in U. Beck, A. Giddens, and S. Lash. *Reflexive Modernization: Politics, Tradition and Aesthetics in the Modern Social Order*. Cambridge: Polity; Stanford, Calif.: Stanford University Press, pp. 56–109.

Giddens, Anthony. 1994b. *Beyond Left and Right: The Future of Radical Politics*. Cambridge: Polity.

Gilbert, Alan. 1990. *Democratic Individuality*. Cambridge: Cambridge University Press.

Gill, Colin. 1985. *Work, Unemployment, and the New Technology*. Cambridge: Polity.

Gill, R. 1996. "Power, Social Transformation and the New Determinism: A Comment on Grint and Woolgar." *Science, Technology & Social Values* 21(3): 347–353.

Gill, R., and K. Grint. 1995. "The Gender-Technology Relation: An Introduction," in K. Grint and R. Gill, *The Gender-Technology Relation: Contemporary Theory and Research*. London: Taylor and Francis, pp. 1–28.

Gilroy, P. 1993. *The Black Atlantic: Modernity and Double Consciousness*. London: Verso.

Gitelman, Lisa. 1999. *Scripts, Grooves, and Writing Machines: Representing Technology in the Edison Era*. Stanford, Calif.: Stanford University Press.

Gleich, A. von. 1988. "Werkzeugcharakter, Eingriffstiefe und Mitproduktivität als zentrale Kriterien der Technikbewertung und Technikwahl," in F. Rauner, ed. *Gestalten: eine neue gesellschaftliche Praxis*. Bonn: Verlag Neue Gesellschaft.

Gleich, A. von. 1991. "Über den Umgang mit Natur: Sanfte Chemie als wissenschaftliches, chemiepolitisches und regionalwirtschaftliches Konzept." *Wechselwirkung* 48: 4–12.

Goddard, Stephen B. 1994. *Getting There: The Epic Struggle between Road and Rail in the American Century*. New York: Basic Books.

Goldstein, Irwin, and Jennifer Berman. 1998. "Vasculogenic Female Sexual Dysfunction: Vaginal Engorgement and Clitoral Erectile Insufficiency Syndromes." *International Journal of Impotence Research* 10 (suppl.): S84.

Gordon, Deborah. 1988. "Tenacious Assumptions in Western Medicine," in M. Lock and D. Gordon, eds. *Biomedicine Examined*. Boston: D. Riedel, pp. 165–196.

Graham, Stephen. 1998. "Spaces of Surveillant Simulation: New Technologies, Digital Representations, and Material Geographies." *Environment and Planning D: Society and Space* 16: 486.

Green, K. 1992. "Creating Demand for Biotechnology: Shaping Technologies and Markets," in R. Coombs, P. Saviotti, and V. Walsh, eds., *Technological Change and Company Strategies*. London: Harcourt Brace Jovanovich, pp. 164–165.

Green, K., and Alan Irwin. 1996. "Clean Technologies," in P. Groenewegen, K. Fischer, E. G. Jenkins, and J. Schot, eds. *The Greening of Industry: Resource Guide and Bibliography*. Washington D.C.: Island Press, pp. 169–194.

Gregory, Richard. 1981. *Mind in Science*. Penguin Books.

Grint, Keith, and R. Gill, eds. 1995. *The Gender-Technology Relation: Contemporary Theory and Research*. London: Taylor and Francis.

Grint, Keith, and Steve Woolgar. 1995. "On Some Failures of Nerve in Constructivist and Feminist Analyses of Technology." *Science, Technology & Human Values* 20(3): 286–310.

Grint, Keith, and Steve Woolgar. 1997. *The Machine at Work*. Cambridge: Polity.

Habermas, Jürgen. 1970. "Technology and Science as 'Ideology'," in *Toward a Rational Society*. Translated by J. Shapiro. Boston: Beacon Press, pp. 62–122.

Habermas, Jürgen. 1983. "Modernity: An Incomplete Project," in Hal Foster, ed. *The Anti-Aesthetic: Essays on Postmodern Culture*. Port Townsend, Wash.: Bay Press, pp. 3–15.

Habermas, Jürgen. 1984–87. *Theory of Communicative Action*. Translated by T. McCarthy. Boston: Beacon Press, 2 vols.

Habermas, Jürgen. 1986. *Autonomy and Solidarity: Interviews*. P. Dews, ed. London: Verso.

Habermas, Jürgen. 1987. *The Philosophical Discourse of Modernity: Twelve Lectures*. Cambridge, Mass.: MIT Press.

Habermas, Jürgen. 1989. *The Structural Transformation of the Public Sphere*. Cambridge, Mass.: MIT Press. Reprinted in 1991.

Habermas, Jürgen. 1996. *Between Facts and Norms*. Cambridge, Mass.: MIT Press.

Hacker, Barton C. 1997. "The Weapons of the West: Military Technology and Modernization in 19th-Century China and Japan," in T. S. Reynolds and S. H. Cutcliffe, eds. *Technology and the West: A Historical Anthology from Technology and Culture*. Chicago: University of Chicago Press.

Hafner, Katie, and Matthew Lyon. 1996. *Where Wizards Stay Up Late: The Origins of the Internet*. New York: Simon and Schuster.

Haggerty, Kevin, and Ericson, Richard. 2000. "The Surveillant Assemblage." *British Journal of Sociology* 54(1): 605–622.

Hajer, Maarten. 1995. *The Politics of Environmental Discourse: Ecological Modernisation and the Policy Process*. Oxford: Clarendon.

Hajer, Maarten. 1996. "Ecological Modernization as Cultural Politics," in S. Lash, B. Szerszynski, and B. Wynne, eds. *Risk, Environment and Modernity: Towards a New Ecology*. London: Sage, pp. 246–268.

Hall, Stuart. 1980. "Encoding and Decoding," in S. Hall, D. Hobson, A. Lowe, and P. Willis, eds. *Culture, Media, Language*. London: Hutchinson, pp. 128–138.

Hamlin, Christopher. 1998. *Public Health and Social Justice in the Age of Chadwick: Britain, 1800–1854*. Cambridge: Cambridge University Press.

Handy, B. 1998. "The Viagra Craze: A Pill to Cure Impotence?" *Time* 151 (May 4): 50–54.

Hanseth, Ole, and Eric Monteiro. 1998. "Understanding Information Infrastructure." University of Oslo, Oslo, Norway. Available at <www.ifi.uio.no/~oleha/Publications/bok.html>.

Hapnes, T., and K. H. Sørensen. 1995. "Competition and Collaboration in Male Shaping of Computing," in K. Grint and R. Gill, eds. *The Gender–Technology Relation: Contemporary Theory and Research*. London: Taylor and Francis, pp. 174–191.

Haraway, Donna. 1991. *Simians, Cyborgs and Women*. London: Routledge.

Hård, Mikael. 1994. *Machines are Frozen Spirit: The Scientification of Refrigeration and Brewing in the 19th Century—A Weberian Interpretation*. Boulder, Col.: Westview; Frankfurt/Main: Campus Verlag.

Hård, Mikael. 1998. "German Regulation: The Integration of Modern Technology into National Culture," in M. Hård and A. Jamison, eds. *The Intellectual Appropriation of Technology: Discourses on Modernity, 1900–1939*. Cambridge, Mass.: MIT Press, pp. 33–67.

Hård, Mikael, and Andrew Jamison, eds. 1998. *The Intellectual Appropriation of Technology: Discourses on Modernity, 1900–1939*. Cambridge, Mass.: MIT Press.

Harding, Sandra. 1996. "Standpoint Epistemology (a Feminist Version): How Social Advantage Creates Epistemic Advantage," in Stephen B. Turner, ed., *Social Theory and Sociology*. Oxford: Blackwell, pp. 146–160.

Harding, Susan. 2000. *The Book of Jerry Falwell*. Princeton, N.J.: Princeton University Press.

Harootunian, Harry. 2000a. *History's Disquiet: Modernity, Cultural Practice, and the Question of Everyday Life*. New York: Columbia University Press.

Harootunian, Harry. 2000b. *Overcome by Modernity: History, Culture, and Community in Interwar Japan*. Princeton, N.J.: Princeton University Press.

Harre, R. 1979. *Social Being*. Oxford: Blackwell.

Hartsock, Nancy. 1987. "Rethinking Modernism: Minority vs. Majority Theories." *Cultural Critique* 14: 15–33.

Harvey, David. 1989. *The Condition of Postmodernity: An Enquiry into the Origins of Cultural Change*. Cambridge: Blackwell.

Hatzichristou, D. G. 1998. "Current Treatment and Future Perspectives for Erectile Dysfunction." *International Journal of Impotence Research* 10 (suppl.1): S3–13.

Hauben, Michael, and Ronda Hauben. 1997. *Netizens: On the History and Impact of Usenet and the Internet.* Los Alamitos, Calif.: IEEE Computer Society Press.

Hauben, Ronda. 1996. *The Netizens' Netbook.* Available at <www.columbia.edu/~hauben/netbook/>

Hausman, B. L. 1995. *Changing Sex: Transsexualism, Technology and the Idea of Gender.* Durham, N.C.: Duke University Press.

Hawkes, Gail. 1996. *A Sociology of Sex and Sexuality.* Buckingham, UK: Open University Press.

Hay, Stephen T. 1970. *Asian Ideas of East and West: Tagore and His Critics in Japan, China, and India.* Cambridge, Mass.: Harvard University Press.

Hecht, Gabrielle. 1998. *The Radiance of France: Nuclear Power and National Identity after World War II.* Cambridge, Mass.: MIT Press.

Heidegger, Martin. 1977. "The Question Concerning Technology," in D. Krell, ed., *Basic Writings.* New York: Harper & Row, pp. 283–318.

Heidegger, Martin. 1998. "Traditional Language and Technological Language." *Journal of Philosophical Research* 23: 129–145 (translated by Wanda Torres Gregory).

Held, David, Anthony McGrew, David Goldblatt, and Jonathan Perraton. 1999. *Global Transformations: Politics, Economics and Culture.* Stanford, Calif.: Stanford University Press.

Hennessy, R. 1993. *Materialist Feminism and the Politics of Discourse.* London: Routledge.

Herf, Jeffrey. 1984. *Reactionary Modernism: Technology, Culture and Politics in Weimar and the Third Reich.* Cambridge: Cambridge University Press.

Herken, Gregg. 1983. *Counsels of War.* New York: Knopf.

Hess, David. 1991. *Spirits and Scientists.* University Park: Pennsylvania State University Press.

Hess, David. 1993. *Science in the New Age.* Madison: University of Wisconsin Press.

Hess, David. 1997. *Can Bacteria Cause Cancer?* New York: New York University Press.

Hess, David. 1999. *Evaluating Alternative Cancer Therapies.* New Brunswick, N.J.: Rutgers University Press.

Higgs, Eric, Andrew Light, and David Strong, eds. 2000. *Technology and the Good Life?* Chicago: University of Chicago Press.

Hitt, J. 2000. "The Second Sexual Revolution." *New York Times Magazine* (February 18): 34–41, 50, 62, 64, 68–69.

Hobsbawm, Eric J. 1952. "The Machine Breakers." *Past and Present* 1: 57–70.

Hobsbawm, Eric J., and George Rudé. 1969. *Captain Swing.* London: Lawrence and Wishart.

Hobday, Michael. 1995. *Innovation in East Asia: The Challenge to Japan.* Aldershot, UK/Brookfield, Vt.: Edward Elgar.

Hoogma, Remco, Rene Kemp, Johan Schot, and Benhard Truffer. 2002. *Experimenting for Sustainable Transport: The Approach of Strategic Niche Management.* London: E&FN Spon.

Horkheimer, Max, and Theodor Adorno. 1972. *Dialectic of Enlightenment.* New York: Seabury.

Hughes, Thomas P. 1983. *Networks of Power: Electrification in Western Society, 1880–1930.* Baltimore, Md.: Johns Hopkins University Press.

Hughes, Thomas P. 1987. "The Evolution of Large Technological Systems," in W. Bijker, T. Pinch, and T. Hughes, eds. *The Social Construction of Technological Systems: New Directions in the Sociology and History of Technology.* Cambridge, Mass.: MIT Press, pp. 51–82.

Hughes, Thomas P. 1989. *American Genesis.* New York: Viking.

Hughes, Thomas P. 1998. *Rescuing Prometheus.* New York: Pantheon.

Hunt, Lynn, ed. 1993. *The Invention of Pornography: Obscenity and the Origins of Modernity, 1500–1800.* New York: Zone Books.

Ihde, Don. 1990. *Technology and the Lifeworld.* Bloomington, Ind.: Indiana University Press.

Ihde, Don. 1993. *Postphenomenology.* Evanston, Ill.: Northwestern University Press.

Ihde, Don. 1999. "Technology and Prognostic Predicaments." *AI and Society* 13: 44–51.

Iliffe, Rob. 2000. "The Masculine Birth of Time: Temporal Frameworks of Early Modern Natural Philosophy." *British Journal of the History of Science* 33: 427–453.

Irwin, A. 1995. *Citizen Science.* London/New York: Routledge.

Irwin, A., and B. Wynne, eds. 1996. *Misunderstanding Science: The Public Reconstruction of Science and Technology.* Cambridge: Cambridge University Press.

Jackson, S. 1999. "Feminist Sociology and Sociological Feminism: Recovering the Social in Feminist Thought." *Sociological Research Online* 4(3): Available at <www.socresonline.org.uk/socresonline/4/3/jackson.html>

Jackson, S., and S. Scott. 1997. "Gut Reactions to Matters of the Heart: Reflections on Rationality, Irrationality and Sexuality." *Sociological Review* 45(4): 551–575.

Jahn, B. 1998. "One Step Forward, Two Steps Back: Critical Theory as the Latest Edition of Liberal Idealism." *Millennium: Journal of International Studies* 27(3): 613–641.

James, C. L. R. 2001. *The Black Jacobins.* Harmondsworth, UK: Penguin.

Jameson, Frederick. 1991. *Postmodernism, or the Cultural Logic of Late Capitalism.* London: Verso.

Jamison, Andrew, Ron Eyerman, Jacqueline Cramer, with Jeppe Laessøe. 1990. *The Making of the New Environmental Consciousness.* Edinburgh: Edinburgh University Press.

Jänicke, M. 1993. "Über ökologische und politische Modernisierungen." *Zeitschrift für Umweltpolitik und Umweltrecht* 2: 159–175.

Jasanoff, Sheila, Gerald Markle, James Petersen, and Trevor Pinch, eds. 1995. *Handbook of Science and Technology Studies.* London: Sage.

Johnson, Allen, and Timothy Earle. 1987. *The Evolution of Human Societies.* Stanford, Calif.: Stanford University Press.

Jørgensen, Ulrik, and Peter Karnøe. 1995. "The Danish Wind-Turbine Story: Technical Solutions to Political Visions?" in A. Rip, T. J. Misa, and J. Schot, eds. *Managing Technology in Society: The Approach of Constructive Technology Assessment.* London: Pinter, pp. 57–82.

Kandal, Terry. 1988. *The Woman Question in Classical Social Theory.* Miami: Florida International University Press.

Kellner, Douglas. 2000. "Crossing the Postmodern Divide with Borgmann, or Adventures in Cyberspace," in E. Higgs, A. Light, and D. Strong, eds. *Technology and the Good Life?* Chicago: University of Chicago Press, pp. 234–255.

Kellner, Douglas. 2001. "Feenberg's *Questioning Technology.*" *Theory, Culture and Society* 18(1): 155–162.

Kemp, René, Arie Rip, and Johan Schot. 2001. "Constructing Transition Paths Through the Management of Niches," in R. Garud and P. Karnøe, eds., *Path Dependence and Creation.* Mahwah, N.J.: Lawrence Erlbaum Associates pp. 269–299.

Kern, Stephen. 1983. *The Culture of Time and Space, 1880–1918.* Cambridge, Mass.: Harvard University Press.

Khan, Haider A. 1996. "Beyond Distributive Justice in the McWorld." Paper presented at the Democratization Conference, Denver, Colorado.

Khan, Haider A. 1997. *Technology, Energy and Development.* Cheltenham, UK: Edward Elgar.

Khan, Haider A. 1998. *Technology, Development and Democracy: Limits of National Innovation Systems in the Age of Postmodernism.* Cheltenham, UK: Edward Elgar.

Kinsey, Alfred C., and the staff of the Institute for Sex Research. 1953. *Sexual Behavior in the Human Female.* Philadelphia: Saunders.

Kline, Ronald R. 2000. *Consumers in the Country: Technology and Social Change in Rural America.* Baltimore, Md.: Johns Hopkins University Press.

Kline, Ronald, and Trevor Pinch. 1996. "Users as Agents of Technological Change: The Social Construction of the Automobile in the Rural United States." *Technology and Culture* 37: 763–795.

Kraybill, Donald B., and Marc Alan Olshan. 1994. *The Amish Struggle with Modernity.* Hanover, N.H.: University Press of New England.

Kuhn, Thomas. 1962. *The Structure of Scientific Revolutions*. Chicago: University of Chicago Press. Second edition 1970.

Kumssa, Asfaw, and Haider Khan, eds. 1996. *Transnational Economics and Regional Economic Development Strategies: Lessons from Five Low-income Developing Countries*. Nagoya, Japan: United Nations Centre for Regional Development.

La Porte, Todd R. 1988. "The United States Air Traffic System: Increasing Reliability in the Midst of Rapid Growth," in R. Mayntz and T. P. Hughes, eds. *The Development of Large Technical Systems*. Boulder, Col.: Westview, pp. 215–244.

La Porte, Todd R., ed. 1991. *Social Responses to Large Technical Systems: Control or Adaptation*. NATO ASI Series D, Behavioural and Social Sciences, vol. 58. Boston: Kluwer Academic.

La Porte, Todd R., and Paula M. Consolini. 1991. "Working in Practice but not in Theory: Theoretical Challenges of 'High-Reliability' Organizations." *Journal of Public Administration Research and Theory* 1: 19–47.

Lamm, S., and G. S. Couzens. 1998. *The Virility Solution: Everything You Need to Know About Viagra, the Potency Pill that Can Restore and Enhance Male Sexuality*. New York: Fireside Books/Simon and Schuster.

Landauer, Thomas K. 1995. *The Trouble with Computers: Usefulness, Usability, and Productivity*. Cambridge, Mass.: MIT Press.

Landes, David S. 1983. *Revolution in Time: Clocks and the Making of the Modern World*. Cambridge, Mass.: Belknap Press.

Lash, Scott. 1999. *Another Modernity: A Different Rationality*. Oxford: Blackwell.

Lash, Scott, and Jonathan Friedman, eds. 1993. *Modernity and Identity*. Oxford: Blackwell.

Lash, Scott, and John Urry. 1994. *Economies of Signs and Space*. London: Sage.

Lash, Scott, Bronislaw Szerszynski, and Brian Wynne, eds. 1996. *Risk, Environment and Modernity: Towards a New Ecology*. London: Sage.

Latour, Bruno. 1984. *Les Microbes: Guerre et Paix, suivi de Irréductions*. Paris: A.M. Métailié.

Latour, Bruno. 1987. *Science in Action*. Cambridge, Mass.: Harvard University Press.

Latour, Bruno. 1991. *Nous n'avons jamais été modernes*. Paris: La Découverte.

Latour, Bruno. 1992. "Where are the Missing Masses? The Sociology of a Few Mundane Artifacts," in W. Bijker and J. Law, eds. *Shaping Technology/Building Sociotechnical Change*. Cambridge, Mass.: MIT Press, pp. 225–258.

Latour, Bruno. 1993. *We Have Never Been Modern*. Cambridge, Mass.: Harvard University Press.

Latour, Bruno. 1994. "Les objets ont-ils une histories? Recontre de Pasteur et de Whitehead dans us bain d'acide lactique," in I. Stengers, ed. *L'Effet Whitehead*. Paris: Vrin.

Latour, Bruno. 1999a. *Politiques de la nature: Comment faire entrer les sciences en démocratie.* Paris: La Découverte.

Latour, Bruno. 1999b. *Pandora's Hope: Essays on the Reality of Science Studies.* Cambridge, Mass.: Harvard University Press.

Latour, Bruno, and Steve Woolgar. 1979. *Laboratory Life: The Social Construction of Scientific Facts.* Princeton, N.J.: Princeton University Press.

Latour, Bruno, and Steve Woolgar. 1986. *Laboratory Life: The Construction of Scientific Facts.* London: Sage.

Law, John, ed. 1991. *A Sociology of Monsters: Essays on Power, Technology and Domination.* London: Routledge.

Law, John. 1994. *Organizing Modernity.* Oxford: Blackwell.

Leff, E. 1995. *Green Production: Toward an Environmental Rationality.* New York: Guilford.

Lente, Dick van. 1998a. "Machines and the Order of the Harbour: The Debate About the Introduction of Grain Unloaders in Rotterdam 1905–1907." *International Review of Social History* 43: 79–109.

Lente, Dick van. 1998b. "Dutch Conflicts: The Intellectual and Practical Appropriation of a Foreign Technology," in M. Hard and A. Jamison, eds. *The Intellectual Appropriation of Technology: Discourses on Modernity, 1900–1939.* Cambridge, Mass.: MIT Press, pp. 189–223.

Lente, Harro van. 1993. *Promising Technologies.* Delft: Eburon.

Lente, Harro van, and Arie Rip. 1998. "Expectations in Technological Developments: An Example of Prospective Structures to be Filled in by Agency," in C. Disco and B. J. R. van der Meulen, eds. *Getting New Technologies Together.* Berlin: Walter de Gruyter, pp. 195–220.

Lerner, Steve. 1997. *Eco-Pioneers.* Cambridge, Mass.: MIT Press.

Levine, David P. 1997. *Self-seeking and the Pursuit of Justice.* Aldershot, UK: Avebury.

Lewis, Tom. 1997. *Divided Highways: Building the Interstate Highways, Transforming American Life.* New York: Viking.

Lie, Merete, and Knut Sørensen, eds. 1996. *Making Technology our Own? Domesticating Technology into Everyday Life.* Oslo: Scandinavian University Press.

Light, Andrew. 2000. "Technology, Democracy and Environmentalism: On Feenberg's *Questioning Technology.*" *Ends and Means: Journal of the Aberdeen Centre for Philosophy, Technology and Society* 4(2): 7–17.

Linebaugh, Peter. 1992. *The London Hanged.* Cambridge: Cambridge University Press.

Lipietz, Alain. 1987. *Mirages and Miracles: The Crisis of Global Fordism.* London: Verso.

Lovins, Amory B. 1977. *Soft Energy Paths: Towards a Durable Peace.* New York: Harper & Row.

Lukes, Steven. 1994. "Methodological Individualism Reconsidered," in M. Martin and L. McIntyre, eds. *Readings in the Philosophy of Social Science*. Cambridge, Mass.: MIT Press, pp. 451–458.

Lundvall, Bengt-Ake. 1992. *National Systems of Innovation: Towards a Theory of Innovation and Interactive Learning*. London: Pinter.

Lyon, David. 1988. *The Information Society: Issues and Illusions*. Cambridge: Polity; New York: Blackwell.

Lyon, David. 1997a. "Cyberspace Sociality: Controversies over Computer-Mediated Communication," in Brian Loader, ed. *The Governance of Cyberspace*. London/New York: Routledge, pp. 23–37.

Lyon, David. 1997b. "Bringing Technology Back in: CITs in Social Theories of Postmodernity," [in Italian] in Mariella Berra, ed. *Ripensare la tecnologia: informatica, informatzione e sviluppo regionale*. Torino: Bollati Boringheieri.

Lyon, David. 1999. *Postmodernity*. Buckingham, UK: Open University Press; Minneapolis: University of Minnesota Press (2nd ed.).

Lyon, David. 2001. *Surveillance Society: Monitoring Everyday Life*. Buckingham, UK: Open University Press.

Lyotard, Jean-François. 1984a. *The Postmodern Condition*. Minneapolis: University of Minnesota Press.

Lyotard, Jean-François. 1984b. "The Unconscious, History and Phrases: Notes on *The Political Unconscious*." *New Orleans Review* (Spring): 73–79.

Lyotard, Jean-François. 1988. *The Differend: Phrases in Dispute*. Translated by Georges van den Abbeele. Minneapolis: University of Minnesota Press.

Lyotard, Jean-François. 1993. *Toward the Postmodern*. R. Harvey and M. Roberts, eds. Atlantic Highlands, N.J.: Humanities Press.

Lyotard, Jean-François, and Jean-Loup Thébaud. 1985. *Just Gaming*. Minneapolis: University of Minnesota Press.

MacKenzie, Donald. 1984. "Marx and the Machine." *Technology and Culture* 25: 473–502.

MacKenzie, Donald, and Judy Wajcman. 1985. "Introduction: The Social Shaping of Technology," in Donald MacKenzie and Judy Wajcman, eds. *The Social Shaping of Technology*, 1st ed. Buckingham, UK: Open University Press.

MacKenzie, Donald, and Judy Wajcman, eds. 1999. *The Social Shaping of Technology*, 2nd ed. Buckingham, UK: Open University Press.

Maines, Rachel P. 1999. *The Technology of Orgasm: "Hysteria," the Vibrator and Women's Sexual Satisfaction*. Baltimore, Md.: Johns Hopkins University Press.

Marcuse, Herbert. 1964. *One-Dimensional Man*. Boston: Beacon Press.

Marquis, Greg. 2000. "The Evolution of Information Technology in Private Security." Paper presented at the National Conference on Information Technology and the Police, Cornwall, Ontario.

Marshall, Barbara L. 1994. *Engendering Modernity: Feminism, Social Theory, and Social Change*. Boston, Mass.: Northeastern University Press.

Marshall, Barbara L. 2000. *Configuring Gender: Explorations in Theory and Politics*. Peterborough, Vt.: Broadview Press.

Marshall, Barbara L. 2002. "Hard Science: Gendered Constructions of Sexual Dysfunction in the 'Viagra Age'." *Sexualities* 5(2): 131–158.

Martin, Emily. 1994. *Flexible Bodies*. Boston: Beacon.

Marvin, Carolyn. 1988. *When Old Technologies Were New: Thinking about Electric Communication in the Late Nineteenth Century*. New York: Oxford University Press.

Marx, Gary T. 1988. *Undercover: Police Surveillance in America*. Berkeley: University of California Press.

Marx, Karl. 1977. *Capital*. vol. 1. New York: Vintage.

Marx, Leo. 1994. "The Idea of 'Technology' and Postmodern Pessimism," in M. R. Smith and Leo Marx, eds. *Does Technology Drive History? The Dilemma of Technological Determinism*. Cambridge, Mass.: MIT Press, pp. 237–258.

Mascia-Lees, F. E., P. Sharpe, and C. Ballerino-Cohen. 1989. "The Postmodernist Turn in Anthropology: Cautions from a Feminist Perspective." *Signs* 15(1): 7–33.

Masters, W. H., and V. E. Johnson. 1966. *Human Sexual Response*. Boston: Little, Brown.

Mauron, Alex. 2001. "Is the Genome the Secular Equivalent of the Soul?" *Science* 291: 831–832.

Mayntz, Renate, and Thomas P. Hughes, eds. 1988. *The Development of Large Technical Systems*. Boulder, Col.: Westview.

Mayr, Otto. 1986. *Authority, Liberty, and Automatic Machinery in Early Modern Europe*. Baltimore, Md.: Johns Hopkins University Press.

McCahill, Mike, and Clive Norris. 1999. "Watching the Workers: Crime, CCTV, and the Workplace," in P. Davies, P. Francis, and V. Jupp, eds. *Invisible Crimes: Their Victims and their Regulation*. London: Macmillan, pp. 209–210.

McCarthy, Thomas. 1991. *Ideals and Illusions: On Reconstruction and Deconstruction in Contemporary Critical Theory*. Cambridge, Mass.: MIT Press.

McGaw, Judith. 1989. "No Passive Victims, No Separate Spheres: A Feminist Perspective on Technology's History," in S. H. Cutcliffe and R. C. Post, eds. *In Context: History and the History of Technology*. Bethlehem, Pa.: Lehigh University Press, pp. 172–191.

McKeown, Thomas. 1976. *The Modern Rise of Population*. New York: Academic Press.

McKeown, Thomas. 1979. *The Role of Medicine*. Princeton, N.J.: Princeton University Press.

McLaren, A. 1999. *Twentieth-Century Sexuality: A History*. Oxford: Blackwell.

Meehan, J., ed. 1995. *Feminists Read Habermas*. London: Routledge.

Melchiode, G., and B. Sloan. 1999. *Beyond Viagra: A Commonsense Guide to Building a Healthy Sexual Relationship for Both Men and Women*. New York: Owl Books (Henry Holt).

Menser, M., and S. Aronowitz. 1996. "On Cultural Studies, Science and Technology," in S. Aronowitz and M. Mesner, eds., *Technoscience and Cyberculture*. London: Routledge, pp. 7–28.

Miller, Daniel. 1994. *Modernity—An Ethnographic Approach: Dualism and Mass Consumption in Trinidad*. Oxford: Berg.

Miller, Daniel, and Don Slater. 2000. *The Internet: An Ethnographic Approach*. London: Berg.

Miller, Daniel, and Don Slater. forthcoming. "Ethnography On and Off Line: Cybercafes in Trinidad," in M. Johnson, ed. *Internet Ethnographies*. Oxford: Berg (in press 2002).

Mills, C. W. 1959. *The Sociological Imagination*. New York: Oxford University Press.

Misa, Thomas J. 1988. "How Machines Make History, and How Historians (and Others) Help Them to Do So." *Science, Technology & Human Values* 13: 308–331.

Misa, Thomas J. 1994. "Retrieving Sociotechnical Change from Technological Determinism," in M. R. Smith and Leo Marx, eds. *Does Technology Drive History? The Dilemma of Technological Determinism*. Cambridge, Mass.: MIT Press, pp. 115–141.

Misa, Thomas J. 1995. *A Nation of Steel: The Making of Modern America, 1865–1925*. Baltimore, Md.: Johns Hopkins University Press.

Mitcham, Carl. 1994. *Thinking through Technology: The Path Between Engineering and Philosophy*. Chicago: University of Chicago Press.

Mokyr, Joel. 1990. *The Lever of Riches*. Oxford: Oxford University Press.

Mol, A. P. J. 1995. *The Refinement of Production: Ecological Modernisation Theory and the Chemical Industry*. Utrecht: Jan van Arkel/International Books.

Mol, A. P. J. 1996. "Ecological Modernisation and Institutional Reflexivity: Environmental Reform in the Late Modern Age." *Environmental Politics* 5(2): 302–323.

Mol, A. P. J. 2000. "The Environmental Movement in an Era of Ecological Modernisation." *Geoforum* 31: 45–56.

Mol, A. P. J., and F. Buttel, eds. 2002. *The Environmental State under Pressure*. Oxford: Elsevier.

Mol, A. P. J., and G. Spaargaren. 1993. "Environment, Modernity and the Risk-Society: The Apocalyptic Horizon of Environmental Reform." *International Sociology* 8(4): 431–459.

Mol, A. P. J., and G. Spaargaren. 2000. "Ecological Modernization Theory in Debate: A Review." *Environmental Politics* 9(1): 17–49.

Mom, Gijs, Marc Dierikx, Adrienne van den Bogaard, and Charley Werff. 1999. *Schiphol: Haven, Station, Knooppunt sinds 1916.* Zutphen, Netherlands: Walburg.

Monahan, Torin. 2000. "Flexible Spaces and Built Pedagogies: Emerging IT Embodiments," Master's thesis, Science and Technology Studies Department, Rensselaer Polytechnic Institute, Troy. N.Y.

Moon, Suzanne M. 1998. "Takeoff or Self-Sufficiency? Ideologies of Development in Indonesia, 1957–1961." *Technology and Culture* 39: 187–212.

Morales, A. 1998. *Erectile Dysfunction: Issues in Current Pharmacotherapy.* London: Martin Dunitz, pp. xv–xvi.

Morgan, Gareth. 1983. "Research Strategies: Modes of Engagement," in G. Morgan, ed. *Beyond Method: Strategies for Social Research.* London: Sage, pp. 19–42.

Morris-Suzuki, Tessa. 1994. *The Technological Transformation of Japan: From the Seventeenth to the Twenty-first Century.* Cambridge: Cambridge University Press.

Morrow, R. A. 1994. *Critical Theory and Methodology.* Thousand Oaks, Calif: Sage.

Moss, Ralph. 1996. *The Cancer Industry.* Brooklyn: Equinox.

Mumford, Lewis. 1934. *Technics and Civilization.* New York: Harcourt, Brace.

Murata, Junichi. 1999. "Interpretation and Design: the Nature of Technology and its Interpretive Flexibility." Paper presented at a seminar at San Diego State University.

Murray, Janet H. 1997. *Hamlet on the Holodeck: The Future of Narrative in Cyberspace.* Cambridge, Mass.: MIT Press.

Nakaoka, Tetsuro. 1999. *Jidousha ga hashitta* (The cars have run). Tokyo: Asahi-Shinbunsha.

Nelkin, Dorothy, ed. 1979. *Controversy: Politics of Technical Decisions.* Beverly Hills, Calif.: Sage.

Nelson, Richard R., and Nathan Rosenberg, eds. 1993. *National Innovation Systems: A Comparative Analysis.* New York/Oxford: Oxford University Press.

Nicholson, Linda. 1994. "Interpreting Gender." *Signs* 20(1): 79–105.

Nishida, Kitaro. 1949a. *Complete Works.* vol. 8. Tokyo: Iwanami.

Nishida, Kitaro. 1949b. *Complete Works.* vol. 9. Tokyo: Iwanami.

Nishida, Kitaro. 1950. *Complete Works.* vol. 12. Tokyo: Iwanami.

Noble, David. 1984. *Forces of Production: A Social History of Industrial Automation.* New York: Knopf.

Nolan, Mary. 1994. *Visions of Modernity: American Business and the Modernization of Germany.* New York: Oxford University Press.

Norberg, Arthur L., and Judy E. O'Neill. 1996. *Transforming Computer Technology: Information Processing for the Pentagon, 1962–1986*. Baltimore, Md.: Johns Hopkins University Press.

Norman, Donald. 1993. *Things that Make Us Smart: Defending Human Attributes in the Age of the Machine*. Reading, Mass.: Addison-Wesley.

Nuvolari, A. 1999. "The 'Machine Breakers' and the Industrial Revolution." Paper prepared for the European Historical Economics Society Summer School, Lund, Sweden.

Nye, David E. 1990. *Electrifying America: Social Meanings of a New Technology, 1880–1940*. Cambridge, Mass.: MIT Press.

Nye, David E. 1997. *Narratives and Spaces: Technology and the Construction of American Culture*. Exeter, UK: University of Exeter Press.

O'Connor, J. 1998. *Natural Causes. Essays in Ecological Marxism*. New York/London: Guilford.

O'Harrow, Robert. 2000. "Firm Tracking Consumers on Web for Drug Companies." *Washington Post* (August 14): A25.

Odaka, Konosuke. 1993. *Shokuninn no Sekai Koujyou no Sekai* (The World of Factories and the World of Artisans). Tokyo: LibroPort.

Ohkubo, Toshimichi. 1988. "Proposal for 'Promoting Enterprise and Developing Products'." (Ohkubo Toshimichi no Shokusan Kougyou ni kansuru Kenngi. *Modern Thoughts of Japan*, vol. 8, The Philosophy of Economics. Tokyo: Iwanami. First published in 1874.

Oldenziel, Ruth. 1999. *Making Technology Masculine: Men, Women and Modern Machines in America, 1870–1945*. Amsterdam: Amsterdam University Press.

Oudshoorn, Nelly, and Trevor Pinch, eds. *How Users Matter: The Co-Construction of Users and Technology* (under review at MIT Press).

Overy, Richard. 1990. "Heralds of Modernity: Cars and Planes from Invention to Necessity," in Mikulas Teich and Roy Porter, eds. *Fin de Siecle and its Legacy*. Cambridge: Cambridge University Press, pp. 54–79.

Paehlke, R. C. 1989. *Environmentalism and the Future of Progressive Politics*. New Haven, Conn.: Yale University Press.

Palmer, Richard. 1969. *Hermeneutics*. Evanston, Ill.: Northwestern University Press.

Paquiet, P., P. Blancher, and C. Zampa. 1996. *Industrie chimique, territoire et société: L'environnement comme catalyseur de nouveaux rapports*. Lyon/Grenoble: Economie et Humanisme/IREPD.

Parnis, D. and J. Du Mont. 1999. "Rape Laws and Rape Processing: The Contradictory Nature of Corroboration." *Canadian Woman Studies* 19(1 and 2): 74–78.

Parnis, D., and J. Du Mont. 2002. "Examining the Standardized Application of Rape Kits: An Exploratory Study of Post-Sexual Assault Professional Practices." *Health Care for Women International* (in press).

Parry, Jonathan, and Maurice Bloch, eds. 1989. *Money and the Morality of Exchange.* Cambridge: Cambridge University Press.

Parsons, Talcott, and Edward Shils, eds. 1951. *Toward a General Theory of Action.* New York: Harper & Row.

Pavitt, Keith. 1984. "Patterns of Technical Change: Towards a Taxonomy and a Theory." *Research Policy* 13(6): 343–373.

Perrow, Charles. 1984. *Normal Accidents: Living with High-Risk Technologies.* New York: Basic Books.

Pickering, Andrew. 1992. *Science as Practice and Culture.* Chicago: University of Chicago Press.

Pickering, Andrew. 1995. *The Mangle of Practice: Time, Agency and Science.* Chicago: University of Chicago Press.

Pinch, Trevor. 1999. "The Social Construction of Technology: A Review," in Robert Fox, ed., *Technological Change: Methods and Themes in the History of Technology.* Amsterdam: Harwood, pp. 17–35.

Pinch, Trevor, and Wiebe Bijker. 1986. "Science, Relativism and the New Sociology of Technology: Reply to Russell." *Social Studies of Science* 16: 347–360.

Pinch, Trevor, and Wiebe Bijker. 1987. "The Social Construction of Facts and Artifacts: Or How the Sociology of Science and the Sociology of Technology Might Benefit Each Other," in W. Bijker, T. Pinch, and T. Hughes, eds., *The Social Construction of Technological Systems: New Directions in the Sociology and History of Technology.* Cambridge, Mass.: MIT Press, pp. 17–50.

Porter, Roy. 2000. *The Creation of the Modern World: The Untold Story of the British Enlightenment.* New York: W.W. Norton.

Porter, Theodore M. 1995. *Trust in Numbers: The Pursuit of Objectivity in Science and Public Life.* Princeton, N.J.: Princeton University Press.

Poster, Mark. 1989. *Critical Theory and Poststructuralism.* Ithaca, N.Y.: Cornell University Press.

Poster, Mark. 1996. "Databases as Discourse," in David Lyon and E. Zureik, eds. *Computers, Surveillance, and Privacy.* Minneapolis: University of Minnesota Press, pp. 175–192.

Potts, Annie. 2000. "The Essence of the Hard On: Hegemonic Masculinity and the Cultural Construction of 'Erectile Dysfunction'." *Men and Masculinities* 3(1): 85–103.

President's Commission on Critical Infrastructure Protection. 1997. *Critical Foundations: Protecting America's Infrastructures.* Washington, D.C.: U.S. Government Printing Office.

Price, David. 1995. "Energy and Human Evolution." *Population and Environment* 16(4): 301–319.

Pursell, Carroll. 1993. "The Rise and Fall of the Appropriate Technology Movement in the United States, 1961–1985." *Technology and Culture* 34: 629–637.

Pursell, Carroll. 1995. *The Machine in America: A Social History of Technology*. Baltimore, Md.: Johns Hopkins University Press.

Rabinbach, Anson. 1990. *The Human Motor: Energy, Fatigue, and the Origins of Modernity*. New York: Basic Books.

Radder, Hans. 1996. *In and About the World: Philosophical Studies of Science and Technology*. Albany: State University of New York Press.

Randall, Adrian. 1991. *Before the Luddites: Custom, Community, and Machinery in the English Woollen Industry, 1776–1809*. Cambridge: Cambridge University Press.

Redmond, Kent C., and Thomas M. Smith. 1980. *Project Whirlwind: The History of a Pioneer Computer*. Boston: Digital Press.

Reed, Sidney G., Richard H. Van Atta, and Seymour J. Deitchman. 1990. *DARPA Technical Accomplishments: An Historical Review of Selected DARPA Projects*. Alexandria, Va.: Institute for Defense Analyses.

Rheingold, Howard. 1993. *The Virtual Community: Homesteading on the Electronic Frontier*. Reading, Mass.: Addison-Wesley.

Rheingold, Howard. 1996. "A Slice of My Life in My Virtual Community," in Peter Ludlow, ed. *High Noon on the Electronic Frontier: Conceptual Issues in Cyberspace*. Cambridge, Mass.: MIT Press. pp. 413–436.

Richardson, J., ed. 1982. *Policy Styles in Western Europe*. London: Allen and Unwin.

Rip, Arie. 1995. "Introduction of New Technology: Making Use of Recent Insights from Sociology and Economics of Technology." *Technology Analysis & Strategic Management* 17(4): 417–431.

Rip, Arie, and Rene Kemp. 1998. "Technological Change," in S. Rayner and E. L. Malone, eds. *Human Choice and Climate Change*. vol. II. Columbus, Ohio: Battelle Press, pp. 327–399.

Rip, Arie, and Barend J. R. van der Meulen. 1996. "The Post-modern Research System." *Science and Public Policy* 23(6): 343–352.

Rip, Arie, Thomas J. Misa, and Johan Schot, eds. 1995. *Managing Technology in Society: The Approach of Constructive Technology Assessment*. London: Pinter.

Rival, L., and D. Slater, 1998. "Sex and Sociality: Comparative Ethnography of Sexual Objectification." *Theory, Culture and Society* 15(3–4): 295–322.

Rochlin, G. I. 1997. *Trapped in the Net: The Unanticipated Consequences of Computerization*. Princeton, N.J.: Princeton University Press.

Rorty, Richard. 1989. *Contingency, Irony, and Solidarity*. Cambridge: Cambridge University Press.

Rose, Mark H. 1995. *Cities of Light and Heat: Domesticating Gas and Electricity in Urban America*. University Park: Pennsylvania State University Press.

Rose, N. 1996. *Inventing Our Selves: Psychology, Power and Personhood*. Cambridge: Cambridge University Press.

Rosen, Paul. 1993. "The Social Construction of Mountain Bikes: Technology and Postmodernity in the Cycle Industry." *Social Studies of Science* 23: 479–513.

Rosenband, Leonard N. 2000. *Papermaking in Eighteenth-Century France: Management, Labor, and Revolution at the Montgolfier Mill, 1761–1805.* Baltimore, Md.: Johns Hopkins University Press.

Rosenberg, Nathan. 1970. "Economic Development and the Transfer of Technology: Some Historical Perspectives." *Technology and Culture* 11: 550–575.

Rostow, Walt Whitman. 1960. *The Stages of Economic Growth: A Non-Communist Manifesto.* Cambridge: Cambridge University Press.

Rothschild, Joan. 1989. "From Sex to Gender in the History of Technology," in S. H. Cutcliffe and R. C. Post, eds. *In Context: History and the History of Technology.* Bethlehem, Pa.: Lehigh University Press, pp. 192–203.

Rothschild, Joan, ed. 1999. *Feminism and Design.* New Brunswick, N.J.: Rutgers University Press.

Rowland, Wade. 1997. *Spirit of the Web: The Age of Information from Telegraph to Internet.* Toronto: Somerville House.

Ruiter, W. de. 1988. "Het Postmoderne Tijdperk." *Wetenschap en samenleving* 40(4): 3–11.

Rule, J. 1986. "Against Innovation? Custom and Resistance in the Workplace, 1700–1850," in T. Harris, ed. *Popular Culture in England, 1500–1850.* New York: St. Martin's Press, pp. 168–188.

Rule, James. 1973. *Private Lives, Public Surveillance.* London: Allen Lane.

Rule, James, D. McAdam, T. Stearns, and D. Uglow. 1983. "Documentary Identification and Mass Surveillance." *Social Problems* 32(1): 222–234.

Russell, Stewart. 1986. "The Social Construction of Artefacts: A Response to Pinch and Bijker." *Social Studies of Science* 16: 331–346.

Sabel, Charles F., and Jonathan Zeitlin. 1985. "Historical Alternatives to Mass Production: Politics, Markets and Technology in Nineteenth-Century Industrialization." *Past & Present* 108: 133–176.

Sabel, Charles F., and Jonathan Zeitlin. 1997. "Stories, Strategies, Structures: Rethinking Historical Alternatives to Mass Production," in C. F. Sabel and J. Zeitlin, eds., *World of Possibilities: Flexibility and Mass Production in Western Industrialization.* Cambridge: Cambridge University Press, pp. 1–33.

Sagan, Scott Douglas. 1993. *The Limits of Safety: Organizations, Accidents, and Nuclear Weapons.* Princeton, N.J.: Princeton University Press.

Said, Edward W. 1995. *Orientalism: Western Conceptions of the Orient.* London: Penguin. Reprint of the 1978 edition, with a new afterword.

Sale, K. 1995. *Rebels Against the Future.* Reading, Mass.: Addison-Wesley.

Sayer, A. 1984. *Method in Social Science.* London: Routledge.

Sayer, A. 2000. "System, Lifeworld and Gender: Associational versus Counterfactual Thinking." *Sociology* 34(4): 707–725.

Sayer, Derek. 1991. *Capitalism and Modernity: An Excursus on Marx and Weber*. London: Routledge.

Scaff, Lawrence A. 1989. *Fleeing the Iron Cage: Culture, Politics, and Modernity in the Thought of Max Weber*. Berkeley: University of California Press.

Schatzberg, Eric. 1999. *Wings of Wood, Wings of Metal: Culture and Technical Choice in American Airplane Materials, 1914–1945*. Princeton, N.J.: Princeton University Press.

Schiller, Herbert. 1981. *Who Knows: Information in the Age of the Fortune 500*. Norwood, N.J.: Ablex.

Schot, Johan. 1991. "Maatschappelijke sturing van technische ontwikkeling: Constructief technology assessment als hedendaags Luddisme." Ph.D dissertation, University of Twente, Enschede, Netherlands.

Schot, Johan. 1995. "Schiphol, Kijkdoos van de Moderniteit," in Vereniging Milieudefensie, ed. *Dossier Schiphol: Over economie en ecologie in Nederland distributieland*. Amsterdam: Vereniging Milieudefensie.

Schot, Johan. 1998. "The Usefulness of Evolutionary Models for Explaining Innovation: The Case of the Netherlands in the Nineteenth Century." *History and Technology* 14: 173–200.

Schot, Johan. 2001. "Towards New Forms of Participatory Technology Development." *Technology Analysis and Strategic Management* 13(1): 39–52.

Schot, Johan, and Adri de la Bruhèze. forthcoming. "The Mediated Design of Products, Consumption and Consumers in the Twentieth Century." In Nelly Oudshoorn and Trevor Pinch, eds. *How Users Matter: The Co-Construction of Users and Technology*. (under review by MIT Press).

Schot, Johan, and Arie Rip. 1998. "The Past and Future of Constructive Technology Assessment." *Technological Forecasting and Social Change* 54: 251–268.

Schumpter, Joseph. 1950. *Capitalism, Socialism, and Democracy*. New York: Harper and Brothers.

Schumpter, Joseph. 1961. *The Theory of Economic Development*. Oxford: Oxford University Press.

Schwarz, Michiel, and Michael Thompson. 1990. *Divided We Stand: Re-defining Politics, Technology and Social Choice*. New York: Harvester Wheatsheaf.

Sclove, Richard. 1995. *Democracy and Technology*. New York: Guilford.

Scott, James C. 1998. *Seeing Like a State*. New Haven, Conn.: Yale University Press.

Scranton, Philip. 1997. *Endless Novelty: Specialty Production and American Industrialization 1865–1925*. Princeton, N.J.: Princeton University Press.

Seely, Bruce E. 1987. *Building the American Highway System: Engineers as Policy Makers*. Philadelphia: Temple University Press.

Segal, Howard P. 1985. *Technological Utopianism in American Culture*. Chicago: University of Chicago Press.

Segaller, Stephen. 1998. *Nerds 2.0.1: A Brief History of the Internet*. New York: TV Books.

Seiden, O. J. 1998. *Viagra: The Virility Breakthrough*. Rocklin, Calif.: Prima.

Sen, Amartya. 1992. *Inequality Reexamined*. Cambridge, Mass.: Harvard University Press.

Sen, Amartya. 1999. *Development as Freedom*. New York: Knopf.

Sherwood, John M. 1985. "Engels, Marx, Malthus and the Machine." *American Historical Review* 90: 837–865.

Sieferle, R. P. 1984. *Fortschrittsfeinde? Opposition gegen Technik und Industrie von der Romantik bis zur Gegenwart*. Munich: Beck Verlag.

Slater, Don. 1997. *Consumer Culture and Modernity*. Cambridge: Polity.

Slater, Don. 1998. "Trading Sexpics on IRC: Embodiment and Authenticity on the Internet." *Body and Society* 4(4): 91–117.

Slater, Don. 2000a. "Consumption without Scarcity: Exchange and Normativity in an Internet Setting," in P. Jackson, M. Lowe, D. Miller, and F. Mort, eds. *Commercial Cultures: Economies, Practices, Spaces*. London: Berg, pp. 123–142.

Slater, Don. 2000b. "Political Discourse and the Politics of Need: Discourses on the Good Life in Cyberspace," in L. Bennett and R. Entman, eds. *Mediated Politics*. Cambridge: Cambridge University Press, pp. 117–140.

Slaton, Amy. 2001. *Reinforced Concrete and the Modernization of American Building, 1900–1930*. Baltimore, Md.: Johns Hopkins University Press.

Slevin, James. 2000. *The Internet and Society*. Cambridge: Polity.

Smalley, Richard E. 2001. "Of Chemistry, Love, and Nanobots." *Scientific American* 285 (September): 76–77.

Smart, Barry. 2000. "Postmodern Social Theory," in Bryan Turner, ed. *The Blackwell Companion to Social Theory*. Oxford: Blackwell, 2nd ed., chap. 14.

Smith, D. 1974. "Women's Perspective as a Radical Critique of Sociology." *Sociological Inquiry* 44: 7–13.

Smith, Dorothy E. 1987. *The Everyday World as Problematic: A Feminist Sociology*. Milton Keynes, UK: Open University Press.

Smith, Dorothy E. 1990. *Texts, Facts, and Feminity: Exploring the Relations of Ruling*. London: Routledge.

Smith, Merritt Roe, and Leo Marx, eds. 1994. *Does Technology Drive History? The Dilemma of Technological Determinism*. Cambridge, Mass.: MIT Press.

Smith, Terry. 1993. *Making the Modern: Industry, Art, and Design in America*. Chicago: University of Chicago Press.

Smits, Martijntje. 1997. "Langdon Winner: Technologie als Schaduwconstitutie," in Hans Achterhuis, ed. *Van Stoommachine tot Cyborg*. Amsterdam: Ambo, pp. 93–115. Translated into English as "Langdon Winner: Technology as a Shadow Constitution," in Hans Achterhuis, ed. *American Philosophy of Technology: The Empirical Turn.* Bloomington, Ind.: Indiana University Press, 2001, pp. 147–169.

Spaargaren, G. 1997. "The Ecological Modernisation of Production and Consumption: Essays in Environmental Sociology." Ph.D. disseration. Wageningen Agricultural University, Wageningen, Netherlands.

Spaargaren, G., and A. P. J. Mol. 1992. "Sociology, Environment and Modernity: Ecological Modernisation as a Theory of Social Change." *Society and Natural Resources* 5: 323–344.

Spinosa, Charles, Fernando Flores, and Hubert Dreyfus. 1997. *Disclosing New Worlds: Entrepreneurship, Democratic Action, and the Cultivation of Solidarity.* Cambridge, Mass.: MIT Press.

Sproull, Lee, and Sara B. Kiesler. 1991. *Connections: New Ways of Working in the Networked Organization.* Cambridge, Mass.: MIT Press.

Standage, Tom. 1998. *The Victorian Internet: The Remarkable Story of the Telegraph and the Nineteenth Century's On-Line Pioneers.* New York: Walker.

Star, Susan Leigh, and Karen Ruhleder. 1996. "Steps Toward an Ecology of Infrastructure: Design and Access for Large Information Spaces." *Information Systems Research* 7: 111–134.

Staudenmaier, John M. 1985. *Technology's Storytellers: Reweaving the Human Fabric.* Cambridge, Mass.: MIT Press.

Staudenmaier, John M. 1989. "The Politics of Successful Technologies," in S. H. Cutcliffe and R. C. Post eds. *In Context: History and the History of Technology.* Bethlehem, Pa.: Lehigh University Press, pp. 150–171.

Staute, J. 1997. *Das Ende der Unternehmenskultur: Firmenalltag im Turbokapitalismus.* Frankfurt: Campus Verlag.

Sterling, Bruce. 1993. "A Short History of the Internet." *Magazine of Fantasy and Science Fiction* (February). Available at <gopher.well.sf.ca.us:70/00/Publications/authors/Sterling/fsf/internet.fsf> (20 Nov. 2001).

Stone, Allurque Rosanne. 1995. *The War of Desire and Technology at the Close of the Mechanical Age.* Cambridge, Mass.: MIT Press.

Strasser, Susan, Charles McGovern, and Matthias Judt, eds. 1998. *Getting and Spending: European and American Consumer Societies in the Twentieth Century.* Cambridge: Cambridge University Press.

Stratton, Jon. 1997. "Cyberspace and the Globalization of Culture," in David Porter, ed. *Internet Culture.* New York: Routledge, pp. 253–276.

Summerton, Jane, ed. 1994. *Changing Large Technical Systems.* Boulder, Col.: Westview.

Suzuki, Jun. 1996. *Meiji no Kikaikougyou* (Machine Industry in Meiji era). Kyoto: Minerva Shobou.

Sydie, R. A. 1987. *Natural Woman, Cultured Man.* Toronto: Methuen.

Sydie, R. A. 1994. "Sex and the Sociological Fathers." *Canadian Review of Sociology and Anthropology* 31(2): 117–138.

Szasz, Thomas, and Marc Hollender. 1956. "The Basic Models of the Doctor-Patient Relationship." *Archives of Internal Medicine* 97: 585–592.

Tainter, Joseph. 1998. *The Collapse of Complex Societies.* New York: Cambridge University Press.

Tellegen, E. 1983. *Milieubeweging.* Utrecht/Antwerpen: Het Spectrum.

Tenner, Edward. 1996. *Why Things Bite Back: Technology and the Revenge of Unintended Consequences.* New York: Knopf.

Thompson, E. P. 1963. *The Making of the English Working Class.* London: Penguin.

Tichi, Cecelia. 1987. *Shifting Gears: Technology, Literature, Culture in Modernist America.* Chapel Hill: University of North Carolina Press.

Tiefer, L. 1996. "The Medicalization of Sexuality: Conceptual, Normative and Professional Issues." *Annual Review of Sex Research* 7: 252–282.

Tobey, Ronald C. 1996. *Technology as Freedom: The New Deal and the Electrical Modernization of the Home.* Berkeley: University of California Press.

Todd, Edmund N. 2001. "Engineering Politics, Technological Fundamentalism, and German Power Technology, 1900–1936," in Michael Thad Allen and Gabrielle Hecht, eds. *Technologies of Power.* Cambridge, Mass.: MIT Press, pp. 145–174.

Toulmin, Stephen. 1990. *Cosmopolis: The Hidden Agenda of Modernity.* New York: Free Press.

Touraine, Alain. 1971. *The Post-Industrial Society: Tomorrow's Social History; Classes, Conflicts and Culture in the Programmed Society.* New York: Wildwood House.

Touraine, Alain. 1995. *Critique of Modernity.* Oxford/Cambridge: Blackwell.

Trommler, Frank. 1995. "The Avant-Garde and Technology: Toward Technological Fundamentalism in Turn-of-the-Century Europe." *Science in Context* 8: 397–416.

Tsunoyama, Sakae. 1984. *Tokei no shakaishi* (The Social History of Clocks). Tokyo: Chuou kouron-Sha.

Tukker, A. 1999. *Frames in the Toxicity Controversy: Risk Assessment and Policy Analysis related to the Dutch Chlorine Debate and Swedish PVC Debate.* Dordrecht: Kluwer.

Turkle, Sherry 1984. *The Second Self: Computers and the Human Spirit.* New York: Simon and Schuster.

Turkle, Sherry. 1995. *Life on the Screen: Identity in the Age of the Internet.* New York: Simon and Schuster.

Turner, Bryan, ed. 1990. *Theories of Modernity and Postmodernity.* London: Sage.

Turner, Bryan. 1993. *Max Weber: From History to Modernity*. London/New York: Routledge.

Ullrich, O. 1979. *Weltniveau: In der Sachgasse der Industriegesellschaft*. Berlin: Rotbuch.

Urometrics. 2000. Press Release: FDA Grants Clearance for EROS-CTD, First Treatment for Female Sexual Dysfunction (May 3).

Van Atta, Richard H., Sidney G. Reed, and Seymour J. Deitchman. 1991. *DARPA Technical Accomplishments: An Historical Review of Selected DARPA Projects*. Alexandria, Va.: Institute for Defense Analyses (vol. 2 of 3).

Van Creveld, Martin. 1985. *Command in War*. Cambridge, Mass.: Harvard University Press.

Van der Ryn, Sim, and Stuart Cowan. 1996. *Ecological Design*. Washington, D.C.: Island Press.

Vaughan, Diane. 1996. *The Challenger Launch Decision: Risky Technology, Culture, and Deviance at NASA*. Chicago: University of Chicago Press.

Verheul, H., and P. Vergragt. 1995. "Social Experiments in the Development of Environmental Technology: A Bottom-up Perspective." *Technology Analysis & Strategic Management* 7: 315–326.

Vig, Norman J. 1988. "Technology, Philosophy, and the State," in Michael E. Kraft and Norman J. Vig, eds. *Technology and Politics*. Durham, N.C.: Duke University Press, pp. 8–32.

Virilio, Paul. 1986. *Speed and Politics: An Essay on Dromology*. Translated by M. Polizzotti. New York: Columbia University Press.

Vogel, D. 1986. *National Styles of Regulation*. Ithaca, N.Y.: Cornell University Press.

Vries, Jan de, and Ad van der Woude. 1996. *The First Modern Economy: Success, Failure, and Perseverance of the Dutch Economy, 1500–1815*. Cambridge: Cambridge University Press.

Waarden, F. van. 1995. "Persistence of National Policy Styles: A Study of their Institutional Foundation," in B. Unger and F. van Waarden, eds. *Convergence or Diversity: Internationalization and Economic Policy Response*. Aldershot, UK: Avebury, pp. 333–372.

Wagner, Peter. 1994. *A Sociology of Modernity: Liberty and Discipline*. London: Routledge.

Wajcman, Judy. 1991. *Feminism Confronts Technology*. Cambridge: Polity. Reprinted 1994.

Wajcman, Judy. 2000. "Reflections on Gender and Technology Studies: In what State is the Art?" *Social Studies of Science* 30(3): 447–464.

Ward, David, and Olivier Zunz. 1997. *The Landscape of Modernity: New York City, 1900–1940*. Baltimore, Md.: Johns Hopkins University Press.

Watson, Peggy. 1997. "(Anti)feminism after Communism," in Ann Oakley and Juliet Mitchell, eds. *Who's Afraid of Feminism?* London: Hamish Hamilton, pp. 144–161.

Weber, K. M., R. Hoogma, B. Lane, and J. Schot. 1999. *Experimenting with Sustainable Transport Innovation: A Workbook for Strategic Niche Management*. Enschede.

Weber, Max. 1958. *The Protestant Ethic and the Spirit of Capitalism*. New York: Scribner's. First published in 1905.

Weber, Max. 1978. *Economy and Society*. Berkeley: University of California Press.

Webster, Frank. 1995. *Theories of the Information Society*. London: Routledge.

Wehler, Hans-Ulrich. 1975. *Modernisierungstheorie und Geschichte*. Gottingen: Vandenhoeck und Ruprecht.

West, Cornell. 1988. "Marxist Theory and the Specificity of Afro-American Oppression," in C. Nelson and L. Grossberg, eds. *Marxism and the Interpretation of Culture*. Chicago: University of Chicago Press, pp. 17–29.

Westrum, Ron. 1991. *Technologies and Society*. Belmont, Calif.: Wadsworth.

White, Lynn, Jr. 1962. *Medieval Technology and Social Change*. Oxford: Oxford University Press.

Whitehead, E. D., and T. Malloy. 1999. *Viagra: the Wonder Drug for Peak Performance*. New York: Dell Books.

Whitney, Charles. 1986. *Francis Bacon and Modernity*. New Haven, Conn.: Yale University Press.

Wingelaar, M., and A. P. J. Mol. 1997. "Industriële en werknemersbelangen," in P. Basset and J. Kastje, eds. *Groen ondernemen: Werken aan milieu-innovaties in bedrijven*. Amsterdam: Wetenschappelijk Bureau Groen-Links, pp. 79–100.

Winner, Langdon. 1977. *Autonomous Technology: Technics-out-of-Control as a Theme in Political Thought*. Cambridge, Mass.: MIT Press.

Winner, Langdon. 1980. "Do Artifacts Have Politics?" *Daedalus* 109: 121–136.

Winner, Langdon. 1986. *The Whale and the Reactor*. Chicago: University of Chicago Press.

Winner, Langdon. 1993. "On Opening the Black Box and Finding It Empty: Social Constructivism and the Philosophy of Technology." *Science, Technology & Human Values* 18: 362–378.

Winner, Langdon. 2001. "Where Technological Determinism Went," in S. H. Cutcliffe and C. Mitcham, eds. *Visions of STS: Counterpoints in Science, Technology, and Society Studies*. Albany: State University of New York Press, pp. 11–17.

Winograd, Terry, and Fernando Flores. 1987. *Understanding Computers and Cognition*. Reading, Mass.: Addison-Wesley.

Wise, M. Norton, ed. 1995. *The Values of Precision*. Princeton, N.J.: Princeton University Press.

Wittgenstein, Ludwig. 1958. *Philosophical Investigations*. New York: Macmillan.

Witz, Anne. 2000. "Whose Body Matters? Feminist Sociology and the Corporeal Turn in Sociology and Feminism." *Body and Society* 6(2): 1–24.

Wooddell, Margaret, and David Hess. 1998. *Women Confront Cancer*. New York: New York University Press.

Woolgar, Steve. 1991. "The Turn to Technology in Social Studies of Science." *Science, Technology and Human Values* 16: 20–50.

Woolgar, Steve. 1996. "Science and Technology Studies and the Renewal of Social Theory," in Stephen Turner, ed. *Social Theory and Sociology*. Oxford: Blackwell, pp. 235–255.

Woolgar, Steve, and K. Grint. 1996. "A Further Decisive Refutation of the Assumption that Political Action Depends on the 'Truth' and a Suggestion that We Need to go Beyond this Level of Debate: A Reply to Rosalind Gill." *Science, Technology & Human Values* 21(3): 354–357.

Wynne, Brian. 1996. "May the Sheep Safely Graze? A Reflexive View of the Expert-Lay Knowledge Divide," in Scott Lash, Bronislaw Szerszynski, and Brian Wynne, eds. *Risk, Environment and Modernity: Towards a New Ecology*. London: Sage, pp. 44–83.

Yates, JoAnne. 1989. *Control through Communication: The Rise of System in American Management*. Baltimore, Md.: Johns Hopkins University Press.

Young, Iris Marion. 1994. "Gender as Seriality: Thinking about Women as a Social Collective." *Signs* 19(3): 713–738.

Young, Stacey. 1997. *Changing the Wor(l)d: Discourse, Politics and the Feminist Movement*. London: Routledge.

Ziman, John, ed. 2000. *Technological Innovation as an Evolutionary Process*. Cambridge: Cambridge University Press.

Zuboff, Shoshana. 1988. *In the Age of the Smart Machine: The Future of Work and Power*. New York: Basic Books.

About the Authors

Philip Brey is associate professor and vice chair of the department of philosophy at the University of Twente, Enschede, the Netherlands. He received his Ph.D. at the University of California, San Diego. He has published widely in the philosophy of technology, with a special emphasis on information and communication technology (ICT). He has been investigating how ICT artifacts and systems can, through their design and embedding in a sociotechnical context, come to function as agents that bring about changes in social and political arrangements and cultural practices, and he has also been studying the ethical dimensions of such changes.

Paul N. Edwards is associate professor of information at the University of Michigan, in Ann Arbor, where he chairs the Program on Science, Technology & Society. His research interests focus on the history, politics, and culture of information technology. Edwards is author of *The Closed World: Computers and the Politics of Discourse in Cold War America* (MIT Press, 1996) and co-editor, with Clark A. Miller, of *Changing the Atmosphere: Expert Knowledge and Environmental Governance* (MIT Press, 2001). He is currently completing a book about climate science and politics since 1950, tentatively titled *The World in a Machine: Computer Models, Data Networks, and Global Atmospheric Politics.*

Andrew Feenberg is professor of philosophy at San Diego State University. He is the author of *Alternative Modernity* (University of California Press, 1995), *Questioning Technology* (Routledge, 1999), and *Transforming Technology* (Oxford University Press, 2002). Feenberg is currently doing research on the Internet and software development for online education under grants from the National Science Foundation and the Fund for the Improvement of Postsecondary Education.

David Hess is a professor in and chair of the science and technology studies department at Rensselaer Polytechnic Institute, Troy, New York. He is the author of ten books and edited volumes in anthropology and science and technology studies, including *Science and Technology in a Multicultural World* (Columbia, 1995) and *Science Studies: An Advanced Introduction* (NYU Press, 1997).

Haider A. Khan holds a permanent faculty position in the Graduate School of International Studies at the University of Denver, Colorado, and has held visiting positions in Tokyo University, Hitotsubashi University, Tilburg University, Liaoning University, University of Costa Rica, and Cornell University. His fields of specialization are economic theory and international and development economics. He has published four books and monographs as a single author, two books as a co-author, and more than thirty articles in professional journals. He was at the Asian Development Bank, Manila, and at the Asian Development Bank Institute, Tokyo, from 1997 to 1999. He will be a fellow at the Center for International Research on the Japanese Economy at the University of Tokyo in 2002.

David Lyon is professor of sociology at Queen's University in Kingston, Ontario. He has been researching the social aspects of communication and information technologies since the early 1980s. His books in the area include *The Information Society: Issues and Illusions* (Polity/Blackwell, 1988), *The Electronic Eye: The Rise of Surveillance Society* (Polity/Blackwell, 1994), *Computers, Surveillance and Privacy* (co-edited with Elia Zureik, Minnesota, 1996), and *Surveillance Society: Monitoring Everyday Life* (Open University Press, 2001). He is currently writing a book on the critique of cyberspace.

Barbara L. Marshall teaches sociology at Trent University in Peterborough, Ontario. She is the author of *Engendering Modernity: Feminism, Social Theory and Social Change* (Polity, 1994) and *Configuring Gender: Explorations in Theory and Politics* (Broadview, 2000), as well as various articles on feminism, social theory, and gender. She is continuing to explore the historical and contemporary intersections of gender, sexuality, technology, and consumer culture, and in a joint project with Anne Witz is conducting an interrogation of the masculine ontology that underpins sociological theories of modernity.

Thomas J. Misa is associate professor of history at Illinois Institute of Technology, Chicago. He is the author of *A Nation of Steel: The Making of Modern America, 1865–1925* (Johns Hopkins University Press, 1995), which was awarded the Society for the History of Technology's Dexter Prize. His forthcoming interpretation of technology and culture since the Renaissance is tentatively entitled *Leonardo to the Internet*.

Arthur P. J. Mol is professor of environmental policy at Wageningen University, the Netherlands. He has published widely on industrial transformations, globalization, and ecological modernization.

Junichi Murata is a professor in the department of history and philosophy of science, University of Tokyo. His main research theme is phenomenological theory of knowledge; he has been working on the philosophy of perception and mind and the philosophy of science and technology. His main publications include *Perception and the Life-World* (University of Tokyo Press, 1995; in Japanese); and "Consciousness and the Mind-Body Problem" in *Cognition, Computation, Consciousness*, edited by M. Ito et al. (Oxford University Press, 1997).

Arie Rip is professor of philosophy at the University of Twente, Enschede, the Netherlands. He is a former secretary of the European Association for the Study of Science and Technology (1981–86) and past president of the Society for Social Studies of Science (1988–89). His research and publications focus on science and technology dynamics and science and technology policy analysis, including constructive technology assessment.

Don Slater is a reader in sociology at the London School of Economics. His research includes work in economic sociology and the relationship beween culture and economy; consumer culture, objects, and technology; and comparative ethnographies of Internet use. Recent publications include *Consumer Culture and Modernity* (Polity, 1997); with Fran Tonkiss, *Market Society: Markets and Modern Social Theory* (Polity,

2000); and with Daniel Miller, *The Internet: An Ethnographic Approach* (Berg, 2000).

Johan Schot is professor of social history of technology at Eindhoven University of Technology and the University of Twente and scientific director of the Foundation for the History of Technology. He is the program leader of the national research program on the history of technology in the Netherlands in the twentieth century. He is co-founder (together with Kurt Fischer) of the Greening of Industry network and project leader of several European Union-funded international projects. He chairs (with Ruth Oldenziel) the European Science Foundation network, Tensions of Europe: Technology in the Making of Twentieth Century Europe. His research work and publications span the history of technology, science and technology studies, innovation and diffusion theory, constructive technology assessment, environmental management, and policy studies.

Index

ACER, 345, 348
Actor-network theory, 52, 60, 87–91, 181, 230, 302n10. *See also* Technology studies; Symmetry principle
Adas, Michael, 26n3
Aesthetic modernism. *See* Modernism, as aesthetic theory
African National Congress, 370
Agency, 59, 90, 121, 123, 340, 360, 364, 365, 370, 372. *See also* Structure-agency problem
AIDS movement, 287, 297
Air Defense Integrated System, 210
Airports, 162, 164, 190. *See also* Schiphol Airport
Alder, Ken, 262
Alternative modernities. *See* Modernity, alternatives/ plurality
Anthropology, 279
Anti-modern theories, 143, 145, 146, 303, 306. *See also* Postmodernism; Poststructuralism; Postcolonialism
Architecture, 36, 45. *See also* Modernism
 International Style, 5, 12
ARPANET, 202, 216–220
Arthur, Brian, 329
Automaton, 246

Bacon, Francis, 2, 6
Ball, Kirstie, 181

Ballistic Missile Early Warning System, 212
Baudrillard, Jean, 44, 57, 176
Bauman, Zygmunt, 45, 55, 308, 372
Beck, Ulrich, 83, 84, 170, 196, 268, 308, 331. *See also* Risk society theory of modernity, 42
Bell, Daniel, 57
Beniger, James, 58, 205, 206, 207, 208, 209, 220
Benton, Ted, 304
Berman, Marshall, 26n2, 37, 103n3
Berners-Lee, Tim, 203, 204
Bigelow, Jacob, 7
Bijker, Wiebe, 59, 61, 92, 328
"Black box," 2, 12, 34, 52, 190, 231, 244, 249, 342
Blühdorn, I., 307
Bogard, William, 176
Boisjoly, Roger, 86
Borgmann, Albert, 74, 98
Brey, Philip, 186

Cable and Wireless, 152
Cailliau, Robert, 203
Calhoun, Craig, 111, 117, 172
Callon, Michel, 53, 60
CAM movement. *See* Complementary and alternative medicine
Capability approach, 335, 337, 338. *See also* Social capabilities

Capitalism, 39, 40, 280, 285, 291, 299, 304
consumer, 166, 179, 181
and environment, 305
Carnival, 146, 150
Castells, Manuel, 35, 37, 43, 61, 68, 187, 208
CERN, 203
Challenger accident, 86, 87
Chemical industry, 313, 316–323
China, 242, 243
Clarke, Adele, 115
Clarke, Roger, 168
Clinical trials, 289
Closed circuit television (CCTV), 170, 171, 178
Cockburn, Cynthia, 116
Co-construction, 26n7, 95, 163, 173, 176, 186, 189
dynamics of, 181, 183, 360, 372
of gender-technology-modernity, 123, 124
as methodology, 3, 10, 16, 122, 129, 360, 363, 364, 365
of modernity and technology, 15, 18, 105, 106, 139, 186, 191, 362, 364, 366, 368, 369
Collins, Harry, 86
Colonialism, 140, 148
Commission on Critical Infrastructure Protection (U.S.), 187
Community, 22, 264, 292, 293, 295, 299, 335
online, 98, 101
Complementary and alternative medicine (CAM), 282–290, 296–297
Computer(s)/computerization, 98–100, 163–167, 169, 178, 207
and communication, 97–99
IBM vs. Apple, 14
Constructive technology assessment, 278n15, 365
Consumer culture, 39, 124
Consumer organizations, 317
Consumption, 165, 172, 179
Control revolution, 39, 205, 207

Cortada, James, 207, 220
Cosmopolitanism, 149, 150
Counter-productivity theorists, 306. *See also* Demodernization
Creativity of technology, 229, 232, 251, 253
Critical theory, 35, 37, 41, 82, 106–113, 116, 130, 132n8, 176
and technology, 56, 119, 120
Culbert, Michael, 287
Cultural ecology, 280
Cultural history, 36
Cultural horizon, 342
Cultural studies, 36, 46, 99
Cultural values, 279, 285, 286, 291
Cyberculture, 99, 142, 146, 147, 155, 216, 219
Cyborgs, 16, 108, 128, 146

Databases, 162, 165, 168, 175, 177, 178
Datatech, 346
Dataveillance, 167, 168, 170, 172
David, Paul, 329
De Stijl, 266
Deconstruction, 61, 109, 114, 186, 327
Delanty, Gerard, 110, 111
Democracy, 279, 286, 300, 337
deep, 340, 356n11
semiotic, 117
and technology, 101, 120, 121, 122, 354
Democratic rationalization, 103n7
Demodernization, 304, 306, 308. *See also* Soft chemistry
Derrida, Jacques, 45
Design process, 220, 273, 274, 275, 276
Development, 11, 328–343, 352–355. *See also* Capability approach; POLIS
Development frame, 351
Device paradigm, 74
De-worlding, 96, 97, 98, 99, 101, 102
Dickens, Peter, 304, 305
Digital convergence, 200
Disclosing New Worlds, 93, 94

Disembedding, 143, 144, 147, 148, 152, 158. *See also* Giddens, Anthony
Douglas, Susan, 13
Dreyfus, Hubert, 93
Durkheim, Emile, 2, 37

Echelon system, 171
Ecological modernization, 22, 23, 304, 305, 309, 310, 311, 312, 321, 323, 324, 325
Ecological reform, 309, 318
Economics of innovation, 341
Economy of signs and spaces, 44
Edwards, Paul, 60
Electronics industry, 344, 345, 346, 347
Ellul, Jacques, 8
Email, 142, 148, 197, 202, 205
"Enframing," 8, 93, 96
Engels, Friedrich, 2, 8
Enlightenment, 6, 12, 36, 39, 41, 45, 107, 108, 111, 131n2, 332
Environmental modernization, 321
Environmental movement, 303, 310, 312, 321, 370
Environmental risk, 284, 294
Environmental state, 312
Erectile dysfunction, 123, 125, 126, 127
Ericson, Richard, 177
Erie Canal, 293
Essentialism, 57, 117, 132n15, 176, 227, 234, 360
Ethnography, 139, 143, 144, 153, 154
Everett, Robert, 214, 215
Everquest, 179
Evolutionary economics, 368
Expert systems, 57, 124, 127
Expositions
  New York (1939), 267
  Osaka (1903), 238
  Vienna (1873), 248
Expressionism, 36

Feenberg, Andrew, 4, 57, 113, 119, 140, 338, 339, 340

Feminist theory, 16, 105, 106, 112, 114, 130, 131n1, 145, 360
Ferguson, Kathy, 114
Fischer, Claude, 201. *See also* User heuristic
Flores, Fernando, 93, 102
Food and Drug Administration, 128
Ford, Henry, 5, 11, 39. *See also* Post-Fordist economy
Forrester, Jay, 214, 215
Forster, E. M., 327
Foucault, Michel, 12, 13, 39, 79, 178. *See also* Panopticism; Power
Frankfurt Institute for Social Research, 110
Fraser, Nancy, 117, 121, 122, 292
Freedom, 13, 24, 157, 330, 335, 339. *See also* Capability approach; Social capabilities
French Revolution, 262
Fuller, Steve, 103n4, 239
Functionalism, 36, 57, 61, 124, 204, 206, 207
Futurism. *See* Italian Futurism

Gemeinschaft, 292
Gender, 39, 48, 53, 64, 66, 106, 197
  and modernity, 11, 27n19, 70n3, 108, 120, 129
  and technology, 105, 106, 109, 115, 117, 120, 129, 130
Genetic engineering, 12, 368
Gesellschaft, 292
Giddens, Anthony, 37, 39, 42, 45, 56, 57, 141, 167, 195, 308, 363
Gitelman, Lisa, 13
Global climate change, 194
Global modernity, 142, 143, 149, 150, 151, 153, 155, 158
Globalization, 26n4, 62, 152, 166, 171, 280, 292, 295, 296, 299, 324, 334
Glocalization, 158, 325
Governmentality, 178. *See also* Foucault, Michel
Graham, Stephen, 178

"Great divide," 2, 76, 77, 359
Grint, Keith, 117
Gropius, Walter, 5
Ground Observer Corps, 210
Guccione, Bob, 123

Habermas, Jürgen, 4, 8, 13, 36, 41,
    45, 56, 80, 81, 82, 83, 110
Hacker, Barton, 240
Haggerty, Kevin, 177
Haraway, Donna, 16, 108, 128
Hård, Mikael, 26n6
Harootunian, Harry, 29n34
Harvey, David, 35, 36, 37, 44, 45,
    46, 69
Hecht, Gabrielle, 26n7, 199
Heidegger, Martin, 8, 56, 57, 93, 98,
    328, 331, 339
Hermeneutics, 76, 91, 92, 95, 235,
    236, 342. *See also* Interpretation
Heterogeneity, 35, 45, 139, 141, 363
History of technology, 46, 48, 49, 50,
    52. *See also* Technology studies
Hobday, Michael, 345, 347
Hsinchu Science Park, 343, 344,
    349, 350
Hughes, Thomas, 5, 21, 198, 208,
    209, 215, 221, 328
Human Genome Project, 127, 368, 369
Hybridity, 128, 146, 230, 284,
    339, 366
Hypersurveillance, 176

IBM, 211, 361
ICQ, 144, 156
Identity, 109, 155, 172, 233, 362
    gender, 11, 122, 123, 197
    modern, 39, 111, 147, 149, 150,
        152, 155, 292, 369
    national, 146, 150, 159, 199, 235
    postmodern, 44, 175
Ideographic explanation, 113
Indonesia, 159, 362
Industrial revolution, 6, 29n34, 33, 39
Industrial society, 33, 37, 40, 42, 43,
    64, 66, 67

Information
    infrastructure, 192, 207, 208, 209
    revolution, 207
    society, 33, 38, 39, 43, 56, 58,
        166, 180
    technology, 13, 43, 44, 55, 58, 156,
        168, 170, 171, 202, 208, 209, 210,
        222, 345, 353
Infrastructural inversion, 192
Infrastructure(s), 60, 185–199, 205,
    207, 212, 218–223, 363, 366, 368
    anti-modern aspects, 204
    definitions of, 186
    and modernity, 191, 209
    and nature, 188, 193–195
Innovation studies, 368
Instrumentalization theory, 95, 98,
    104n10, 119
Interlevel analysis, 67
International Style. *See* Architecture
Internet, 83, 98, 100, 162, 171, 179
    and ethnography, 143
    Network Working Group, 219
    and Trinidad, 142–153
Interpretation, 51, 91, 92, 93, 99,
    114, 117. *See also* Hermeneutics
Interpretative flexibility, 94, 227, 228,
    230, 231, 234, 235, 244, 342
"Iron cage," 41, 165. *See also*
    Weber, Max
Italian Futurism, 2, 5, 28n28, 266, 363

James, C.L.R., 141
Jameson, Frederick, 44, 45
Japan
    and mechanical clocks, 244–246
    modernization of, 236–240

Kellner, Douglas, 27n23, 104n10
Kuhn, Thomas, 48, 73, 76, 77, 78
Kyoto protocol, 317

Latour, Bruno, 2, 47, 53, 60, 78, 79,
    84–91, 99, 100, 128, 186, 327,
    333, 340
Law, John, 60, 88

Le Corbusier, 5
Level of abstraction, 56, 63–67
Levels of analysis, 116
Lifeworld, 81, 126, 292. *See also*
  System/lifeworld distinction
  of technology, 99, 339
Light, Andrew, 133n18
Load factor, 221
Louise, Charlotte, 284
Luddites, 259, 260, 261, 264, 272
Lyotard, Jean-François, 45, 46

Machine tax, 260
Macro scale, 197, 204, 206–209,
  218–223
Macro-level analysis, 34, 35, 38, 50,
  59–63, 66–68
Macronix, 350
Marcuse, Herbert, 98
Marshall, Barb, 162
Marx, Gary, 168
Marx, Karl, 2, 7- 9, 37, 39, 56, 57,
  73–76, 83, 105
Marx, Leo, 11, 27n20
Marxist theory, 40, 56
  and environment, 304, 305
Massachusetts Institute of
  Technology, 7, 211, 367
Mayr, Otto, 6
Mechanical clock, 241–246
Media studies, 13, 156
Medical pluralism, 294
Medical technology, 280
Meiji Restoration (1868), 236,
  238–240, 244, 249
Merton, Robert K., 48
Meso level, 35, 50, 68, 186, 197, 198,
  213, 216–223, 279, 364
Methodology, 14, 15, 91, 92, 259,
  359. *See also* Co-construction;
  Essentialism; Interpretation; Micro-
  macro problem; Mutual orientation
  comparative approach, 154, 155
  critical theory, 112–115, 119, 121, 130
  genealogy, 114
  interlevel analysis, 67, 68

multiscalar approach, 186, 215, 218,
  222, 224
Metropolitanism, 142
Micro scale, 197, 201–204, 207, 216,
  218, 222
Microelectronics Technology Inc., 350
Micro-level analysis, 60, 63, 66
Micro-macro problem, 35, 62, 63, 65,
  116. *See also* Meso level; Structure-
  agency problem
Mies van der Rohe, Ludwig, 5, 12
Miller, Daniel, 141
Mills, C. Wright, 110, 112
Minitel, 100
Misa, Thomas, 61, 192, 197, 204, 364
Miss Universe competition, 150
Modernism. *See also* Architecture
  as aesthetic theory, 6, 36
  as ideology of modernity, 45
Modernist
  keywords, 7, 367
  settlement, 186, 189, 190, 196, 221
  technology politics, 20, 257, 258,
    263, 269–273, 332
Modernity
  alternatives/plurality, 20, 29n34, 84,
    154, 251, 331, 365, 367
  definitions of, 2, 5, 35, 39, 107, 141,
    186, 257, 280
  and gender, 11, 27n19, 70n3, 111
  "late," 38, 41, 62, 66, 279, 287, 288
  and surveillance technologies,
    163–169
  and technology, 4, 33
Modernity theory/studies, 34–41, 62,
  69, 89, 141, 147, 153, 186, 222,
  223, 228, 359, 360
  treatment of technology, 55–58,
    73–75, 79, 80, 84, 96, 97, 143, 221
Modernization, 6, 19, 20, 37, 42,
  140, 147, 151, 152, 157, 246, 280,
  286–294, 298–308, 328, 330, 337,
  354, 355
  definitions of, 257, 290
  in Japan, 228, 236–250
  theory, 11, 76, 83

Morgan, Gareth, 112
Morris-Suzuki, Tessa, 249
Morton Thiokol, 86
Mosaic, 204, 219
Mumford, Lewis, 6, 8, 241
Murata, Junichi, 342
Muschamp, Herbert, 5
Mutual orientation, 213, 215, 216, 218, 222, 225n11

Nakaoka, Tetsurou, 238
Nanotechnology, 12, 369
National Center for Supercomputing Applications (U.S.), 204
National innovation systems, 339–343, 352, 354, 347, 351–355, 368
NATO Air Defense Ground Environment (NADGE), 212
Natural disasters, 195
and infrastructures, 193
Natural paints, 321. *See also* Demodernization
Naturalism, 36
Navy Special Devices Center, 215
Negotiation spaces, 273
Neolithic revolution, 281
Netherlands, 263, 270
Network society, 166, 174, 180, 210, 220, 222
New York World's Fair of 1939, 267
NIS. *See* National innovation systems
Nishida, Kitaro, 228, 232, 233, 234, 235, 253
Nomothetic explanation, 113
Nongovernmental organizations (NGOs), 23, 267, 270, 294, 299, 301, 310, 312, 317, 319, 334
Nuclear Sarajevos, 213
Nuclear waste storage, 195
Nussbaum, Martha, 335

O'Connor, James, 304, 305
Objectification, 155
Odaka, Konosuke, 249
Ogborn, Miles, 7

Oldenziel, Ruth, 11, 27n19
Online education, 98, 100, 101
Ontological designing, 102
Original equipment manufacturing (OEM), 347, 348
Ormrod, Susan, 116
Orwell, George, 164, 169
Osaka Exposition (1903), 238
"Otherness of technology," 229, 230, 232

Panopticism, 13, 61, 174–176, 204, 210, 221, 222
"Parliament of things," 339, 353
Path dependence, 24, 329, 351, 353
Pavitt, Keith, 322
Perry, Matthew C., 236
Philosophy of science, 48, 76–79
Philosophy of technology, 47, 276n2
Pinch, Trevor, 28n25, 59, 86, 92, 328
Pluralism, 293
POLIS. *See* Positive feedback loop innovation structure
Political economy, 279, 291, 296, 298, 299
Pornography, 39, 144, 158
Positive feedback loop innovation structure (POLIS), 341, 343, 344, 347, 351–355
Postcolonialism, 145
Poster, Mark, 175, 176
Post-Fordist economy, 43, 65
Posthumanism, 99, 108, 146
Postmodernism, 12, 14, 45, 46, 145, 337
definitions of, 44, 131n2
and environment, 303, 307
and technology, 13
Postmodernity, 38, 43, 44, 45, 220, 252
definitions of, 44, 45, 107, 165, 173
and surveillance, 165, 173–175, 179–182
Poststructuralism, 13, 14, 145
definitions of, 131n2

Power, 4, 39, 43, 87, 109, 115, 116, 129, 154, 164, 167, 169, 171, 178, 200, 204, 210, 220, 221, 222, 273, 332, 334, 365
Premodern societies, 291
President's Commission on Critical Infrastructure Protection, 219
Protected entity, 298
Public participation, 270, 272
Public understanding, 268
Purification, 84, 339

Radder, Hans, 88
Railroads, 185, 198, 205, 206
RAND, 217
Rationality, 5, 8, 75, 77, 85, 95, 145, 284, 333
  bounded, 333
  communicative or social, 41
  ecological, 310
  purposive or instrumental, 41, 56, 57, 73, 74, 81, 82, 92, 96, 110, 119, 126, 228, 238, 244
Rationalization, 2, 9, 28n33, 40, 56, 64, 66, 73, 76, 79, 80, 81, 102, 124. *See also* Computerization
Reductionism, 223
Reflexive modernity/modernization, 21–24, 38, 42, 83, 267–271, 304, 308–310, 331
Refractive reflexivity, 332, 333
Regulation, 285, 299
Reification, 57, 103n2
Renaissance, 7, 33, 36
Reproductive sciences, 115
Resistance, 88–91, 141, 267, 296, 370
  to technology, 258, 259, 263
Responsible Care, 316
"Reverse determination," 228, 233–234, 251, 253
Risk management, 169, 170, 177
Risk society, 38, 43, 84, 170, 196, 268, 282, 285, 308, 309, 319, 331, 333. *See also* Beck, Ulrich
Rochlin, Gene, 190, 220

Roman aqueducts, 195
Romanticism, 262, 276, 285
Rorty, Richard, 45
Rosen, Paul, 69, 252
Rosenberg, Nathan, 247, 248
Rotterdam harbor, 263
Royal Dutch/Shell Company, 371
Ruhleder, Karen, 189
Rule, James, 167

SAGE (Semi-Automatic Ground Environment), 210, 211, 216
Said, Edward, 355n4, 370
Sampo Corporation, 345
Scales
  of force, 186, 192–194, 220, 222
  of social organization, 186, 190, 192, 196, 197, 204, 205, 208, 216–223
  of time, 194
Schiphol Airport, 1, 21, 25, 269, 271, 272, 362
Schnaiberg, Allan, 304
Schumacher, E. F., 268
Scientific revolution, 6, 76–79
Sclove, Richard, 53
Sen, Amartya, 335
September 11th (2001), 28n28, 162, 168, 169
Sexual Assault Evidence Kit (SAEK), 120
Sexual dysfunction, 123, 124
Sexual technologies, 106, 123–130
Slater, Don, 122
Smith, Dorothy, 362
Social capabilities, 331, 339–343, 351–354. *See also* Capability approach; Freedom
Social construction of technology (SCOT), 10, 201, 227, 329. *See also* Interpretative flexibility; Micro scale; Stabilization
Social constructivism, 51, 61, 85, 186, 197, 218, 222, 223
Social shaping of technology. *See* Technology studies

Social theory. *See* Modernity theory
Social worlds approach, 115
Sociology of scientific knowledge, 48
Sociotechnical landscape, 366
Sociotechnical network, 52
Sociotechnology, 47
Soft chemistry, 320, 321. *See also*
    Demodernization
Sombart, Werner, 265, 266
Sonderweg, 265
Sony World, 361
South Africa, 370
South Korea, 338, 342, 343, 345
    chaebol, 347
Space of flows, 187
Spinosa, Charles, 93
Stabilization, 52, 53, 252, 329
Standpoint epistemologies, 134n25
Star, Leigh, 189
Staudenmaier, John, 47, 49
Strategic Air Command, 212
Structuration, 195. *See also* Giddens,
    Anthony
Structure-agency problem, 62, 111,
    116. *See also* Micro-macro problem
STS. *See* Technology studies
Subpolitics, 290, 312. *See also* Beck,
    Ulrich
Surrealism, 36
Surveillance technologies, 162, 163,
    172, 174, 177, 202, 204, 219. *See
    also* Closed circuit television;
    Dataveillance; Hypersurveillance
    and modernity, 33, 42, 161
    and postmodernity, 165, 166,
    179–181
    and surveillant assemblage,
    162, 172
Sustainability, 281, 307
Suzuki, Jun, 249
Swift, Jonathan, 327
Symmetry principle, 74, 85, 88, 89,
    100, 230. *See also* Actor-network
    theory
System/lifeworld distinction, 4, 41,
    81–83, 92, 111, 116–122, 126,

129, 130, 132n14, 135n31. *See also*
    Lifeworld

Taiwan, 338, 342–355
Taiwan Semiconductor
    Manufacturing Corp., 350
Tanaka, Hisashige, 246
Tatung, 345, 347
Taylor, Frederick, 5, 165
Technical code, 119, 353
Technological determinism, 9, 28n24,
    50, 57, 58, 61, 71n17, 123, 129,
    183, 231, 237, 245, 248, 338, 340.
    See also TINA
    and postmodernism, 13
Technological enthusiasm, 267
Technological failure, 190
Technological fundamentalism, 6
Technological modernism, 368, 369
Technological shaping of society, 10, 52
Technological systems, 3, 4, 12, 18,
    47, 185, 196, 198, 201, 328, 329,
    335, 351
Technology
    alternative, 20, 96, 101, 268, 296,
    300, 321
    definitions of, 7, 185, 266, 279
    iconic character, 367
    modern, 12, 33, 39, 56, 96,
    227–229, 240, 245, 250, 251, 253,
    266, 362
    traditional, 244–247, 250–253
Technology studies, 8, 14, 15, 19,
    33–35, 46–50, 53, 54, 62, 105,
    107, 113, 116, 130, 140, 186, 197,
    201, 222, 227, 228, 283, 294, 359,
    360, 365. *See also* Actor network
    theory; History of technology
    momentum, 198, 221
    pluralism, 293, 294, 296
    regime, 353, 357n17
    theoretical claims, 50, 79
    treatment of modernity, 58, 59, 62, 73
    user heuristic, 201, 202, 232
    variation and selection model, 51
Tenner, Edward, 193, 232

"Terminal subject," 98, 101
Theories of modernity. *See* Modernity
  theory/studies
Throwaway society, 319
TINA ("There Is No Alternative"), 371
Totalitarianism, 164
Touraine, Alain, 55
Trinidad, 141, 142, 146, 148–159
  and Internet, 142, 144
  and "virtuality," 145
Turkle, Sherry, 14, 53, 99

Ullrich, Otto, 306
United Fiber Optic Communications
  Inc., 350
United Microelectronic Corporation,
  345
Universalism, 285, 286, 289
Usenet, 202
User heuristic, 201, 202, 232
Utilitarianism, 335

Van Doesburg, Theo, 266
Van Lente, Dick, 367
Venezuela, 149
Venturi, Robert, 12
Viagra, 123, 124, 127
Vienna exposition (1873), 248
Virtual/virtuality, 143, 144, 145
  communities, 203, 204
  infrastructures, 220

Wagner, Peter, 37
Wajcman, Judy, 53, 105, 129, 183
Weber, Max, 2, 8, 37, 40, 80, 166
Westmoreland, William, 212
Whirlwind, 214, 215
Williams, Raymond, 7
Winbond Electronics Corp., 350
Winner, Langdon, 28n25, 53, 191,
  275, 365
Winograd, Terry, 102
Woolgar, Steve, 117
World Trade Center, 28n28, 162
World Wide Military Command
  Control System (WWMCCS), 212

World Wide Web (WWW), 185, 200,
  203, 219, 220
Wynne, Brian, 332